COLD GAS DYNAMIC SPRAY

COLD GAS DYNAMIC SPRAY

Roman Gr. Maev
Volf Leshchynsky

CRC Press
Taylor & Francis Group
Boca Raton London New York

CRC Press is an imprint of the
Taylor & Francis Group, an **informa** business

MATLAB® is a trademark of The MathWorks, Inc. and is used with permission. The MathWorks does not warrant the accuracy of the text or exercises in this book. This book's use or discussion of MATLAB® software or related products does not constitute endorsement or sponsorship by The MathWorks of a particular pedagogical approach or particular use of the MATLAB® software.

CRC Press
Taylor & Francis Group
6000 Broken Sound Parkway NW, Suite 300
Boca Raton, FL 33487-2742

© 2016 by Taylor & Francis Group, LLC
CRC Press is an imprint of Taylor & Francis Group, an Informa business

No claim to original U.S. Government works

Printed on acid-free paper
Version Date: 20160224

International Standard Book Number-13: 978-1-4665-8442-6 (Paperback)

**Visit the Taylor & Francis Web site at
http://www.taylorandfrancis.com**

**and the CRC Press Web site at
http://www.crcpress.com**

Contents

Preface

In the past 10 years, cold gas dynamic spray (or cold spray—CS) technology and CS equipment have made remarkable progress. CS methods are now recognized and respected as a competitive technology and as an alternative to other thermal spray and laser cladding or conventional welding techniques. To take advantage of the performance capabilities of the CS process, leading industrial companies such as General Electric (Fairfield, CT), Ford (Dearborn, MI), and Boeing (Chicago, IL) are gradually implementing this coating technique into their engineering processes. As CS equipment and technology have considerably expanded the application possibilities of thermal spray techniques, it is beneficial for students and engineers to describe and analyze the main trends and developments behind this method. Furthermore, it is advantageous to present a comprehensive assessment of CS technology as it fits within the framework of thermal spray manufacturing processes.

Advances are reflected in (1) a conventional approach, in which the important steps are grit blasting and cold spraying and (2) hybrid CS-sintering technology that offers integrity of microstructure, compositional homogeneity, and mechanical property levels equal to (and frequently better than) those of the wrought counterpart.

This book first of all addresses graduate students, postdoctoral fellows, and engineers in practice who are familiar with the conventional approach to applied physical metallurgy, powder metallurgy, and thermal spraying technologies. The authors hope that it may also prove useful to mechanical engineering undergraduates studying metallurgy as an ancillary subject. To students for whom metallurgy, and, in particular, thermal spraying processes, is a principal subject, this book can offer a helpful approach to certain sections of the physical background of cold spraying, metallurgical, and mechanical principles involved. In the text, the basic principles of CS technology are covered. This book is primarily intended for those engineers who have acquired this knowledge either by studying thermal spraying and powder metallurgy as subjects in an engineering course or by careful reading of Chapter 2.

In recent years, the widened scope of thermal spraying technologies has been reviewed in number of publications. Therefore, a brief survey of thermal spray methods in Chapter 1 serves as a sufficient introduction to the thermal spraying methods in general.

The authors trust that the discussion on the basic principles of plasticity theory in Chapter 2 will enable readers to follow the study of the CS processes without being at a disadvantage, if they have not studied the plasticity theory as an independent subject.

Chapter 3 presents a description of the CS equipment, the nozzle design, and the geometry of a CS gun. To illustrate the main aspects of determination of CS system parameters, the numerical simulation and experimental data are presented in Section 3.2.2.

In this book, emphasis is placed on the proper development of CS coatings as well as their structure and properties, which are of paramount importance to the success of CS. Broadly, Chapter 4 deals with the microstructural and mechanical properties of CS metals and alloys. Processes such as bonding, hardening, and softening of the coating materials during CS and coating structure formation with following heat treatment (sintering) are dealt with from a theoretical as well as from a practical aspect.

The use of CS technology is reasonable in cases where conventional thermal spraying technologies cannot be successfully applied. A detailed analysis of aircraft component repair is presented in Chapter 6; the methodology and accompanying technical data for the transition of CS technology into the aerospace industry and several case studies are presented. The case studies show the tremendous impact that CS has had for a few select applications. Three aerospace materials—magnesium, aluminum, and titanium—are presented while attempting to include as much technical data substantiating the advantages and benefits gained by the transition of the process.

Economic aspects of CS applications are analyzed in Chapter 7. The costs from application to application can vary significantly, but the cost of each is dependent on the same set of operating parameters. Once defined, these parameters can be used to accurately calculate the cost of a finished product or can be used to predict the cost of a potential product. The methods presented in Chapter 7 assume that a complete product has not yet been fabricated, and cost estimates may be calculated by students based on the geometry and composition of the desired product.

The frequent references to appropriate sources and book chapters make it easy for the reader to look up the metallurgical and mechanical principles governing any particular process under consideration.

Roman Gr. Maev
and
Volf Leshchynsky
University of Windsor

MATLAB® is a registered trademark of The MathWorks, Inc. For product information, please contact:

The MathWorks, Inc.
3 Apple Hill Drive
Natick, MA 01760-2098 USA
Tel: +1 508 647 7000
Fax: +1 508 647 7001
E-mail: info@mathworks.com
Web: www.mathworks.com

Acknowledgments

The editors express their appreciation to the many individuals within government agencies, companies, and universities who have pushed CS technology forward during the past 20 years. Their projects and public dissemination of the technical results contributed immeasurably to the emergence of CS technologies as a strong and growing manufacturing process. We express our kind appreciation to all contributors of the book, and also to Professor A. Papyrin, Dr. E. Irrisou, Dr. J.-G. Legoux, Dr. J. Vlcek, Dr. J. Kocimiski, Dr. Ol. Bielousova, Dr. Ev. Leshchinsky, Dr. H. Weinert, Dr. J. Borowski, and Dr. J. Wengland for their discussion and comments on some aspects of the book as well as for providing some of the data. Thanks are extended to Suong Manchini, Sabina Baraoniciu, Sarah Beneteau, Mike Lee, and Ninel Nyestyerova for their help during manuscript preparation. Finally, we acknowledge the support, tolerance, and understanding of the authors' families throughout this endeavor.

Editors

Roman Gr. Maev was born in Moscow, Russia. He received his MSc degree in theoretical nuclear physics from the Moscow Physical Engineering Institute followed by a PhD on the theory of semiconductors from the Physical P.N. Lebedev Institute of the USSR Academy of Sciences. In 1990, he received a fellowship from the Gore–Chernomyrdin Commission, as part of which he successfully completed a course project for the Scientific Business Management Fund for one semester of study at Harvard Business School (Boston, MA). In 2001, he received a DSc degree from the Russian Academy of Sciences, and in 2005, he received a full professor diploma in physics from the Government of the Russian Federation.

Dr. Maev is the founding director-general of the Institute for Diagnostic Imaging Research—a multidisciplinary, collaborative research and innovation consortium with one of its research interests in nanotechnology. He is also a full faculty professor in the Department of Physics at the University of Windsor, Ontario, Canada, and in 2007 was granted the title *University Professor Distinguished*. Dr. Maev is a fellow of IEEE, BI NDT, and the Engineering Academy of Russian Federation.

The diverse range of disciplines encompassed by Dr. Maev includes theoretical fundamentals of physical acoustics, experimental research in ultrasonic and nonlinear acoustical imaging, and the nanostructural properties of advanced materials and their analysis. He is the author of four monographs, editor and coeditor of 10 books, has published over 477 articles in leading international journals, and holds 31 international patents.

Volf Leshchynsky received his PhD from the Technical Physics Institute of Byelorussian Academy of Sciences in 1968, and his DSc from the Institute of Metallurgy of the same University in 1989. He then held a chair as professor of the Metal Forming Institute, Poznan, Poland, and became head of the Metal Forming Department at the East Ukraine University. From 2004 onward, he was a visiting professor at the Physics Department of the University of Windsor, Ontario, Canada. He is currently a leading researcher at the Institute for Diagnostic Imaging Research of the University of Windsor, where he conducts research in new powder metallurgy, nanostructuring, and cold spray techniques. Dr. Leshchynsky's research interests include the fundamentals of nanotechnology and nanostructural materials, and the development and application of powder spraying and powder metallurgy. He is the author of more than 170 scientific papers, 1 monograph, and holds 35 patents.

Contributors

Chris Berghorn
Flame-Spray Industries
Port Washington, New York

Victor K. Champagne, Jr.
US Army Research Laboratory
Aberdeen Proving Ground, Maryland

Victor K. Champagne, III
Department of Mechanical Engineering
University of Massachusetts
Amherst, Massachusetts

Alberto Colella
MBN Nanomaterialia SpA
Vascon Di Carbonera (TV), Italy

Dmitry Dzhurinskiy
Institute for Diagnostic Imaging Research
University of Windsor
Windsor, Ontario, Canada

Hirotaka Fukanuma
Plasma Giken Co., Ltd
Saitama, Japan

Dennis Helfritch
TKC Global LLC
Herndon, Virginia

and

US Army Research Laboratory
Aberdeen Proving Ground, Maryland

Keith Kowalsky
Flame-Spray Industries
Port Washington, New York

Volf Leshchynsky
Institute for Diagnostic Imaging Research
University of Windsor
Windsor, Ontario, Canada

Roman Gr. Maev
Institute for Diagnostic Imaging Research
University of Windsor
Windsor, Ontario, Canada

Paolo Matteazzi
MBN Nanomaterialia SpA
Vascon Di Carbonera (TV), Italy

Kazuhiko Sakaki
Department of Mechanical Systems and
    Engineering
Faculty of Engineering
Shinshu University
Wakasato, Nagano, Japan

Wolfram Scharff
IFU GmbH Diagnostic Systems
Lichtenau, Germany

Thomas Van Steenkiste
Flame-Spray Industries
Port Washington, New York

Emil Strumban
Institute for Diagnostic Imaging Research
University of Windsor
Windsor, Ontario, Canada

Sergey Titov
Institute for Diagnostic Imaging Research
University of Windsor
Windsor, Ontario, Canada

1 Introduction

Roman Gr. Maev

CONTENTS

1.1 OVERVIEW OF THERMAL SPRAYING TECHNOLOGIES

1.1.1 COATING CHARACTERIZATION

Today numerous surface treatment procedures are used to modify the surface properties of various materials without altering their bulk characteristics. These techniques are used within industrial environments to improve resistance to corrosion, wear, fatigue, and heat. To properly analyze and, in turn, distinguish between these methods, it is first beneficial to differentiate between thin and thick layers.

Thin coatings have been developed more on a considerably wider scale. These include techniques such as CVD or PVD (chemical or physical vapor deposition), plasma spraying, ion-assisted deposition, magnetron-sputtered deposition, and chemical deposition under a laser beam. Their thicknesses, being generally less than a few micrometers, have no direct effect on the mechanical characteristics of the treated material; they do not alter such aspects as yield strength or ultimate tensile strength.

On the other hand, thick coatings, including cladding and welding, have thicknesses varying from a few hundred micrometers to several centimeters. They generally not only serve as protective layers, but also add mechanical strength. Most of the problems encountered when welding or cladding materials concern the heat-affected zone (HAZ) and result from the heat generated by a laser (Kalla, 1996) or electron beam (Yilbas et al., 1998), or alternatively as a result of capacitive discharge (Simmons and Wilson, 1996). In processes characterized by high energy densities such as plasma arc, electron beam welding, and laser beam welding (Tjong et al., 1995; Sun and Karppi, 1996), the HAZ is considerably smaller than with more conventional processes such as gas metal arc and submerged arc welding (Cullison, 2001; Kannatey and Asibu, 1997). This is primarily due to very short interaction times. Even with a small HAZ, incompatibilities may still exist due to differences in the physical properties of the materials such as fusion points and thermal expansion coefficients (Henderson, 1976; Conzone et al., 1997). This may ultimately induce cracking

1

(Wang, 2000; Conzone et al., 1997), high residual stresses (Wang, 2000; Conzone et al., 1997), or brittle intermetallic compounds (Wang, 2000).

Thermally sprayed coatings are used extensively within a wide range of industrial applications. These techniques generally involve spraying molten powder or wire feedstock that has been melted using oxy-fuel combustion or an electric arc (plasma). The molten particles are subsequently accelerated toward a properly prepared (often metallic) substrate. Solidification occurs at rates similar to those obtained in rapid solidification technology, resulting in deposits that have ultrafine grain structure, with similar microstructure characteristics.

Thermal spray is a highly flexible process that is able to generate dense, tenaciously bonded coatings with low porosity. It is commonly used to shield surfaces from wear, high temperatures, and chemical attack, along with environmental corrosion protection in infrastructure maintenance engineering. Typical materials that are sprayed in this method include most metal alloys and ceramics. In fact, virtually any material can be thermally sprayed if it does not decompose prior to melting.

Various thermal spraying methods, such as flame spraying (including high-velocity air-fuel and high-velocity oxy-fuel [HVOF] thermal spray devices), plasma spraying, and electric arc spraying, have been used to coat both metallic and nonmetallic surfaces. A general overview of existing spraying techniques is given next.

1.1.2 FLAME SPRAYING

Although *flame spraying* was one of the earliest thermal spray processes to be developed, its usefulness endures even today. In this method, metals, ceramics, or cement-like materials are deposited onto a substrate by means of combustion reaction. The flame spray device includes a combustion chamber that receives a mixture of fuel (e.g., propylene or propane) and oxidant (e.g., oxygen or air) in the form of a high-temperature, high-pressure gas stream. This combustion stream is initially directed into a flow nozzle where the spray material (e.g., a powder, a solid rod or wire) is introduced and at least partially melted and *atomized*. It is then propelled in this form to the target substrate. Depending on the design, particle streams may be accelerated up to supersonic or hypersonic velocities. This supersonic particle stream may be generated by a single or two stage combustion devices, or alternatively by those that produce steady-state, continuous detonations.

In HVOF techniques, propylene, propane, or hydrogen (depending on user requirements) is combined with oxygen in a proprietary siphon system in the front portion of the Diamond Jet gun. The thoroughly mixed gases are then ejected from a nozzle and ignited externally from the gun. Meanwhile, the coating material (in powdered form) is fed axially through the gun using nitrogen as a carrier gas. Thus, the ignited gases form a circular flame configuration that surrounds and uniformly heats the powdered spray material as it exits the gun and is propelled to the specimen surface.

It should be noted that due to the high kinetic energy transferred to the particles by the HVOF process, the coating material generally does not need to be fully melted. Instead, the powder particles are heated to a molten state and flatten plastically as they impact the surface. For this reason, coatings are formed having more predictable chemistries that are very homogeneous and have fine granular structure.

Through this use of efficient and controlled thermal output with a high kinetic energy, HVOF processes produce dense, very low porosity coatings that exhibit very high bond strengths (exceeding 12,000 PSI or 83 MPa), have low oxidization, and yield extremely fine finishes, even without further processing. The resultant coatings exhibit low residual internal stresses and can be sprayed to thicknesses that are not normally associated with dense, thermal sprayed coatings.

1.1.3 ARC WIRE SPRAYING

Arc wire spraying uses two metallic wires as the coating feedstock. These wires are electrically charged with opposing polarity and fed into the arc gun at matched, controlled speeds. When the wires are brought together at the nozzle, their opposing charges create enough heat to continuously melt their tips. Compressed air is used to atomize the subsequently molten material and, in turn, accelerates it toward the surface, forming the coating. In arc wire spraying, the weight of the coating that can be deposited per unit of time is a function of the electrical power (amperage) of the system, as well as of both the density and melting point of the wire. New arc wire spray coating systems have unique, advanced mechanisms such as synchronized *push–pull* wire feed motors that ensure consistent wire delivery.

1.1.4 PLASMA SPRAYING

Plasma spraying is perhaps the most flexible of all thermal spray processes in that it can create sufficient energy to melt any material. Further, because it uses powder as the coating feedstock, the number of coating materials that can be used in the plasma spray process is almost unlimited. In this method, the plasma gun consists of a cathode (electrode) and an anode (nozzle) with a small gap forming a chamber between the two. As gases are passed through the chamber, DC power is applied to the cathode and, in turn, arcs across to the anode. The powerful arc is sufficient to strip the gases of their electrons, forming a state of matter known as plasma. As the coating material is introduced into this gas plume it is rapidly melted, forming a particle–plasma stream which is accelerated up to a hypersonic velocity and propelled toward the substrate.

In this process, various mixtures of nitrogen, argon, and helium (usually 2 of the 4) are used in combination with the electrode current to control the amount of energy produced by a plasma system. Because the flow of each of the gases and the applied current can be accurately regulated, repeatable and predictable coatings can be obtained. In addition, the point and angle at which the material is injected into the plume, as well as the distance between the gun and the target can be precisely controlled. This provides a high degree of flexibility in developing appropriate spray parameters for materials having a very wide range of melting temperatures.

1.1.5 RAPID PROTOTYPING

In materials processing, specific parts are manufactured to particular specifications from a variety of materials. For viable product realization in the marketplace, a spectrum of issues regarding performance, cost, and environmental impact must constantly be addressed. Novel processing methodologies, as well as the innovation of new materials, provide opportunities for product development, improvements in product performance, reductions in cost, and minimization of environmental impact through the complete life cycle (Zhang et al., 2001). One of these innovations, namely additive manufacturing, has allowed for product modification through controlled consolidation, layering, and/or coating, in turn leading to superior product attributes. In fact, the idea of fabricating components through the addition of material led to the establishment of the rapid prototyping industry approximately 15 years ago (Zhang et al., 2003).

Rapid prototyping was initially considered a tool for the rapid development of 3D models and prototype parts, used almost exclusively to verify the design feasibility rather than for functional application. Currently available commercial rapid prototyping techniques, such as stereolithography, selective laser sintering, laminated object manufacturing, fused deposition manufacturing, and 3D printing, are only able to produce prototypes using wax, plastic, nylon, paper, polycarbonate materials, and so on. There are, however, many difficulties that must be addressed when attempting to replicate these processes techniques using materials with high melting temperature such as metal. These include poor material strength, high porosity, oxidation, warping, and *step* effects. There is, however, a great interest in further developing current materials additive manufacturing (MAM) techniques or alternatively to create new methods that are capable of directly producing metal parts.

The MAM techniques that can be used to directly manufacture metal parts can be divided into two groups: (1) 3D cladding (Mazumder et al., 1997; Milewski et al., 1998) and (2) 3D welding

(Schmidt et al., 1990). In 3D cladding, a laser beam creates a weld pool into which powder is injected and melted. The substrate is then scanned by a laser–powder system to trace a cross section that, upon solidification, forms the cross-section of a part. Consecutive layers are then additively deposited, thereby producing a 3D component. An example of this technique, known as *laser engineering net shaping*, was developed by the Sandia National Laboratory. In this method, metal components are fabricated directly from CAD solid models, thus further reducing the lead times for metal part fabrication. A similar process by the name of *directed light fabrication* is under development at Los Alamos National Laboratory.

1.1.6 PLASMA DEPOSITION MANUFACTURING

Plasma deposition manufacturing (PDM) is a newly developed direct metal fabrication process that belongs to the category of 3D welding (Zhang et al., 2003). In this technique, a supply head delivers a well-defined flow rate of metal powder, which is deposited in a molten pool formed by controlled plasma heating. In this way, full strength parts with the single or multifarious materials can be built up, layer by layer, by melting and rapidly solidifying the feed material into a particular shape.

The PDM process has many advantages over traditional material subtractive technologies: (1) conventional metal melting and casting techniques frequently require considerable time and expenditure with respect to mold production, particularly in low-volume manufacturing. PDM, by contrast, allows for the direct production of individual parts in a single step. Also, the manufacturing materials may be varied either between parts or within a single component. (2) Most MAM techniques are limited to the production of parts composed of plastic, wax, or paper; thus, the required engineering properties cannot always be obtained. The PDM process, on the other hand, allows for the direct fabrication of metal or end-use-material components. Further, the deposition rate and particular materials used in PDM can be adjusted through a wide range, making this technology extremely flexible. (3) PDM can be used as an effective remanufacturing tool, helping to increase product life cycle.

There is, however, a primary disadvantage to the PDM process in that the effects of local melting due to the high temperatures cannot be avoided. For this reason, it seems reasonable (and quite advantageous) to apply the principles of PDM to gas dynamic spraying (GDS). In this case, the solid-state bonding mechanism will allow the deposited material to retain its original structure.

The most efficient among the thermal spraying methods is plasma deposition, combining a high-temperature gas jet with the high productivity of a coating process. Unfortunately, while plasma spraying can produce high-quality coatings, this method suffers from several minor limitations. First, the necessary apparatus is relatively complex and expensive. Second, although plasma spray processing is a well-established method for forming thick coatings, in coatings thinner than 100 μm it is difficult to control material properties. Also, in spite of the plasma jet velocity being about 1000–2000 m/s, and the particles being accelerated up to 50–200 m/s (and even up to 350 m/s in extreme cases), the acceleration process is not yet efficient enough. The primary imperfection associated with plasma spraying, however, is its high processing temperature, making a number of substrate materials unfit for treatment due to local heating, oxidation, and thermal deformations. Further, the high-temperature jet that is incident on the product surface intensifies chemical and thermal processes, causing phase transformations and the appearance of oversaturated and nonstoichiometric structures that bring about structural changes in the substrate material. Also, the high cooling rates result in the hardening of these heated materials. And finally, the liberation of gases during crystallization brings about both porosity and the appearance of microcracks.

In short, thermal spraying methods have the following disadvantages: (1) a high level of thermal and dynamic effects on the surface being coated; (2) substantial changes in the properties of the material to which the coating is being applied (i.e., electrical and heat conductance, etc.); (3) changes in the material structure result from the chemical and thermal effects of the plasma jet and the hardening of overheated melts; (4) ineffective powder particle accelerations due to the low density of the plasma; (5) intensive evaporation of fine powder fractions having sizes ranging from 1 to 20 µm; (6) insufficient cost-per-mass ratio of the applied material; and (7) reduced equipment portability due to the high temperature requirements of the method.

1.1.7 EXPLOSIVE CLADDING

The processes that are best able to avoid the problems associated with thermal changes appear to be spot welding (Turgutlu et al., 1995, 1997), impact welding (Date et al., 1999), or explosive cladding (EC). These methods allow for flyer plate velocities that exceed plasma and detonation gun spraying—both of which result in poor bonding and increased surface roughness (Li and Ohmori, 1996). Explosive bonding is considered to be a solid-state welding process because melting is not required at the interface of the two materials in order for bonding to occur (Zimmerly et al., 1994). Moreover, as there is little to no diffusion between the two metals or alloys, each one retains its intrinsic properties. This process makes it possible to join materials having greatly differing melting points without the need for intermetallic compounds—even those having the historic reputation of being incompatible (e.g., titanium and stainless steel, nickel, aluminum, etc.) (Gerland et al., 2000). This is contrary to what occurs in other techniques such as laser cladding (Kalla, 1996; Yilbas et al., 1998).

EC is a solid-state fusion welding process in which the joining of two metals is accomplished through the application of a shock pressure that results from the detonation of an explosive pack. Material interaction at the mating surfaces during EC produces a *surface jetting* effect, ultimately resulting in a strong bond being formed between the metals (Raghukandan, 2003). As it turns out, an effect analogous to this *surface jetting* is the main feature of cold spraying (CS) processes, suggesting that the EC and CS are of the same fundamental nature. One should further note that the impact velocities of EC vary in the range 1400–3900 m/s, whereas the particle velocities of CS are known to have lesser values ranging from 500 to 1600 m/s for different carrier gases. Therefore, the intensity of *surface jetting* or jet flow in the CS process seems to be less than that of EC. Unlike in EC, however, the particle impact velocity in CS may be precisely controlled; it is through this control that the structure and material properties of the deposited layers may be managed.

1.2 GENERAL CHARACTERIZATION OF CS METHOD

Cold gas dynamic spray (CS) is a rapidly emerging coating technology, in which spray particles in a solid state are deposited on a substrate via a high-velocity impact, at temperatures lower than the melting point of the powder material. As shown by McCune et al. (2000), the methodology for generating surface coatings from metallic particles in solid form was first patented by Rocheville (1963). This process employs a supersonic nozzle into which solid particles of feedstock material are introduced and accelerated toward a substrate, either forming a thin surface coating, or being directly embedded as isolated particles in small depressions on the surface. An alternative technique for producing thick deposits of various metals from *cold* jets was reported by Alkhimov et al. (1990). In this method, solid particles ranging in size from 10 to 50 µm are introduced by means of a pressurized powder feeder into the high-pressure, high-temperature chamber of a converging-diverging (de Laval) nozzle and are subsequently accelerated into a supersonic stream by the gas (Figure 1.1a). Here, as the gas has been electrically preheated prior to its introduction into the nozzle, the term *cold spraying* appears to contradict with the nature of this deposition process.

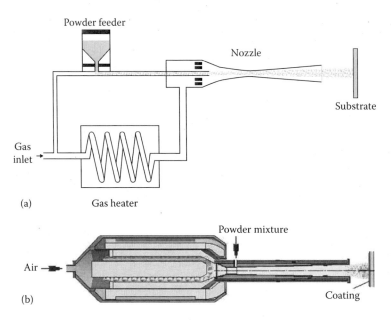

FIGURE 1.1 GDS operating systems. It accelerates micron-sized particles to high velocities by entraining the particles in the flow of a supersonic nozzle—(a) axial injection (From Assadi H. et al., Bonding mechanism in cold gas spraying. *Acta Mater.*, 51, 4379–4394, 2003. With permission.) and (b) radial injection. (From Maev R.G. and Leshchynsky V. *Introduction to Low Pressure Gas Dynamic Spray: Physics & Technology.* Wiley-VCH Verlag GmbH, Weinheim, Germany, 2007. With permission.)

However, this process is more appropriately referred to as *cold spray* because in most cases the particle impact temperature is relatively low.

In deposition, spray materials experience only minor changes in microstructure and little oxidation of decomposition. Most metals such as Cu-, Al-, Ni-, Ti-, and Ni-based alloys can be deposited by cold spray (Papyrin, 2001); even cermets (Karthikeyan et al., 2001) and ceramics (Karthikeyan et al., 2000) can be embedded into a substrate to form a thin layer coating in cold spray.

The most important parameter in the cold spray process is the particle velocity prior to impact on the substrate (Alkhimov et al., 1990). For a given material, there exists a critical particle velocity that must be achieved. Only particles whose velocities exceed this value can be effectively deposited, in turn producing the desired coating. Conversely, particles that have not reached this threshold velocity contribute to the erosion of the substrate (Alkhimov et al., 1990). Naturally, critical particle velocities vary according to the particular spray material that is chosen. In fact, critical velocities are dependent on both the size of the particles and their size distribution of particles (Alkhimov et al., 1990; Van Steenkiste et al., 1999) and the particular substrate material (Gilmore et al., 1999). For example, it has been reported that the critical particle velocities of Cu, Fe, Ni, and Al are approximately 560–580, 620–640, 620–640, and 680–700 m/s, respectively (Alkhimov et al., 1996).

The particle velocity, on the other hand, is determined by the nature, operating pressure, and temperature of accelerating gas, as well as by nozzle design (Dykhuizen and Smith 1998; Gilmore et al., 1999). Material parameters such as density, particle size, and morphology influence the acceleration and subsequent deposition behavior of the particles (Gilmore et al., 1999; Van Steenkiste et al., 1999). It has been demonstrated that a dense coating can indeed be deposited through CS if these spray conditions contribute to particle velocities that exceed the critical particle velocity (McCune et al., 1996). However, even in the case of deposition efficiencies greater than 80%, Karthikeyan et al. (2000) reported that porous titanium coatings may be formed.

A unique feature of cold GDS is its ability to generate a wide range of deposition layer thicknesses ranging from tens of microns up to a centimeter. In this regard, this process extends beyond the concept of *coating* a substrate, providing a means for developing three-dimensional structures (McCune et al., 1996). In fact, in a recent paper by Gabel and Tapphorn (1997) a material-forming process is discussed that utilizes a low-temperature gas stream to accelerate solid particles into a jet. Further, Bhagat et al. (1997) reported the use of cold spray processes for the production of nickel–bronze layers that may be employed in wear resistance applications.

The acceleration of powder particles in GDS is realized within the CS nozzle. Here compressed gases (having inlet pressures up to 30 bar) flow through a converging/diverging nozzle, thereby developing supersonic velocities. The powder particles are metered into the gas flow immediately upstream of the converging section of the nozzle where they are accelerated by the rapidly expanding gas. It is perhaps obvious that compressed gases that are preheated are able to achieve higher gas flow velocities in the nozzle. However, even with preheat temperatures as high as 900 K, as used by Dykhuizen et al. (1998), the gas rapidly cools as it expands in the diverging section of the nozzle. Hence, the duration of particle interaction with the hot gas is relatively brief, and the temperatures of the solid particles remain substantially below that of the gas (Dykhuizen et al., 1998). Thus, the primary function of the carrier gas is to only accelerate the particles to above the critical velocity.

The actual mechanism by which the solid-state particles deform and bond has not yet been fully characterized. It seems possible that plastic deformation may disrupt thin surface films, such as oxides, and provide intimate conformal contact under high local pressure, thus permitting bonding to occur (Dykhuizen et al., 1998). Though unproven, this hypothesis is consistent with the fact that a wide range of ductile materials, such as metals and polymers, have been cold-spray deposited. Experiments with materials, such as ceramics, on the other hand, have proven unsuccessful unless sprayed along with a ductile powder material. If indeed this is the case, the minimum critical velocity corresponds to the kinetic energy that is necessary to bring about powder material plastic flow during deposition. Calculations by Schmidt et al. (2003) indicate that typical kinetic energies found at impact are less than that which is required to melt the particles. Micrographs of cold-sprayed materials (Dykhuizen et al., 1998) also reveal that the deposition mechanism is primarily, if not entirely, a solid-state process.

Because of the fact that sufficient particle velocity is essential for successful cold spray deposition, it is critical to fully understand the relative influence of process variables (such as gas inlet pressure, temperature, and nozzle geometry) on particle velocity. For this reason, discussions of the basic governing equations, computational results, and comparisons with experimental datum will take place primarily with respect to the relationship between these variables. It is hoped that this form of analysis will bring about a deeper understanding of the CS process, and that this book may ultimately serve to widen the study and application of CS.

1.3   IMPACT FEATURES OF CS

1.3.1   Main Features

It is perhaps obvious that particle impact is the main feature that governs the consolidation process in CS. In fact, particle impact phenomena have been studied for many years for a wide range of particle velocities (Klinkov et al., 2005). The nature of layer formation in the cold spray process, on the other hand, is far from being fully understood. Currently, extensive theories explaining the processes of cold spray deposition are being developed (Karthikeyan, 2005). At this point in time, however, the actual mechanism by which the powder particles are consolidated, deformed, and bonded with the substrate is not well understood, particularly with respect to CS processes having small particle velocities or low pressures. For this reason, it is yet necessary to closely scrutinize the impact interactions of the particles in hopes of demonstrating how these main features influence the consolidation process.

The two main phenomena that may be used to characterize the impact are (1) the interaction of a single particle with the surface of the substrate, or a previously bonded particle and (2) the shock compression of the powder layer (Dykhuizen et al., 1999). Although the individual particle approach is being intensively examined and modeled, the shock compression model requires further scrutiny. The shock-compression processes that are realized by CS seem to produce highly activated powder mixtures. This occurs through defect generation and grain size reduction via fracturing and/or the formation of subgrain structures during interparticle sliding, severe particle deformation, and consolidation. These effects result in significantly increased mass transport rates and enhanced chemical reactivity in the powders, creating new paths for the movement of point defects at the interface. In particular, the interaction of the powder mixture components can be accelerated by the presence of dislocations, shorter diffusion distances, more intimately cleansed surface contacts, and higher packing densities, all of which favor solid-state diffusion reactions.

Microparticle impact with surfaces continues to be studied because of its involvement in many important processes such as surface contamination, powder transport, and coating application. Although a significant number of studies have been conducted within the past 30 years, an overall understanding of the deposition process is yet to be realized. In fact, although the fate of microparticles after impact with a surface can be satisfactorily explained by several models, none of them have been entirely validated by experimental data. This is because these models require information that is not easy to quantify, such as the amount of elastic and plastic deformation, the effective inertial mass of the substrate, and the bulk and surface properties of the material (Li et al., 1999). Ideally, these data should be based not only on the properties of the bulk material, but also on those of the microparticles. Presently, however, this information is difficult to obtain. For this reason, it is perhaps useful to develop models that instead establish the general relationships between these variables.

1.3.2 REBOUND AND EROSION PROCESSES

In principle, particles impinging onto the specimen can rebound from, stick to, or penetrate into a bulk substrate. Oftentimes, the impact of a particle on a surface leads to the deformation or destruction of both the particle and the solid body. Klinkov et al. (2005) have classified the results of particle impact on plane surfaces. This classification is based on two important dimensional parameters that may be used to characterize particle impact: the impact velocity, v_p, and the diameter, d_p, of impinging particles (Figure 1.2). These impact parameters for certain CS processes may be identified through observation of the results published in many papers. From these works, one can readily see that the regions of the each CS process (high-pressure, low-pressure, or kinetic CS) strongly depend on both v_p—and, in turn, on the kinetic energy of the particle—and d_p. Further, it should be noted that all of the CS processes occur with particle velocities ranging from 10^2 to 10^3 m/s.

The indentation and erosion of metallic plates due to the impact of particles has been observed for a wide range of particle sizes and velocities (Zukas, 1990). Maximum erosion for a ductile target, however, has been shown to occur at incidence angles of approximately 70°–80° to the normal. It is generally believed that this indentation and erosion takes place as a result of a combination of plowing and cutting actions. In this regime, plastic flow and friction are the key mechanisms of deformation and energy dissipation. To gain insights into the fundamental mechanisms underlying solid-particle erosion, Molinari and Ortiz (2002) performed detailed finite-element simulations modeling the impact of spherical particles on metallic plates. It was found that the primary process governing this erosion is adiabatic shear band formation due to plastic deformation and friction, heat generation, and thermal softening.

Upon examining the dependence of rebound velocity on impact velocity (Figure 1.3), three particle velocity zones are clearly visible (Molinari and Ortiz, 2002). The nondimensional

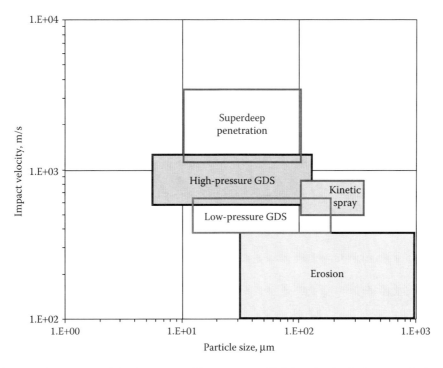

FIGURE 1.2 High-pressure, kinetic spray, and low-pressure GDS location in the impact velocity–particle size map.

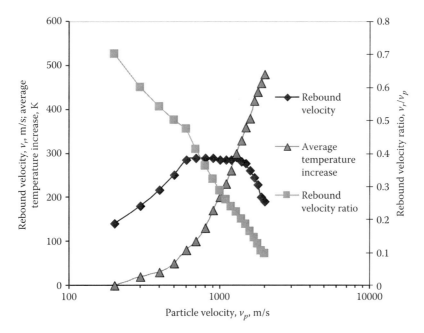

FIGURE 1.3 Normal incidence impact: rebound velocity, rebound velocity ratio, and an average temperature increase versus particle impact velocity. (After Molinari and Ortiz, 2002. With permission.)

velocity parameter, v_r/v_p, and the average particle temperature increase function, ΔT_r (in K), on the other hand, may only be divided into two regimes. In fact, a logarithmic approximation of the ratio v_r/v_p gives $v_r/v_p = -0.21 \ln(v_p) + 1.8$ for the low particle velocities (up to 600 m/s), and $v_r/v_p = -0.29 \ln(v_p) + 2.3$ for higher particle velocities (above 600 m/s). The large difference between the coefficients of these equations suggests that there are two distinct mechanisms of energy dissipation that take place for particles in these velocity ranges. This behavior can also be detected when measuring the average particle temperature increase, ΔT_r. Its specific dependence on particle velocity can be seen in Figure 1.3, where these same low and high impact velocity groups can be observed.

The logarithmic approximations for the temperatures in these velocity groups are $\Delta T_r = 67.2 \ln(v_p) - 361.4$ K for low impact velocities, and $\Delta T_r = 376.6 \ln(v_p) - 2391.9$ K for high impact velocities. These trends appear to be related to a change in the particle impact mechanism from nearly elastic to plastic; this change occurs at the critical velocity of 600 m/s. A simulation of the change in contact forces by Li et al. (1999) is shown in Figure 1.4 for low particle velocities (normal velocity of 5.2 m/s for the molybdenum surface). From this simulation, it is abundantly clear that the Hertzian dissipation force is relatively small, implying nearly perfect elastic impact. It is further observed that the adhesion dissipation force is prevalent during the rebound phase, and is thus primarily responsible for the total energy lost during the impact.

Because of plastic deformation, energy losses result in a decrease in the velocities of rebound particles. In this regime, the deformation occurs too rapidly for effective heat conduction, resulting in thermal softening and adiabatic shearing, and ultimately reducing the bearing capacity of the material. The sharp nature of this transition is suggestive of a critical phenomenon and is typical of shear localization instabilities (Molinari and Ortiz, 2002).

The region of superdeep penetration (SDP) phenomena coincides with particle velocities ranging from 1.2×10^3 to 2.5×10^3 m/s. The main characteristic of SDP is the penetration of a small fraction of particles (~0.1%) to a specified depth, 10 d_p, within the substrate. This occurs for particle velocities exceeding 10^3 m/s, provided that the hardness of the particles is

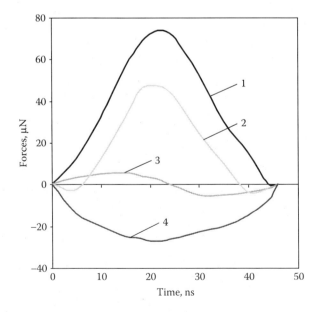

FIGURE 1.4 Contact force variation with contact time. 1—Hertzian force, 2—total force, 3—adhesion dissipation force, and 4—adhesion force. (After Li et al., 1999. With permission.)

greater than substrate (Kiselev and Kiselev, 2002). Finally, an area characterized by hypervelocity impact ($v_p > 2 \times 10^3$ m/s) lies beyond consideration (see Figure 1.2).

1.3.3 BONDING PHENOMENON

In the region where CS processes are realized, particles impact and bond with the surface for specifically defined particle velocities. Many authors (Papyrin, 2001; Alkhimov et al., 2001; Kreye and Stoltenhoff, 2000) consider cold spray phenomenon to be strictly governed by the interaction of particles with the substrate. However, there is a distinct lack of literature describing and evaluating collective particle behavior in CS. As discussed in the previous section, the interaction of incoming particles with the substrate can result in particles that rebound from (rather than adhere to) the surface. To fully understand the CS process, then, it is necessary to characterize the effect of preliminary impacts that do not result in material adhesion. Klinkov et al. (2005) suggests that these impacts lead to the activation of the surface. Succeeding particles impinging on an activated surface area are then more likely to bond to the substrate. Presently, however, proper characterization of this surface activation with respect to, for example, the nature and quantity of preliminary impacts is difficult at best. For this reason, its physical mechanism is not fully understood.

As described earlier, in CS high particle velocities are obtained through accelerating an expanding gas stream to velocities in the range of 500–1200 m/s by means of a converging diverging de Laval-type nozzle. In this process, the gas is heated only to increase the particle velocity and to increase particle deformation upon impact. Thus, powder layers form as a result of the kinetic energy of the particles as they hit the substrate; the bonding of particles in CS is a result of extensive plastic deformation and related phenomena at the interface. In general, for the particle to be bonded, its kinetic energy must be transformed into heat and strain energy within both the particle and substrate. For this reason, inelastic collision processes that involve the plastic deformation of the particle and substrate are the main phenomenon causing particle consolidation. Such plastic deformation can be roughly approximated by the flattening of spherical incident particles into a plate-like structure having aspect ratios between 3:1 and 5:1 (Van Steenkiste et al., 2002). In order for sufficient plastic deformation to take place, the contact stress at the interface must exceed the yield stress of the particle. Thus, the powders that are typically employed in most CS processes consist of metals with relatively high ductility. Hence, the fundamental theories of plasticity provide the scientific basis for particle impact analysis. Moreover, the localization of deformation plays a very important role in this phenomenon, and will be appropriately pursued in the next chapters.

The bonding mechanisms in CS can be compared to those in processes such as explosive welding and powder compaction. In EC, an area of the interface undergoes a severe deformation that is characterized by adiabatic shear bands, highly elongated grains, recrystallized grains, ruptured oxide films, and even resolidified microstructures (Hammerschmidt and Kreye, 1981). In explosive powder compaction, dense powder materials can be produced through specifically chosen combinations of shock, pressure, and duration (Grujicic et al., 2004). As with EC, successful bonding in powder compaction is dependent on the critical conditions for extensive plastic deformation at either the particle–particle or particle–substrate interfaces (Nesterenko, 1995). However, Assadi et al. (2003) point out that despite the similarities that exist among these processes, it has not yet become clear as to what extent the theories describing EC or shock wave powder compaction can be applied to CS.

In light of the above review, these two approaches must then be considered together, with their principles coupled together to achieve a deeper understanding the CS process. In particular, to develop an adequate theory describing low-pressure CS (LPCS) one must consider (i) one-particle impact, (ii) multiparticle impact, and (iii) powder ensemble behavior during the impact.

1.4   APPLICATION PROSPECTS OF CS METHOD

As shown in Section 1.1, the advanced thermal spray technologies such as HVOF, wire arc or flame spraying, plasma spraying, and detonation spraying are widely used in the automotive industry, among many others. As an example, according to Barbezat (2006), a large portion of engine and transmission components are coated using various processes. Although some of these applications have already reached maturity, many others are relatively new, representing an important market for the future of spraying technologies and, in particular, CS. The range of materials that may be coated, as well as the quality of CS deposits, depends to a large extent on the range of sufficient particle velocities that may be attained, as well as the mechanical characteristics of both the particles and substrate. From this viewpoint, it is beneficial to compare the capabilities of high-pressure CS (HPCS) and LPCS processes in the context of their industrial applications.

As discussed by Villafuerte (2005), the application of helium as a processing gas for high pressure gas dynamic spray provides the highest possible supersonic flow velocities. This technology, then, is capable of successfully depositing the widest range of materials in an oxygen-free environment. Unfortunately, helium is both relatively expensive and requires the use of special gas-handling and recovery/recycling equipment. Nitrogen, on the other hand, is considerably less expensive and equally capable of providing an oxygen-free environment. The supersonic velocities obtained with nitrogen, however, are lower than with helium, limiting the range of coatings that can be produced. Air in fact offers the least expensive alternative but has the narrowest range of known applications, being generally limited to those processes that are not extremely sensitive to the presence of oxygen at low temperatures.

In the case of LPCS, the particle velocities are measured to be in the range of 400–600 m/s that are only sufficient for the successful bonding of soft metals such as Zn, Al, and Sn. It will be shown in Chapters 2 and 4 that localized plastic deformation at the particle–substrate interface is necessary for the kinetic energy transformation. For this reason, successful powders and substrates for cold spray are mostly metals with relatively high plasticity. Nevertheless, as discussed in Chapters 2 and 3, there exist clear-cut distinctions between the behavior of single and collective particle groups when impacting a surface. It is for this reason that it is indeed possible to successfully spray powder mixtures consisting

TABLE 1.1
The Development of CS Applications

Desired Properties	Applications	Examples	CS Technology Comments
Mechanical properties	Metal restoration and sealing	Engine blocks, castings, molds and dies, weld seams, autobody repair, HVAC equipment, refrigeration equipment, heat exchangers, and cryogenic equipment	1. The application of soft metals (Al, Zn, Sn) that is preferable with LPCS 2. Need of portable mobile LPCS equipment
	Wear-resistant, low-friction coatings	Die components, valves, and piston rod guides with a low-friction coefficient	1. The application of wear resistant compositions (WC-Co, NiCr alloys) is possible with HPCS 2. Some components with hardness about 200HB may be deposited by LPCS
	Friction coatings	Increase friction of rolls for papermaking	Composite friction coating may be made by LPCS
	Antistick properties	Deposits impregnated with release agents such as polytetrafluoroethylene or silicone	The soft metal based coatings with high porosity are being made by LPCS

(Continued)

TABLE 1.1 (*Continued*)
The Development of CS Applications

Desired Properties	Applications	Examples	CS Technology Comments
Thermal and electrical properties	Thermal barriers	Aluminum piston heads, manifolds, disc brakes, and aircraft engine components	Al-Ti based coatings and composite coating are being made by HPCS and LPCS
	Thermal dissipation	Copper or aluminum coatings on heat sinks for microelectronics	The Cu and Al based coating is being made by HPCS with a high deposition efficiency
	Soldering priming	Microelectronics components and printed circuit boards	The soldering Ag and Sn composite coating is being made by HPCS and LPCS
	Electrically conductive coatings	Copper or aluminum patches on metal and ceramic or polymeric components (McCune, 2003)	The Cu and Al based coating is being made by HPCS with a high deposition efficiency
	Dielectric coatings	Ceramic coatings for aerospace, automotive, and electronic packaging	The ceramic-based composite coating is being made by HPCS
Corrosion properties and biocompatibility	Localized corrosion protection	Zinc or aluminum deposits on affected helms, weldments, or other joints in which the original protective layer on the base material has been affected by the manufacturing process	1. The application of soft metals (Al, Zn, Sn) that is preferable with LPCS 2. Need of portable mobile LPGDS equipment
	Biomedical	Biocompatible/bioactive materials on orthopedic implants, prostheses and dental implants. Porous coatings of these materials on load-bearing implant devices facilitate implant fixation and bone in growth, replacing the need for cements and screws.	GDS of hydroxyapatite-based coating allows one to retain the nanostructure of hydroxyapatite as compared with HVOF or plasma straying
	Decorative coatings	Metal and ceramic components	The portable mobile LPCS equipment is preferable
Technology modification	Rapid prototyping and near net manufacturing	Cold spraying can produce a well-defined footprint. Fabrication of parts with custom metal matrix composite structure or gradient structures.	CS allows us to create the more flexible and cheaper technology for thin wall components

of various metals and ceramics onto both metallic and ceramic substrates alike. In fact, the presence of a metal–ceramic powder mixture allows sufficient particle deformation to take place at particle velocities that are significantly lower than that required for the bonding of pure metals. Therefore, the range of materials that can be successfully sprayed by LPCS may be expanded to include not only pure metals such as Al, Zn, Cu, Fe, Ni, and Ti, but also composites such as $Al-Al_2O_3$, $Al-B_4C$, $Al-Zn-Al_2O_3$, Cu-Pb-Sn Cu-W, Cu-Zn-TiC, and Ni-SiC.

To emphasize the benefits of CS, some specific applications are described in Chapter 6. To facilitate this task, it is perhaps useful to classify these CS technologies into four groups according to their mechanical, thermal and electrical, corrosion and biocompatibility, and technological parameters. Table 1.1 summarizes known and possible applications for cold spray technology in accordance with this classification.

2 Theoretical Description of Cold Spray Process

Volf Leshchynsky

CONTENTS

2.1 PARTICLE ACCELERATION

2.1.1 GENERAL RELATIONSHIPS

Widespread adoption of a cold spray (CS) technology requires the application of a variety of coating materials, each one specifically suited to particular applications. For this reason, to determine the dependence of the microstructure of spray coatings on the operating conditions of the gas dynamic spray system is of great practical interest. To obtain high-quality coatings, the spray technology parameters must be selected carefully, and due to the large variety of process parameters, much trial and error goes into optimizing the process for each specific coating and substrate combination. Consequently, a complete model of the CS coating process should be developed. This model should include the following three distinct subprocesses:

1. Spraying (controlled by parameters such as particle size, temperature, velocity, and morphology of powder laden jet)
2. Particle impact, deformation, and bonding
3. Coating microstructure formation, including shock compaction, and adiabatic shear bands formation

In the CS process, particles, in metallic or nonmetallic form, are fed into a carrier gas where they are heated, accelerated, and propelled at a high velocity onto the surface to be coated. Particles land on the solid surface where they deform and agglomerate to create a thin layer. It is well known that the introduction of new coating materials is both time consuming and costly. To date, CS parameters are being optimized merely on the basis of trial and error, from which empirical relationships are derived. Because a very large number of parameters are involved (e.g., gas type and temperature; particle size and velocity; substrate material and temperature), this approach is laborious and expensive, and must be repeated for each material being sprayed. The development of theoretical or computational models that can predict the microstructure of coatings could greatly reduce the cost of this development. Ideally, these models will allow operators to customize coating properties to meet the requirements of particular applications without having to do extensive experimentation. Indeed, these models will also lead to improved designs of CS systems that optimize the technological parameters of CS processes. Indeed, the modeling of low-pressure CS process will promote the widespread adoption of a cold gas dynamic spray technology in expanded industrial environments.

As suggested earlier, a complete description of CS includes modeling three distinct processes: particle acceleration (the main variables being particle size, density, gas type, pressure and temperature), particle impact deformation and bonding (the main variables being particle velocity, density, temperature, particles distribution in the gas-powder jet, and the stress–strain state at an interface); and the powder coating microstructure formation including adiabatic shear banding, particle consolidation, and mechanical alloying (the main variables being shock compaction parameters of different powder mixtures, characteristics of a heat generation due to impact, diffusion constants, etc.). Specifically, this model consists of several complementary parts and should include submodels that simulate: (a) gas flow and particle acceleration; (b) impact, deformation, and bonding of particles on the substrate; and (c) shock compaction (densification) and consolidation of powder layers.

2.1.1.1 The Governing Equations of Single-Phase Turbulent Flow

There are many industrial processes similar to gas dynamic spraying that involve jets imping-ing on a solid surface. Such examples include explosive welding, jet mixing systems, water jet cutting, and shrouding systems. Many of these are discussed extensively in books by Schlichting (1979), Abramovich (1963), and Hinze (1975). For the specific problem of turbulent free jets and compound jets consisting of more than one flow stream, the flow structure is well understood, with close agreement being demonstrated between theory and practice by Rajaratnam (1976). However, present literature does not include much work on compound jets containing flowing powder.

One feature of turbulent jet flow is that momentum, heat, and mass are transferred at rates much greater than those of laminar flow (where molecular transport processes take place via viscosity and diffusion). The conservation equations used for turbulent flows are obtained from those of laminar flows using a time-averaging procedure commonly known as Reynolds averaging (Rajaratnam, 1976). When the flow is turbulent, the characteristic variables of pressure, velocity, and temperature may vary with both space and time.

The mathematical model used in simulation is based on the Navier–Stokes system of differen-tial equations with the Reynolds method of averaging the time-dependent equations, together with the standard k–ε turbulence model. The time-averaged, governing equations for turbulent flow are expressed as follows (Rajaratnam, 1976).

- Conservation of mass:

$$\frac{\partial}{\partial x_j}\left(\rho_g v_j\right)=0 \tag{2.1}$$

where ρ_g is the density of the gas and v_j is the velocity vector in the jth direction.

- Conservation of momentum:

$$\frac{\partial}{\partial x_j}\left(\rho_g v_i v_j\right)=-\frac{\partial p}{\partial x_i}+\frac{\partial \tau_{ij}}{\partial x_j}+\rho_g g_i \tag{2.2}$$

where τ_{ij} is the stress and g_i is the gravitational acceleration.

The stress is given by

$$\tau_{ij}=\left[\left(\mu+\mu_t\right)\left(\frac{\partial v_i}{\partial x_j}+\frac{\partial v_j}{\partial x_i}\right)\right]-\frac{2}{3}\mu_t\frac{\partial v_i}{\partial x_i}\delta_{ij} \tag{2.3}$$

where μ is the molecular viscosity and $\delta_{ij}=1$ for $i=j$; otherwise, $\delta_{ij}=0$. Also, μ_t is the turbulent viscosity given by

$$\mu_t=\rho_g C_\mu \frac{k_t^2}{\varepsilon_t} \tag{2.4}$$

where $C_\mu=0.09$ is a constant, k_t is the kinetic energy of turbulence, and ε_t is the dissipa-tion of kinetic energy of turbulence, which will be defined in the k–ε turbulence model to follow.

2.1.1.2 The k–ε Model for Turbulent Flows

The modeling of turbulent flows requires appropriate methods to describe the effects of the unstable fluctuation of the velocity and scalar quantities within the basic conservation equations. Because numerical simulations of the turbulence have been thoroughly studied, the scope of this description is not concerned with model development, but focuses only on adopting a reliable turbulence model for jet flow problems.

The standard k–ε model of turbulence is generally used to close the system of conservation equations described in (2.1)–(2.4). Two additional conservation conditions include the kinetic energy of turbulence k_t and its dissipation ε_t. In accordance with FLUENT[*], these are

$$\frac{\partial}{\partial x_i}\left(\rho_g v_i k\right) = \frac{\partial}{\partial x_i}\left(\frac{\mu_t}{\sigma_k}\frac{\partial k}{\partial x_i}\right) + G_k + G_b - \rho_g \varepsilon_t \tag{2.5}$$

$$\text{and } \frac{\partial}{\partial x_i}\left(\rho_g v_i \varepsilon_t\right) = \frac{\partial}{\partial x_i}\left(\frac{\mu_t}{\sigma_\varepsilon}\frac{\partial \varepsilon_t}{\partial x_i}\right) + C_{1\varepsilon}\frac{\varepsilon_t}{k_t}\left(G_k + G_b\right) - C_{2\varepsilon}\rho_g\frac{\varepsilon_t^2}{k_t} \tag{2.6}$$

Here,

$$G_k = \mu_t \left(\frac{\partial v_j}{\partial x_i} + \frac{\partial v_i}{\partial x_j}\right)\frac{\partial v_i}{\partial x_j} \tag{2.7}$$

$$G_b = -g_i \frac{\eta_t}{\rho_g \sigma_h}\frac{\partial \rho_g}{\partial x_i} \tag{2.8}$$

with $C_{1\varepsilon}$, $C_{2\varepsilon}$, σ_k, and σ_ε being the empirical constants, σ_h the turbulent Prandtl number, $\mu_t C_p/k_t$, G_k the rate of production of kinetic energy of turbulence, and G_b the generation of turbulence due to buoyancy (the FLUENT users guide).

2.1.1.3 Particle Dynamics in Gas Flow

The effective characterization of the CS particle-laden jet provides a means for determining the drag force per unit particle mass, drag coefficients, and the effect of friction on the particle velocity for various particle-to-gas mass flow ratios. Naturally, these parameters must be defined with respect to particle flow, as the behavior of particles suspended in turbulent flow depends on the properties of both the particles and the flow. Turbulent dispersion of both the particles and the carrier fluid can be handled via energy analysis for definite particle size distributions. The momentum transfer during the interaction between the gas and particle phases has been thoroughly investigated and many criteria have been isolated. However, due to the inherent computational complexity that exists in particle and gas flow analysis, Streeter (1961) suggests that the uncertainty of current prediction models is still large. In fact, to define the properties of a gas–particle mixture, the volume must be large enough to contain sufficient particles to achieve a stationary average. For this reason, the particles cannot be treated as a continuum in a flow system of comparable dimensions.

A two-phase flow problem involving a dispersed second phase may be solved through the addition of a second transport equation. The trajectory of a dispersed phase particle is solved by integrating the force balance on the particle in a Lagrangian reference frame (following the particle coordinate). Usually, investigators treat a solid phase either as a discrete system (Triesch and Bohnet, 2001) or as a continuum (Gidaspow, 1994). For computing dilute flows, researchers (e.g., Oesterle and Petitjean, 1993; Triesch and Bohnet, 2001) often use a discrete method where separate particle trajectories are calculated through a Lagrangian approach, where as particle–particle collisions are treated as a random process using the Monte Carlo algorithm. It is assumed that a

---

[*] FLUENT (1996) is a software package that models turbulent gas-particle flow. It utilizes a finite difference numerical algorithm based on a specified volume control approach.

particle moves in a cloud formed by other particles. The motion through this cloud is accompanied by collisions with particles forming the cloud. The continuum approach of Gidaspow (1994) and Louge et al. (1991) is based on kinetic theory. According to this theory, particles are considered to be chaotically moving gas molecules, and this granular gas is characterized by its granular pressure, temperature, conductivity, and the flow viscosity. Here, the flow viscosity (which is calculated using the theory of inhomogeneous gases) accounts for the stresses caused by the chaotic motion of the particles.

The force balance equation equates the particle inertia with the force acting on the particle, and can be written as

$$\frac{dv_{p,i}}{dr} = f_D(v_{\infty,i} - v_{p,i}) + \frac{g_i(\rho_p - \rho_\infty)}{\rho_p} + F_i \tag{2.9}$$

where $f_D(v_{\infty,i} - v_{p,i})$ is the drag force per unit particle mass and

$$f_D = \frac{18\mu}{\rho_p d_p^2} \frac{C_D \, Re}{24} \tag{2.10}$$

where Re stands for the relative Reynolds number, which is defined as

$$Re = \frac{\rho_\infty d_p \left| v_{p,i} - v_{\infty,i} \right|}{\mu} \tag{2.11}$$

The drag coefficient, C_D, is a function of the relative Reynolds number, and is of the following general form:

$$C_D = \alpha_1 + \frac{\alpha_2}{Re} + \frac{\alpha_3}{Re^2} \tag{2.12}$$

where α_1, α_2, and α_3 are constants that apply over several ranges of Re. Integration with (2.9) yields the velocity of the particle at each point along the trajectory, with the trajectory itself predicted by

$$\frac{dx_i}{dr} = v_{p,i} \tag{2.13}$$

To adequately predict the trajectories of the dispersed phase, the basic force balance (2.9) must be solved in each coordinate direction. However, this equation incorporates additional forces F_i in the particle balance that can be important under special circumstances such as interparticle friction and friction between the particles and the walls. Thus, to solve the system, it is necessary to incorporate the friction force equations.

Because the friction of the particles is determined by the shear stress at the wall, gas–particle flow may be treated as a continuous flow and may be described in terms of fluid dynamics.

Experimental studies of a gas flow containing monodisperse particles in a round tube by Tsuji et al. (1984) revealed a considerable effect of friction on the particle velocity for various particle-to-gas mass flow ratios. It is therefore of great importance to account for the effects of friction at the gas–particle jet both inside and outside the nozzle.

To describe the supersonic gas-powder flow behavior in a de Laval nozzle in CS processes, the fundamental results of computational fluid dynamics (CFD) were used. Further, the analyses of gas and particle dynamics were undertaken using the principles of kinetic theory that were put forth by Gidaspow (1987) and Louge et al. (1991) and further developed by Eskin et al. (2003).

2.1.2 Gas Flow and Particle Acceleration

A single stage of the gas flow and particle acceleration may be divided into the following three periods:

1. Gas and powder flow and mixing
2. Movement and acceleration of particles in the divergent part of the nozzle
3. Movement of particles in free jet area (including rebound from the substrate)

Each of these processes has its own features and may be characterized by several fundamental parameters. The analysis of these characteristics through physical and numerical modeling allows for the optimization of these values. As shown in Figure 2.1, the gas and particle velocities change at the each period of acceleration. One can also see that there is a significant difference between gas and particle velocities due to the variation of the drag force.

As shown in Chapter 1, the volume solid content of the gas–particle jet in a high-pressure CS process is in the range of 10^{-5}–10^{-6}, while the volume content in low-pressure CS system is in the range of 10^{-4}–10^{-5}. For this reason, parameters of gas-powder flow vary considerably and are strongly dependent on the particular CS parameters. For example, in low-pressure systems, interparticle friction plays an increasingly significant role. However, the lack of modeling and experimental data does not permit the exact evaluation of such specific parameter dependences.

Researchers have historically calculated nozzle accelerations without taking into account particulate friction (Rudinger, 1980). This can in fact be justified if the solid content is low and/or the ratio of the nozzle's length to its diameter is relatively small. Acceleration calculations for the case of high-volume solid content were made by Eskin and Voropaev (2004). Here, the authors describe in detail how two frictional models may be applied to acceleration nozzles. In the first method, the friction factors for solids are employed. Using a method often employed in engineering calculations of pneumatic conveying, a fluid-like behavior of the particulate phase is assumed. Unfortunately, the application of this method leads to discrepancies between calculated and measured pressure gradients that have been shown to reach and even exceed 40%. In the second model, the frictional forces between solids is taken into account by assuming that the particles acquire radial velocity components due to noncentral collisions with particles of various sizes (Eskin et al., 1999, 2001, 2003). This radial

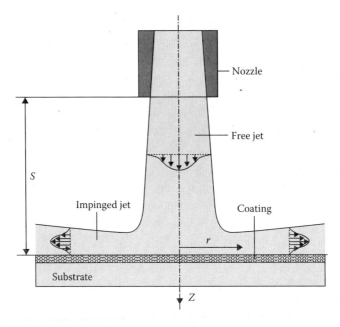

FIGURE 2.1 Particle-laden jet scheme.

motion of particles leads to particle–wall collisions that give rise to partial losses in the particle axial momentum components. This momentum loss is considered to account for particulate friction. The major disadvantage of this model is that it does not take into account collisions between particles of equal size. However, when this method is being applied to CS processes using powders consisting of varying particle size and content, this disadvantage is assumed to be minimal.

Eskin and Voropaev (2001) provide a detailed analysis of flow acceleration within a jet mill nozzle. Perhaps the most interesting finding of this investigation is that the ratio between the energy transferred from the gas to the particles and that which acts to accelerate the mixture is dependent on both particle size and solids loading. In fact, the authors demonstrate through thermodynamic analysis that this ratio very rapidly increases with a decrease in particle size, and significantly increases with an increase in solids loading. The most important results of particle acceleration processes modeling in de Laval nozzles will be discussed next from this point of view.

2.1.2.1   CFD Basics

Numerical simulations are often very helpful in analyzing the CS process. As such, a short review of numerical simulation results is presented by Jen et al. (2005). Numerous CS characteristics such as nozzle geometry, processing parameters, and spray particle parameters have been examined using the CFD code (Champagne, 2007). Through this analysis, it has been found that the primary processing parameters that influence the particle velocity are the carrier gas type, operating pressure, and temperature. As for nozzle geometry, the expansion ratio and divergent section length of the spray gun nozzle also have a significant effect on particle velocity. Moreover, the density, size, and morphology of powder play a notable role. The effects of these main parameters are sufficiently summarized by an equation that may be obtained through the nonlinear regression of the simulated results for particle velocity (Jen et al., 2005). Li et al. (2005) have conducted numerical simulations for two-phase gas–solid flows. Here, the airflow field is obtained by solving three-dimensional Navier–Stokes equations with a standard k–ε turbulence model and nonequilibrium wall function. The second phase, the coating powder, is thought to consist of spherical particles that are dispersed within the continuous phase (the air). In addition to solving the transport equations for the air, the authors calculate the trajectories of the particles by solving the particle motion equations using a Lagrangian method.

As shown previously, a high-pressure CS process is characterized by low particle–gas flow ratios. It is on this basis that several CFD analyses of gas and particle dynamics in CS processes have been performed by McCune et al. (1996), Dykhuizen and Smith (1998), Dykhuizen and Neiser (2003), and Sakaki and Shimizi (2001). The gas flow model was first introduced by Dykhuizen and Smith (1998). Dykhuizen et al. (1989, 1999, 2003) modeled the internal and external flow of a supersonic nozzle in the CS process. Over the years, to optimize system and nozzle design, many studies have focused on the behavior of flying particles. In this case, it is assumed that the gas flow conditions can be calculated without considering feedback from the powder flux.

The effect of Mach number (speed of sound) on particle velocity has been investigated with respect to an isentropic gas flow model. In this work, Dykhuizen and Neiser (2003) assume that the gas flow is isentropic (adiabatic and frictionless) and one dimensional. Thus, heat and friction losses are not considered. Under these simplified conditions, the changes of state are functions of the local Mach number and the isentropic coefficient or specific heat-capacity ratio γ (Stoltenhoff et al., 2002):

$$\frac{p_g}{p_0} = f(M,\gamma); \quad \frac{\rho}{\rho_0} = f(M,\gamma); \quad \frac{T}{T_0} = f(M,\gamma) \tag{2.14}$$

The subscript 0 represents the upstream stagnation condition of the gas.

In the isentropic gas flow model, sonic conditions for flow gas may only be obtained for sufficient stagnation gas pressures (Dykhuizen et al., 1999). After the sonic condition is obtained, increasing the pressure does not result in an effective increase in gas velocity upstream of the throat. In this model, the gas velocity v_g can be expressed as

$$v_g = M v_s = M \sqrt{\frac{\gamma R T}{M_w}} \tag{2.15}$$

where M is local Mach number, v_s is the speed of sound, and γ is the ratio of specific heats. For monatomic gases γ is 1.66, and for diatomic gases γ is typically 1.4. Also, R is the gas constant (J/kmol K), T is gas temperature, and M_w is the molecular weight of the gas. The local Mach number M only depends on the inner form of the nozzle, the calculation of which is presented in detail by Dykhuizen (1998, 1999, 2003). Typical CS geometry includes a converging–diverging nozzle, which is shown schematically in Figure 2.2.

In one-dimensional numerical models concerning particle acceleration, the particle velocity is calculated by defining the drag force on a single particle in fluid flow on the basis of the following equation:

$$m \frac{dv_p}{dt} = m v_p \frac{dv_p}{dx} = \frac{C_D A_p \rho_g (v_g - v_p)^2}{2} \tag{2.16}$$

It can be seen from (2.16) that the gas pressure does not affect the gas velocity directly. However, the drag coefficient C_D is a function of Re and the particle Mach number M_p, that is, $C_D = f(M_p, \mathrm{Re}_p)$. Re is in turn a function of the gas density ρ_g that itself depends on gas pressure

$$p_g = \rho_g \cdot R \cdot T$$

Alkhimov et al. (2001) obtained an empirical equation that includes the particle velocity, the gas velocity, and the gas pressure for nitrogen (which is often used as a processing gas):

$$v_p = \frac{v_g}{1 + 0.85 \sqrt{d_p / x} \sqrt{\rho_s v_g^2 / p_0}} \tag{2.17}$$

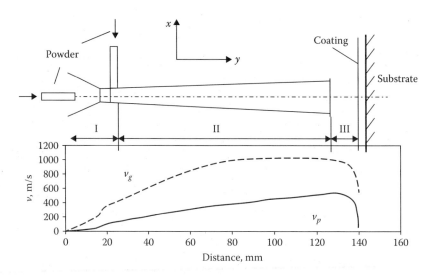

FIGURE 2.2 The gas and particle velocities change during each phase of acceleration.

Here, v_p is the particle velocity, p_0 is the nitrogen supply pressure measured at the entrance of the nozzle, ρ_s is the particle density, d_p is the particle diameter, and x is the axial position. If the gas velocity is replaced with (2.15), the correlation between the particle velocity, the gas temperature, and pressure can be rewritten as

$$v_p = \frac{1}{1/M\sqrt{(M_{N_2}/\gamma RT) + 0.85\sqrt{d_p/x\sqrt{\rho_s/p_0}}}} \tag{2.18}$$

where M_{N_2} is the molecular weight of the nitrogen gas.

The magnitude of particle acceleration and particle velocity can be obtained from the isentropic gas flow model and the one-dimensional numerical particle acceleration model. Unfortunately, some complex phenomena such as shock waves, flow separation, and the velocity distribution of in-flight particles cannot be adequately analyzed. Even more, there are little experimental data available about the velocity distribution of in-flight particles. In the paper of Wu et al. (2005), a professional velocity measuring device is used, and experimental velocity measurement results are presented. The strong particle velocity dependence on pressure and velocity distribution within the nozzle suggests that it is necessary to account for gas pressure and nozzle geometry along with gas type.

Nozzle geometry is of utmost importance in both cold gas dynamic spraying and high-velocity oxygen-fuel (HVOF) processes. For this reason, GDS research has in recent years been focused on nozzle modeling and design. For example, Sakaki and Shimizu (2001) carried out both numerical simulation and experiments to investigate the effect of the entrance geometry of the gun nozzle on the HVOF process. (The results of this study are described in Chapter 3. They may be easily applied to the CS method.) A similar research project concerning wedge-shaped supersonic nozzles was also conducted by Alkhimov et al. (2001). Here, the authors show that for a particular nozzle and particle type, it is possible to adapt the CS parameters in such a way as to maximize the particle velocity that may be achieved.

Dykhuizen et al. (1998, 1999, 2003) employed the CFD program FLUENT to simulate the use of nitrogen and helium to accelerate copper particles having diameters varying from 5 to 25 μm and inlet conditions of $P_0 = 2.5$ MPa and $T_0 = 673$ K. They found that particle velocity is a controlling factor that can determine the properties of bonding with the substrate. The carrier gas helium was found to be 2.5 times faster than nitrogen in the supersonic nozzle. Furthermore, smaller particles were shown to travel faster than larger particles, and an increase in either gas pressure or temperature contributed to an increased particle velocity. The acceleration of particles in de Laval nozzles at the air pressures in the range of 0.2–1.0 MPa is analyzed by Eskin et al. (1999, 2001, 2003) and Eskin and Voropaev (2004). These results may be used for analysis of the low-pressure CS process.

2.1.2.2 An Engineering Model with Particle Friction

In accordance with Eskin et al. (2003), the force acting on ith fraction of particles may be treated as the sum of three forces:

$$f_i = f_{Di}^{(g)} + f_i^{(c)} + f_i^{(fr)} \tag{2.19}$$

Here, $f_{Di}^{(g)}$ is the drag force, which is defined by the (2.10), and $f_i^{(c)}$ is the force arising from the collisions of particles having different sizes (particle rotation is neglected). Thus,

$$f_i^{(c)} = m_i \sum_{j=1}^{n} \frac{\langle \Delta v_{ij} \rangle}{t_{ij}} \tag{2.20}$$

where

$$\langle \Delta v_{ij} \rangle = \frac{m_j}{m_i + m_j} (v_j - v_i) |v_j - v_i| \frac{1 + k_n}{2} \tag{2.21}$$

is the change in the axial velocity v_i as a result of a collision of a particle within the ith fraction with one from of the jth fraction (averaged over all the possible directions of eccentric collisions). k_n accounts for the change in particle velocity due to the inelasticity of collisions, and

$$t_{ij} = \frac{4m_j}{\rho_g (d_i + d_j)^2 \rho_p \varepsilon_j |v_j - v_i|} \tag{2.22}$$

is the mean free time between interparticle collisions.

Having developed the mathematical basis for particle–particle friction, it is now beneficial to derive an equation with which to calculate the forces arising from particle-on-wall friction $f_i^{(fr)}$. These forces are believed to be related to the radial motion of particles that results from the eccentric collisions between particles in the ith and jth fractions. The radial velocity v_{rij} of a particle within the ith fraction after collision with a particle of the jth fraction is

$$v_{rij} = \frac{m_i}{m_i + m_j} \frac{(1 + k_n)}{4} |v_j - v_i| \tag{2.23}$$

where, according to Jenkins (1992), k_n may be used to represent the loss of kinetic energy due to the inelasticity of collisions, $\Delta E = 1 - k_n^2$. For most powders $k_n = 0.4$–0.5; that is, 75%–84% of the energy is lost because of inelastic collisions.

The equation for the radial projection of the drag force is

$$m_i \frac{v_{rij} - v_{rij}^{(f)}}{t_{ij}} = F_{ij}^{(r)} \tag{2.24}$$

where the radial projection of the drag force can be calculated by the equation

$$F_{ij}^{(r)} = \frac{1}{8} \rho_g C_{Di} \pi d_i^2 |v_g - v_i| v_{rij}^{(m)} \tag{2.25}$$

The mean radial velocity of a particle moving in a polydisperse medium can be calculated using the averaging relationship

$$v_{ri} = t_i \sum_{f=1}^{n} \frac{v_{rij}}{t_{ij}}, \quad t_i = \frac{1}{\sum_{j=1}^{n} (1/t_{ij})} \tag{2.26}$$

Here, $v_{rij}^{(m)}$ is the mean radial velocity of a particle of the ith fraction in the time interval t_{ij}. (Note that t_{ij} represents the time from the instant it collides with a particle of the jth fraction until it collides against the nozzle wall.) On the basis of the assumption that the radial velocity decreases linearly with time, the mean radial velocity of the particles in the time interval t_{ij} can be defined as $v_{rij}^{(m)} = 0.5(v_{rij} + v_{rij}^{(f)})$, where $v_{rij}^{(f)}$ is the radial velocity of a particle of the ith fraction that has collided with a particle of the jth fraction at the instant it is colliding against the wall. Finally, t_i is the mean free time of the particle in the polydisperse medium.

Based on these parameters, Eskin et al. (2003) obtained the following equation for the average radial velocity of the particles:

$$v_{ri}^{(m)} = \frac{v_{ri} + v_{ri}^{(f)}}{2} = v_{ri} \frac{1}{1 + \chi_i} \tag{2.27}$$

where

$$\chi_i = \frac{3}{8} C_{Di} \frac{\rho_g |v_g - v_i| t_i}{\rho_s d_i} \tag{2.28}$$

is the mean drag coefficient for the radial motion of the particles in the time internal t_i.

The force arising from particle-on-wall friction $f_i^{(fr)}$ may be determined by using the *colliding layer* estimation. Any particle of this layer is assumed to collide against the wall if its velocity vector (after a collision with another particle) is directed toward the wall. The thickness of this layer for particles of the ith fraction is

$$h_i = v_{ri}^{(m)} t_i \tag{2.29}$$

The friction force equation is given by

$$f_i^{(fr)} = -\frac{1}{2} \psi \frac{m_i v_i}{t_i} \left[1 - \left(\frac{2h_i}{D} - 1 \right)^2 \right] \tag{2.30}$$

where the coefficient ψ ($0 \leq \psi \leq 1$) takes into account that the velocity loss due to the wall collision:

$$\delta u_i = -\psi u_i \tag{2.31}$$

The model developed by Eskin and Voropaev (2003) enables one to both calculate the loss of energy due to friction for nozzles of various designs and estimate the scaling factor. Although this model is developed for rarefied flows, it is consistent with kinetic theory. However, in this case, mean radial particle velocities replace particulate medium temperatures as the cause of the chaotic particle motion. Indeed, it is clear from the analysis presented above that chaotic motion is caused by interparticle collisions, which are induced by the polydispersity of the flow. For this reason the application of polydispersive mixtures in GDS processes may contribute to an overall improvement in technology.

2.1.3 CALCULATED DATA AND DISCUSSION

2.1.3.1 Simulation of Gas-Particle Flow in the Nozzle

Consider for a moment the gas–particle flow parameter data that have been presented by various authors based on the models described above. A comparison of the deposition characteristics of high-pressure and low-pressure CS techniques is shown in Figure 2.3. Here, a considerable difference in the solid volume concentration of the gas-powder jet (10^{-4}–10^{-6}) can be observed for each CS system. The specific influence that particle content has on particle velocity in a powder laden jet, however, has not yet been estimated for real GDS conditions. On the other hand, some simulation results for heavily laden gas–particle jets have been obtained that take into account both particle collisions and wall friction effects. They are presented in a series of papers by Eskin and Voropaev (1999, 2001, 2003). In this case, the solid volume fraction of the gas–particle jet is in the range 10^{-3}–10^{-1}. The simulation results from Eskin and Voropaev (1999), along with the experimental data of Mebtoul et al. (1996), have been used to construct a plot of exit particle velocity as a function of solid volume concentration in the gas–particle jet (Figure 2.4a). One can see that the dependency $v_p = f(\varepsilon)$ fits the power approximation (Figure 2.4b) with a veracity of 0.98–0.99.

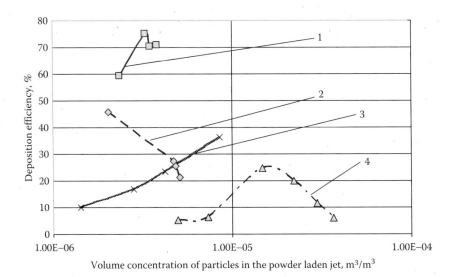

FIGURE 2.3 Deposition efficiency of high-pressure CS (1,2,3) and low-pressure CS (4) processes. 1: Copper, particle size 63–106 μm; 2: Copper, particle size <45 μm; 3: Aluminum, particle size 45 μm; 4: Al, air temperature 550°C, air pressure 7 bar, particle size 45 μm. (1,2: After Dykhuizen and Smith, 1998. With permission; 3: After Papyrin A., Cold spray technology. *Adv. Mater. Process.*, 159, 49–51, 2001. With permission.)

Some velocity measurements for particles of size 120 μm were made by Mebtoul et al. (1996) in an air jet having a stagnation pressure of 0.6 MPa. Here, the jet was formed within a de Laval nozzle with diameter of throat of 3 mm and divergence angle 1°. These experimental data clearly show that for a dilute jet ($\varepsilon < 3 \times 10^{-4}$) the particle velocity is independent of jet concentration at a given gas flow rate (Figure 2.4a). Above a solid volume fraction of 3×10^{-4}, however, particle–particle interactions can no longer be neglected and the velocity may be seen to decrease. It is obvious that this concentration threshold depends on particle size and, as shown in Figure 2.4b, shifts to lower concentrations for smaller particles. One can see that the extrapolation of power approximations up to lower concentrations leads to the reasonably accurate prediction of particle velocities. Thus, simulations of heavily laden gas–particle jets using an engineering model of particulate friction—as developed by Eskin and Voropaev (2004)—permits the computation of gas–particle jet velocity parameters for a wide range of particle concentrations.

Various one-dimensional simulations of gas–particle jets and the dependency of particle velocity on the normalized axial distance parameter (x/D_{throat}) have been undertaken over the years. A comparison of these works by various authors is shown in Figure 2.5. Analysis of these results show that the numerical simulations of Van Steenkiste (1999) and Sakaki and Shimizi (2001) yield higher values of particle velocity than the more detailed simulation procedure developed by Jen et al. (2005). The method presented by Jen et al. (2005) also allows the analysis of some more complex phenomena such as shock waves, flow separation, and the effects of wall friction. As a result, the axial particle velocity at the nozzle exit is affected by frictional forces within the divergent section of the nozzle, causing the velocity of the carrier gas to decrease considerably. The results of Eskin et al. (1999) are similar to Sakaki and Shimizi (2001) for axial particle exit velocities. This agreement suggests that it is indeed possible to approximate particle behavior within the gas-powder jet using an *engineering model*.

2.1.3.2 Influence of Gas Pressure

Previous experimental results have demonstrated the relationship between the variation of the gas inlet pressure and the velocity of the particles at the exit of the nozzle, mass flow rate of the gas, the solid/gas mass flow ratio, and other parameters of the gas-powder jet. As suggested by

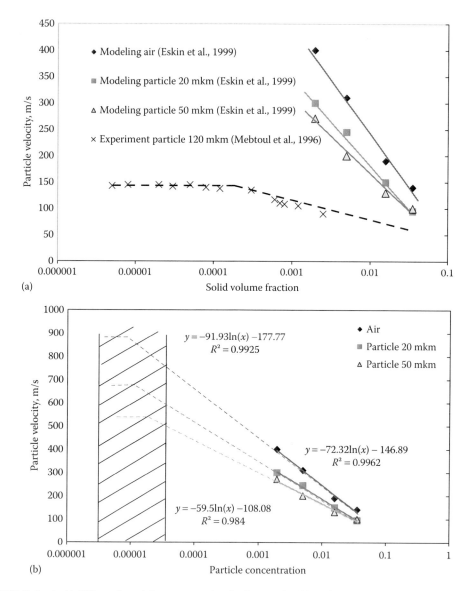

FIGURE 2.4 (a, b) Effect of particle concentration in the nozzle. (Data from Eskin D. et al., Simulation of jet milling. *Powder Technol.*, 105, 257–265, 1999; Mebtoul M. et al., High velocity impact of particles on a target—An experimental study. *J. Mineral Process.*, 44–45, 77–91, 1996.)

Jodoin et al. (2005), however, it is the nozzle mass flow rate that should be chosen as a global examination parameter because

1. It is not an input variable of the model but rather a result from it
2. It can be monitored precisely within the experimental setup
3. It contains all the relevant physics of the process

In fact, the mass flow rate may be predicted through one-dimensional isentropic analysis by means of the relationship,

$$\dot{m} = \frac{p_0}{\sqrt{T_0}} A^* \sqrt{\frac{\gamma}{R}} \left(1 + \frac{\gamma - 1}{2} \right)^{(\gamma - 1/2 - 2\gamma)} \tag{2.32}$$

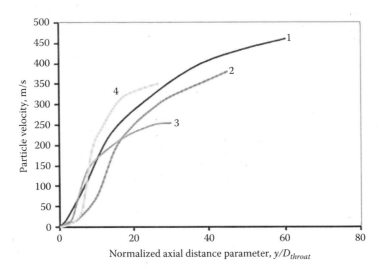

FIGURE 2.5 Numerical simulation results of the different authors: (1) Sakaki and Shimizi (2001); (2) Van Steenkiste et al. (1999); (3) Jen et al. (2005); (4) Eskin et al. (1999). In most of these studies cases, copper powder with particle size of 50 μm was used; Eskin et al. (1999) used cement powder with density 3 g/cm³; nitrogen (Sakaki and Shimizi 2001; Jen et al., 2005) and air (Van Steenkiste et al., 1999; Eskin et al., 1999) were used as carrier gases with inlet pressures of 2.0 and 0.6 Mpa and temperatures of 800, 750, and 350 K. The data of Eskin et al. (1999) were extrapolated to a solid volume concentration of about 1×10^{-5}. (Data from Sakaki and Shimizi, 2001; Van Steenkiste T.H. et al., Kinetic spray coatings. *Surf. Coat. Technol.*, 111, 62–71, 1999; Jen T.C. et al., Numerical investigations on cold gas dynamic spray process with nano- and microsize particles. *Int. J. Heat Mass Transfer*, 48, 4384–4396, 2005; Eskin D. et al., Simulation of jet milling. *Powder Technol.*, 105, 257–265, 1999. With permission.)

The experimental results of Jodoin et al. (2005) (grey and black stars) and curves calculated by (2.32) are shown in Figure 2.6 for nitrogen and helium at a temperature of 750 K. In both cases, due to the complex nature of gas-powder flow inside the nozzle (including friction and shock wave effects), (2.32) permits only an approximation of the mass flow rate values. In fact, the experimental data reveal that the mass flow rate dependence is particularly nonlinear for relatively low stagnation pressures (up to 1 MPa). This appears to stand in contradiction with (2.32) where the mass flow rate scales proportionally with the stagnation pressure. Certainly, this issue requires further study.

Results of modeling the particle velocity dependence on gas pressure are presented in Figure 2.7. Also shown in this figure are the experimental data for a wide range of gas pressures (for nitrogen), as generated by Sakaki and Shimizi (2001). There are two areas of gas pressure within which the axial particle velocity dependences differ considerably. Also, for relatively low gas pressures, the difference between gas and particle velocity is lower than for high gas pressures. This may be attributed to the drag coefficient dependence on the pressure, as was shown in previous analysis (see Equation 2.16). The particle velocity calculation of (2.17) (in accordance with Alkhimov et al., 2001) clearly shows that at a high gas pressure, there is only a small discrepancy between the modeling results of Sakaki and Shimizi (2001) and Alkhimov et al. (2001) (see Figure 2.7). Supersonic flow theory predicts that particle velocity should vary as the log of the stagnation pressure (Dykhuizen and Smith, 1998). Figure 2.7 shows the experimental data of Gilmore et al. (1999) that are well fit by a logarithmic curve for high gas pressures. However, the dependences of particle velocity on the gas pressure for relatively low pressures have not yet been studied in detail.

The main conclusion that may be drawn from the above discussion is that variations in gas pressure up to 1 MPa result in a great increase in particle velocity. Thus, there appears to be considerable opportunity in this region for particles to reach velocities that approach 500–550 m/s. Of course, the maximum velocity that may be attained is ultimately dependent on particle size and density.

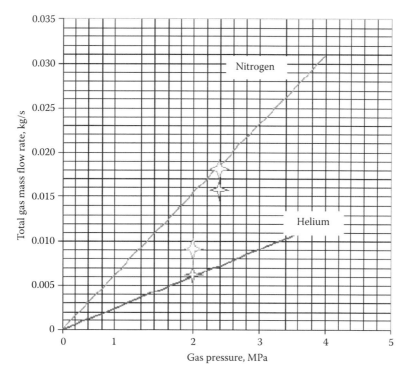

FIGURE 2.6 Pressure dependence of mass flow rate. (After Jodoin et al., Cold spray modeling and validation using an optical diagnostic method. *Surf. Coat. Technol.*, 2005. With permission.)

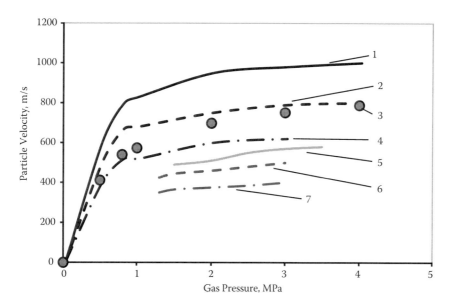

FIGURE 2.7 Modeling of gas pressure influence. (1) Nitrogen, 750K (after Sakaki and Shimizu 2001); (2) 10μm bronze particles, Nitrogen 750K (after Sakaki and Shimizu 2001); (3) 10μm bronze particles (Nitrogen, 750K) results of calculation with Equation (2.17); (4) 20μm bronze particles, Nitrogen 750K (after Sakaki and Shimizu 2001); (5) 15μm Cu particles, Nitrogen 593K (after Stoltenhof et al. 2002); (6) 22μm Cu particles, air 573K (after Gilmore et al. 1999); and (7) 22μm Cu particles, air 298K (after Gilmore et al. 1999. With permission.)

2.1.3.3 Effects of Particle Concentration

An effect of jet particle concentration on the process characteristics of high-pressure GDS has been studied by Gilmore et al. (1999) and Van Steenkiste (1999). The results of this modeling suggest that above the mass powder/gas flow ratio of 0.03%, the mean particle velocity decreases linearly with an increase in the powder feeding rate. Gilmore et al. (2005) believe that this decrease in particle velocity might be due to the increased mass that must be accelerated by the gas flow. Unfortunately, this effect cannot be examined in detail because the increase in powder feeding rate leads to an overburdening of the nozzle with powder (Van Steenkiste et al., 1999). Nevertheless, the particle concentration greatly influences particles interaction and, in turn, the acceleration process. Controlling the powder concentration becomes even more important when powder is radially injected directly into the divergent portion of nozzle (Figure 2.2a). In this case, heavily laden particle jets may be formed.

Some calculations regarding an Al powder-laden jet within radial injection CS (Figure 2.8) show that variations in the powder feeding rate strongly affect average interparticle distances. In fact, increasing the powder feeding rate M_s from 0.1 to 6 g/s results in interparticle distances that range from 2.12 to 0.035 mm. As a result, the role of interparticle and particle–wall collisions is enhanced by many times. Particle concentration also influences particle behavior during formation of the powder layer. Experimental results for the determination of the CS deposition rate for the case of radial injection low pressure are shown in Figure 2.9. Here, it can be seen that an increase in particle volume concentration leads to optimal results, increasing the buildup capacity (thickness per pass) by five or more times.

2.1.3.4 Effects of Nozzle Wall Friction

As described previously, particle accelerations and particle velocities are generally calculated on the basis of isentropic flow using one-dimensional numerical particle acceleration models. The effects of wall friction, however, have not yet been explored in detail. For this reason, there are little experimental data available regarding the velocity distribution of in-flight particles within the CS process (Wu et al., 2005). Nevertheless, this problem is being investigated deeply using theories of multiphase flow in both numerical and physical modeling. The effect of nozzle walls on a high-speed, two-phase gas–particle flow can be defined by interparticle and particle–wall collisions, changes in particle concentrations in the near wall region, axial particle velocities to compare dissimilar experimental conditions directly, these dependences must be characterized in normalized, nondimensional coordinates. To do this, x/b is considered to be the normalized channel height; the particle velocity v_p is normalized through division by the average gas velocity, that is, $v_p/v_{g\text{-average}}$. A distribution of

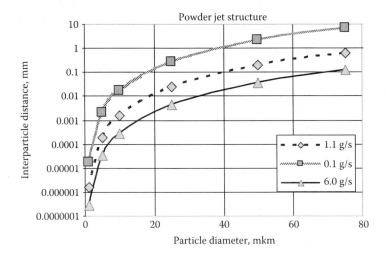

FIGURE 2.8 An interparticle distance in Al powder jet $D_{\text{throat}} = 3.5$ mm, air consumption $= 0.022$ m^3/s $= 1.3$ m^3/min.

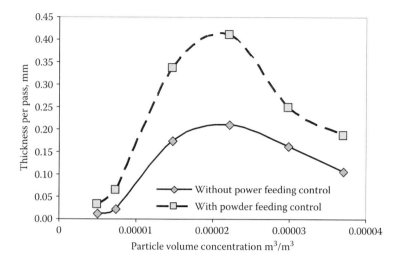

FIGURE 2.9 Buildup effectiveness of the radial injection low-pressure CS of Al–Zn–Al$_2$O$_3$ powder mixture. Air pressure—6 bar; air temperature—300°C; air flow rate—0.022 m^3/s; stand-off distance—10 mm; travel speed—100 mm/s.

this velocity parameter $v_p/v_{g\text{-average}}$ through the nozzle channel up exit from the nozzle (calculated via particle velocity measurements by Wu et al. (2005) is shown in Figure 2.10. Here, they are compared with the modeling results of Volkov et al. (2005). The experimental data of the air velocity distribution presented by Kussin and Sommerfeld (2002) are also shown in this figure for illustration purposes.

The profiles of the particle velocities in comparison with the single-phase air velocity show that an increase in particle size results in a decrease in the particle velocity gradient. It can be seen in Figure 2.10 that in areas near the nozzle wall, the particles travel at velocities that exceed the air flow due to transverse dispersion and the slip boundary conditions. The increase in particle size

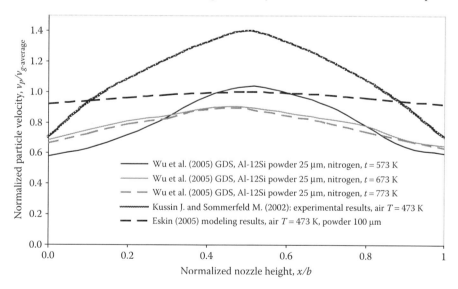

FIGURE 2.10 Effect of nozzle wall friction on the gas and particle velocity distribution: experimental (Kussin J. and Sommerfeld M., Experimental studies on particle behaviour and turbulence modification in horizontal channel flow with different wall roughness. *Exp. Fluids*, 33, 143–159, 2002.; Wu et al., Measurement of particle velocity and characterization of deposition in aluminium alloy kinetic spraying process. *Appl. Surf. Sci.*, 252, 1368–1377, 2005); and modeling (Eskin, D., Modeling dilute gas-particle flows in horizontal channels with different Li W-Y wall roughness. *Chem. Eng. Sci.*, 60, 655–663, 2005) results comparison.

diminishes the response of the particles to the gas flow, in turn enhancing transverse dispersion due to wall collisions. As a result, the streamline particle velocity profiles exhibit less dramatic variation. The experimental results of Wu et al. (2005) reveal that the effect of gas temperature on the profiles of the particle velocities is similar to those observed with the change in particle size (Figure 2.10). Perhaps this is because the increase in gas temperature causes a subsequent increase in the frequency of wall collisions. This, in turn, increases momentum losses due to particle–wall collisions and, therefore, the particle velocity distribution through the nozzle becomes more uniform. Hence, the particle size distribution, particle concentration, and gas-powder jet temperature are the main parameters that influence the velocity profile and the loss of kinetic energy of the particles.

The longitudinal distribution of the radial particle velocity parameter and the specific frequency of particle–wall collisions are shown in Figure 2.11 for a nozzle with $D_{throat} = 17$ mm (Eskin et al., 2003). As can be seen, an increase in particle size results in a decrease in the frequency of particle–wall collisions, as well as the radial particle velocity. It is perhaps obvious that the loss of kinetic energy should be considerably higher for the nozzles used in low-pressure CS with $D_{throat} = 2.5–3$ mm. This is because particles in these nozzles need to travel a shorter distance to collide against the nozzle wall. In fact, the thickness of this *colliding layer* h_i (Equation 2.27) is dependent on the particle velocity. The estimation of the *colliding layer* thickness h_i for a typical CS nozzle reveals a value of $h_i = 0.08–0.2$ mm. Thus, the nozzle wall friction during CS results in considerable losses in particle kinetic energy within the nozzle wall area. This diminishes the deposition efficiency of CS processes because energy losses in the nozzle wall area permit the particles from achieving the critical velocity threshold. A numerical simulation of gas–particle flow inside a straight cylindrical nozzle has been conducted by Jen et al. (2005). Results of this study indicate that the friction in this section results in a significant decrease in both gas and particle velocity, confirming the assumption presented above.

2.1.4 Free Jet Characterization

2.1.4.1 Shock Wave Features of the Jet

In the research discussed above, the analysis of gas–particle flow was focused on supersonic flow inside the nozzle. As mentioned, the most important parameters having influence on the dynamics of the jet gas are temperature, particle velocity, and the concentration of particles in the powder laden jet. In actuality, the kinematics of particle movement is strongly affected by particle flow

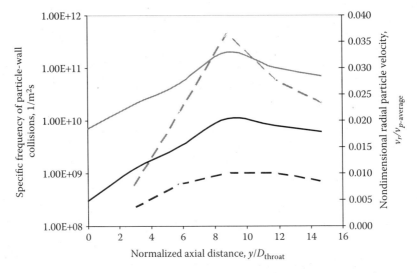

FIGURE 2.11   Axial distribution of the radial particle velocity parameter and specific frequency of particle-against-wall collisions. (After Eskin et al., Effect of particle size distribution on wall friction in high-velocity gas-dynamic apparatuses. *Theor. Found. Chem. Eng.*, 37, 122–130, 2003. With permission.)

outside the nozzle as well (area III, Figure 2.2). As such, the simulation of gas flow before and after exiting the nozzle has been completed by Jen et al. (2005). These datas reveal that compression waves occur both inside and outside the nozzle. Moreover, the multiple reflections of the shock waves result in velocity fluctuations. These shock fluctuations are formed due to the pressure difference between the gas jet and the ambient atmosphere at the nozzle exit.

As previously shown, the calculated values of h_i for a CS nozzle with $D_{throat} = 3.0$ mm are in the range of 0.08–0.2 mm. Thus, within the nozzle the ratio of h_i/D_i is found to be approximately 0.03–0.07. This suggests that it is the nozzle wall friction that leads to particle flow localization. However, the localization effect seems not to be visible when measuring particle velocity. For this reason, the particle velocity results shown in Figure 2.10 reveal smooth curves without sharp peaks.

Jen et al. (2005) have analyzed the principle features of the free jet structure. They state that there are three primary zones between the nozzle exit and the substrate: (1) the Mach disk—located close to the nozzle exit, (2) a second weaker Mach disc, and (3) a Bow shock formed on the substrate. One can see that the pressure in the area enclosed by the Bow shock and the substrate is much higher and vortex formation is observed. The existence of this Bow shock precludes small particles (<0.5 μm) from bonding to substrate.

2.1.4.2 An Engineering Model of the Free Jet

Numerical simulation is a powerful tool for studying CS processes both inside and outside of nozzle. There are, however, significant influences on the particle laden gas jet that may be characterized by simple relationships that are based on an engineering model of the free jet. This description of the nozzle free jet was first introduced by Kosarev et al. (2003); though relatively simple, it has proven quite useful in the calculation and analysis of real GDS processes. In particular, this method of analysis is based on the assumption that the profiles of velocity and dynamic pressure (ρv^2) are self-similar throughout the three regions of jet flow between the nozzle and the substrate.

To estimate the relationship between the velocity and temperature difference profiles, a normalized jet thickness parameter must be introduced: x/δ_M. Here, δ_M is a jet thickness along x-axis, which is defined as the distance from the jet axis to the point where $M^2(\delta_M) = 0.5 M^2_m$. The normalized velocity parameter is $\varphi_{M =} (M/M_m)^2$, where M_m is the axial Mach number. Using this parameter the data obtained by Kosarev et al. (2003) may fit by the function

$$\varphi_M = \exp[-(0.83x/\delta_M)^2] \tag{2.33}$$

A similar equation can be used to approximate the stagnation temperature (T_0) at each point of the free jet.

It is commonly accepted in jet theory that the profiles of the stagnation temperature difference ($\Delta T_0 = T_0 - T_a$) are self-similar to the velocity profile. This suggests the relationship:

$$\left(T_0 - T_a\right)/\left(T_{0m} - T_a\right) = \Delta T_0/\Delta T_{0m} = \left(v_g/v_m\right)^{\sigma} \tag{2.34}$$

where T_a is the ambient temperature, T_{0m} is the axial stagnation temperature, and σ is a ratio between the velocity profile thickness (δ_v) and the temperature profile thickness (δ_T):

$$\delta_v^2 = \sigma \delta_T^2 \tag{2.35}$$

More specifically, δ_T is the temperature profile thickness along the x-axis, which is defined as the distance from the jet axis to the point where $\Delta T_0 (\delta_T) = 0.5 \Delta T_{0m}$.

The experimental results shown in Figure 2.12 reveal the similarity of the Mach and stagnation temperature profiles, allowing the gas temperature and velocity distribution to be described by 2.12. The ratio σ must be specifically defined for individual nozzle geometries; the type (axial or radial) and point of particle injection into the nozzle also seem to influence σ.

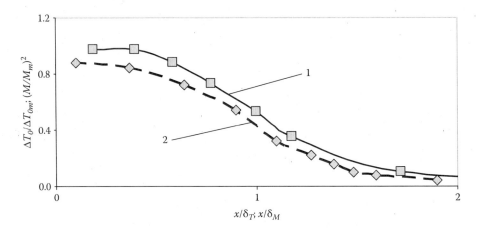

FIGURE 2.12 A distribution of the normalized stagnation temperature difference and normalized M^2 in free jet cross-section. 1—$\Delta T_0/\Delta T_{0m} = f(x/\delta_T)$; 2—$(M/M_m)^2 = f(x/\delta_M)$. (After Kosarev et al., On some aspects of gas dynamics of the cold spray process. *J. Therm. Spray Technol.*, 12(2), 265–281, 2003. With permission.)

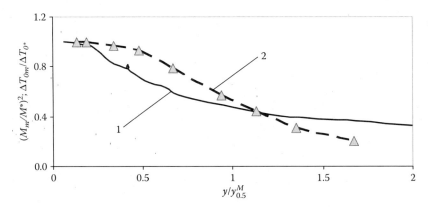

FIGURE 2.13 A distribution of the normalized stagnation temperature difference and normalized M^2 versus the longitudinal coordinate. 1: $1 - \Delta T_{0m}/\Delta T_{0*} = f(y/y_{0.5})$; 2: $(M_m/M^*)^2 = f(y/y_{0.5})$. (After Kosarev et al., On some aspects of gas dynamics of the cold spray process. *J. Therm. Spray Technol.*, 12(2), 265–281, 2003. With permission.)

A y-axis (longitudinal) distribution of axial jet characteristics M_m and ΔT_{0m} are shown in Figure 2.13 for the normalized parameters $(M_m/M_*)^2$ and $\Delta T_{0m}/\Delta T_{0*}$, where M_* and ΔT_{0*} are the parameters at the nozzle exit. The experimental data may fit to the curve

$$\left(\frac{M_m}{M_*}\right)^2 = \left[1 + 3\left(y/y_{0.5}^M\right)^4\right]^{-0.5} \tag{2.36}$$

where y is the longitudinal coordinate in the jet, and $y_{0.5}^M$ is the coordinate where $M_m^2(y_{0.5}^M) = 0.5M_*^2$. The approximation of the longitudinal distribution of ΔT_{0m} (Figure 2.13) is consistent with the equation

$$\Delta T_{0m} = \Delta T_{0*}\Big/\left[1 + \left(2^8 - 1\right)\bar{y}^4\right]^{1/8} \tag{2.37}$$

where

$$\bar{y} = y/y_{0.5}^T, \, y_{0.5}^T \approx 2y_{0.5}^M \tag{2.38}$$

Further, Kosarev et al. (2003) show that the relationship between the axial stagnation difference and the axial value of M^2 should be close to the form

$$\Delta T_{0m} / \Delta T_{0*} = (M_m^2 / M_*^2)^{0.25} \tag{2.39}$$

One of the fundamental problems of jet theory is the determination of the jet thickness, or for a symmetric jet, the jet diameter as a function of the longitudinal coordinate. As stated by Abramovich (1963), a linear increase in thickness is observed both in the initial and primary regions of the jet, though with different proportionality coefficients. For the case of a plane jet (i.e., one in which its expansion is ignored in the direction perpendicular to flow), Kosarev et al. (2003) approximates this thickness with a nonlinear function:

$$\delta_x = h \left[1 + 3 \left(y / y_{0.5}^M \right)^4 \right]^{0.5} \tag{2.40}$$

Experimental data, on the other hand, are more accurately fit with (Figure 2.14):

$$\delta_x = 0.75 h \left[1 + 3 (y / y_{0.5}^M)^4 \right]^{0.4} \tag{2.41}$$

Therefore, the semiempirical Equations 2.33–2.41 permit characterization of the gas flow parameters of the free jet. Consequently, this facilitates a means by which it is possible to define their influence on the structure formation process.

2.1.4.3 Particle Collisions

The effect of high-speed, two-phase gas–particle jet impingement on a solid body is of significant interest in many applications such as aircraft and turbine design and industrial processing. The presence of particles in the free stream—even if their concentration is very low—modifies the flow properties as compared to those of a pure gas (Volkov et al., 2005). The mechanics of aerosols, primarily developed by Fuchs (1997), suggest that for given body shape and flow parameters, there is a critical value of particle radius r_{p*} above which bonding will take place. However, despite exceeding this critical size, on impingement, particles may either bounce off or stick to the body surface. If the particle concentration in the free jet is sufficient, the reflected particles can collide with the incident ones, resulting in the formation of a near-wall layer in which particles move chaotically, colliding with each other. Theoretical estimations show that for coarse-grained particles (the radii of which are much more than r_{p*}) collisions between incident and reflected particles during flow over a blunt body can play a noticeable role, particularly if particle volume fractions are minimal (at $\alpha_{p\infty} \sim 10^{-6}$, Tsirkunov, 2001).

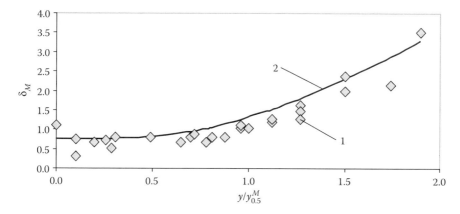

FIGURE 2.14 A dependence of jet thickness versus the longitudinal coordinate. 1: experimental results and 2: Equation 2.41. (After Kosarev et al., On some aspects of gas dynamics of the cold spray process. *J. Therm. Spray Technol.*, 12, 265–281, 2003. With permission.)

The importance of collision between particles in two-phase gas–particle jets has been demonstrated in studies of flow in pipes and channels, and also in the analysis of particle dispersion in homogeneous isotropic turbulence (Tanaka and Tsuji, 1991; Oesterle and Petitjean, 1993; Sommerfeld and Kussin, 2003). The effect of interparticle collisions on wall erosion and heat transfer by particles in an impinging particle-laden jet has been studied numerically with the use of the direct simulation Monte Carlo method by Kitron et al. (1981). Volkov et al. (2005) further clarified the role of interparticle collisions and two-way coupling effects in a dusty-gas flow over a blunt body, particularly inside the boundary layer. This paper provides detail descriptions of various models of gas and dispersed phase flows, including gas–particle interactions, particle–particle collisions, and particle–wall impact interactions. The fine flow structure of each phase is studied with respect to the free stream particle volume fraction $\alpha_{p\infty}$ and the particle radius r_p.

The computational flow model of Volkov et al. (2005) represents a combination of a CFD method for the carrier gas and a Monte Carlo method for the *gas* of particles. The numerical results illustrate many important features that are commonly observed in high-speed two-phase flows over blunt bodies. Of particular interest in this work is the calculated particle flow structure in the area of particle impingement. The particle phase flow patterns were found to be quite different for fine particles ($r < r_p$) and coarse-grained particles ($r > r_p$). Figure 2.15 shows the distribution of the volume fraction for particles that collide with a blunt body of radius R.

The role of interparticle collisions is shown to be negligible for fine particles (<1 μm); whereas, the effect of collisions on the behavior of larger particles is more significant as the particle content $\alpha_{p\infty}$ increases. This results in a sharp increase in the particle concentration near the surface of substrate. In fact, some particle concentration peaks are seen only for low particle concentrations (curve 1). The longitudinal distribution of $\alpha_p/\alpha_{p\infty}$ changes qualitatively when $\alpha_{p\infty}$ increases from 10^{-6} to 3×10^{-5}. Unfortunately, numerical simulation data are available only for particles of size <1 μm. Nonetheless, these results demonstrate the strong longitudinal dependence of the concentration profile at the near surface field.

2.1.4.4 Concluding Remarks

Chapter 2.1 provides analytical equations that can be used to estimate the gas dynamics of the CS process. It is shown how the spray particle velocity depends on particle size and density, gas stagnation pressure, total gas temperature, particle collisions, and other factors. Additionally, it is demonstrated that the particle velocity is the main parameter controlling CS process. The supersonic

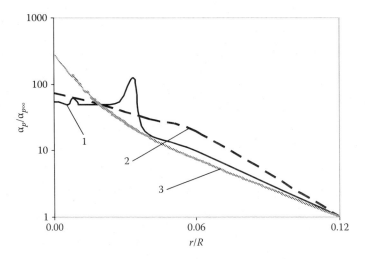

FIGURE 2.15 Distribution of a relative particle volume fraction $\acute{\alpha}_p = \alpha_p/\alpha_{p\infty}$ for particles with $r_p = 1$ μm. 1—$\alpha_{p\infty} = 3 \times 10^{-6}$, 2—$\alpha_{p\infty} = 10^{-5}$, 3—$\alpha_{p\infty} = 3 \times 10^{-5}$.

particle velocities are required for deposition of various metals. The preferable way is to raise the total gas temperature or reduce the gas molecular weight to increase the gas velocities.

2.2 PARTICLE IMPACT

2.2.1 PARTICLE CONSOLIDATION AT COLD SPRAYING

2.2.1.1 General Description

It is a well known fact that coatings in CS are formed due to significant plastic deformation in both the sprayed particles and the substrate, resulting from the impact of accelerated particles. In fact, these high impact pressures bring about direct contact between pure metals through the breakup and subsequent dispersion of a thin oxide film on both the particle and substrate surfaces, in turn leading to the creation of strong metallic bonds (Dykhuizen et al., 2003). On the other hand, the intensive deformation of the particle and substrate lead to increased strain and temperature at the localized contact areas. Several numerical simulation studies have indicated that the temperature at the contact zone, caused by the adiabatic shear during deposition, can reach values near the melting point of the sprayed materials, resulting in a thin melted layer (Alkhimov et al., 2000; Grujicic et al., 2003). Based on the metallic jetting observed with the impact of cold sprayed $TiAl_6V_4$ particles, Vlcek et al. (2003) suggested that this jetting may be attributed to melting at the impact interface. However, the occurrence of jetting may be a result of localized deformation under dynamic loading at the interface surface (Walter, 1992). It should be noted however, that strain localization may not be limited to the interface. Adiabatic shear bands may also occur in bulk particles (Walter, 1987) as well as in powder granular media (Hu and Molinari, 2004). The lack of strain localization analysis—particularly with respect to its influence on the structure of the formed coatings—limits our present understanding of the deposition mechanism in the CS process.

Because the common approach to CS coating formation analysis is related to the interaction between a single particle and the substrate, it is rather important to study adiabatic shear bonding processes in powder layers, taking into account the structure of the granular media, that is, particle ensembles. In fact, if the air-particle jet consists of particles of various sizes, the nature of the particle interaction and coating formation depends on the behavior of particle ensembles on impingement (Kochs et al., 1975). In the present approach, we attempt to combine the theories of fluid mechanics that are relevant to solid particles suspended in a supersonic gas stream with the theory of particle consolidation during the spraying process.

In current CS technology, it is of fundamental importance to create a very high speed gas that can accelerate the injected powders to velocities exceeding the critical velocity (about 500–800 m/s). This is normally achieved by expanding helium or nitrogen in a Laval nozzle. The pressure ratio for this expansion is in the order of 23. Because of its light weight, helium is a much easier gas with which to work. It is however very expensive, making the use of this process within an industrial environment economically impractical.

The critical gas velocity that must be achieved in CS is material dependent, being affected by particle size and density, elastic–plastic properties, and powder composition (Stoltenhoff et al., 2002; Grujicic et al., 2003). In fact, by using relatively large particles in CS processes using air, the minimum kinetic energy necessary for deformation and bonding with low particle critical velocities may be easily obtained (Air GDS processes were first referred to as Kinetic Spraying by Van Steenkiste et al., 1999). For critical velocities with Mach numbers 1.2–2, the air-pressure ratio is calculated to be in the range of 2.4–7.8. Thus, to achieve critical particle velocities using an air stream, it is necessary to use a CS system with minimal air pressure parameters of about 5–8 bars. Thus, the purpose of CS analysis is to bring about a more thorough understanding of the effect the primary parameters have on structure formation when using various powder materials. These parameters are believed to be gas temperature, particle velocity, particle concentration in the powder laden jet, and powder composition. The CS experimental results were analyzed with respect to the effects of air temperature T_g, powder jet composition, and stand-off-distance x_d. The air temperature influences the air and particle velocity, which are to be as high as possible. However, optimization of particle interaction and bonding during the spraying process may be achieved through the intensification of particle deformation and interparticle sliding. These processes are controlled through the composition of the powder impinging the substrate, particle sizes, and the incidence angle. In an effort to better understand the process, the following dependences are to be examined.

1. Deposition efficiency versus air temperature (Figure 2.16).
2. Average single pass thickness of deposited layer versus air temperature (Figure 2.17).

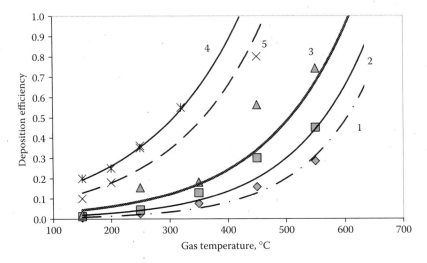

FIGURE 2.16 Deposition efficiency for powder mixtures: 1: Powder mixture with weight concentration $0.25Al_2O_3 + 0.5Al + 0.25Zn$; 2: Powder mixture with weight concentration $0.5Ti + 0.25Al + 0.25Zn$; 3: Powder mixture with weight concentration $0.5W + 0.25Al + 0.25Zn$; 4: Cu. (After Borchers et al., Microstructural and macroscopic properties of cold sprayed copper coatings. *J. Appl. Phys.*, 93, 10064–10070, 2003. With permission.); 5: Ti. (After Li and Li, Deposition characteristics of titanium coating in cold spraying. *Surf. Coat. Technol.* 167, 278–283, 2003. With permission.)

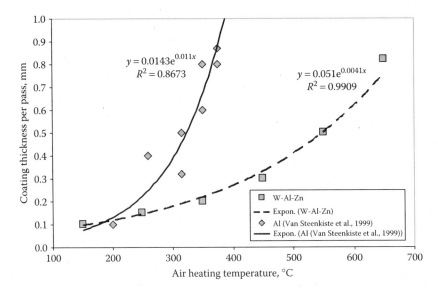

FIGURE 2.17 Average single pass thickness—air temperature dependence. Experimental data and exponent approximations for Al (Data from Van Steenkiste T.H. et al., Kinetic spray coatings. *Surf. Coat. Technol.*, 111, 62–71, 1999. With permission.) and Al-based powder mixture with weight concentration $0.5W + 0.25Al + 0.25Zn$.

2.2.1.2 Deposition Efficiency and Buildup Parameter

Experimental results demonstrate various features common to powder mixtures sprayed at low air pressures: the deposition efficiency χ_D of CS is found to be low for temperatures ranging from 100°C to 500°C. However, a sharp increase can be observed at temperatures higher than 600°C, becoming comparable with the values of χ_D for Ti and Cu (Stoltenhoff et al., 2002). It is important to note that nitrogen was applied for CS of Ti (Borchers et al., 2003; Li et al., 2003). It is known that the nitrogen and helium propellant gases allow higher gas and particle velocities to be achieved, as compared with that of air (Dykhuizen et al., 1998). In fact, particle velocities reach critical values at temperature of 300°C for nitrogen. The deposition efficiency of Ti is above 0.6 at this temperature (Figure 2.16) (Li et al., 2003). For air CS at a low pressure of 0.5 MPa, a χ_D value of ~0.6 is realized at air temperatures of 650°C. Thus, we can assume that the critical velocity threshold of 500 m/s (Dykhuizen et al., 1998) is achieved at this temperature.

The dependence of the deposition efficiency on temperature is accurately described by an exponential approximation for all powders and powder mixtures (Figure 2.16). Simulation veracity is in the range of $R^2 = 0.94$–0.99. The deposition efficiency equation is approximated by

$$\chi_D = \chi_{Do} e^{-Q_D/RT_0} \tag{2.42}$$

where χ_{Do} is an exponent coefficient of (2.42). Q is an activation constant of the CS process and is fixed for definite powder mixtures.

Approximations of the experimental data for studied powder mixtures are shown in Table 2.1 for three types of powder mixtures.

Analysis of the constants in (2.42) shows that the values of χ_{Do} do not vary for powder mixtures 1–3, while the values of Q_D are slightly different. This means that critical particle velocity, defined by the gas temperature, is the same for all powder mixtures studied because they consist of soft Al and Zn particles that are deformed during impact with substrate during coating formation. The values of the activation constants of (2.42) reveal the kinetic parameters of the powder buildup process, which depend on both the type and volume concentration of the larger and heavier particles. Thus, the deposition efficiency of the powder mixture 3 (see Table 2.1) is high due to the higher kinetic energy of the heavy tungsten particles, resulting in the conversion of plastic work into heat and stored energy within the formed CS coating.

TABLE 2.1
Parameters of the Exponential Approximations of Equations 2.42 and 2.43

Number	Powder Composition by Weight	Exponent Coefficient of Equation 2.42 χ_{Do}	Activation Constant of Equation 2.42 Q_D (kJ/mol)	Exponent Coefficient of Equation 2.43 t_o	Activation Constant of Equation 2.43 Q_t (kJ/mol)
1	25%Al$_2$O$_3$+50%Al+25%Zn	21.95	29.07	3.19	12.9
2	50%Ti+25%Al+25%Zn	20.35	26.05	4.97	12.92
3	50%W+25%Al+25%Zn	21.62	23.4	6.93	12.83
4	Copper (nitrogen)[a]	6.48	12.45	–	–
5	Titanium (nitrogen)[b]	16.5	37.5	–	–
6	Aluminum (Air)[c]	–	–	164.2	28.47

[a] Borchers, C. et al., Microstructural and macroscopic properties of cold sprayed copper coatings. *J. Appl. Phys.*, 93(2), 10064–10070. 2003.

[b] Li and Li. Deposition characteristics of titanium coating in cold spraying. *Surf. Coat. Technol.*, 167, 278–283. 2003.

[c] Van Steenkiste et al., Kinetic spray coatings. *Surf. Coat. Technol.*, 111, 62–71. 1999.

The average single pass thickness of a CS-coated layer or buildup parameter t is one of the main parameters characterizing the capability for CS buildup. Its dependence on air temperature is shown in Figure 2.17. The semiempirical relationship that gives a good approximation of the experimental data is in fact found to be similar to (2.42):

$$t = t_o e^{-Q_t/RT_0} \tag{2.43}$$

The parameters t and Q_t of Equation 2.43 are shown in Table 2.1. It can be seen from these data that the activation constant Q_t does not depend on the air temperature and, consequently, on particle velocity. At the same time, the parameter t increases by about 2 times showing the influence of the powder composition on the buildup process. The comparison of the activation constants in Equation 2.2 reveals the same order of magnitude. These expressions, describing the deposition efficiency and average single pass thickness parameters by means of an exponential function, correctly reflect the behavior of certain physical parameters. For deeper understanding of the physical mechanism behind CS, it is necessary to develop new phenomenological models that include the interaction of particles during the deposition of powder materials.

2.2.1.3 Arrhenius Approximations

Based on the experimental data presented in the previous chart, it is important to discuss three primary issues: (1) the basis for the application of Arrhenius' law in the phenomenological analysis of CS processes, (2) the behavior of particles during CS, and (3) the characteristics of shear localization and mechanical alloying in CS coating structure formation.

An expression for the deposition efficiency and average single pass coating thickness may be obtained using the Arrhenius equations (2.42 and 2.43). This appears to be reasonable because, according to Equation 2.44 (Dykhuizen et al., 1998), the particle velocity within the powder jet is a function of the gas temperature:

$$V_p = (M-1)\sqrt{\frac{\gamma_c RT_0}{1+\left[(\gamma_c-1)/2\right]M^2}} \tag{2.44}$$

Here, T_0 is the initial gas temperature (in convergent part of nozzle), M is Mach number, and γ_c is the ratio of specific heats or isentropic coefficient (Stoltenhoff et al., 2002). For monatomic gases, γ is 1.66, and for diatomic gases γ is typically 1.4. (Air is typically modeled as a diatomic gas because it is primarily composed of nitrogen and oxygen.) The modeling results and analytic equations based on isentropic gas flow model that were developed by Dykhuizen et al. (1998) allow for a rough estimation of the particle velocity. This analytical equation (2.44) demonstrates the dependence of particle velocity on gas temperature. As Dykhuizen et al. (1998) describes, this equation may be used to approximate the particle velocity only when an optimal nozzle shape is used. Further, this relationship does not take into account particle parameters. It is in fact quite difficult to obtain analytical solutions using equations whose main variables are the drag coefficient, particle density, and particle size. Equation 2.44, however, provides a means for the preliminary analysis of the dependence of particle velocity on the temperature and type of the propellant gas (RT_0).

Contact time, t_c, is formally described as the time during which the impact particle decelerates from V_p to $V_p = 0$. Mathematically, this may be approximated as

$$t_c \approx \frac{2\varepsilon_p d_p}{V_p} \tag{2.45}$$

The t_c can be approximated by a simplified form of (2.45):

$$t_c \approx \frac{d}{V_p} \tag{2.46}$$

where d is the diameter of the incoming particle. The severe deformation of particles due to impact is described by the shear strain γ that may be approximated using microstructure analysis. The shear strain rate may be calculated as

$$\dot{\gamma} \approx \frac{\gamma}{t_c} \qquad (2.47)$$

One can find RT_0 from Equation 2.42 as

$$RT_0 = f(\gamma_c, M) V_p^2, \text{ where } f(\gamma_c, M) = \frac{1 + (\gamma_c - 1/2) M^2}{(M-1)^2 \gamma_C}$$

Finally, Equation 2.43 may be modified by taking into account (2.46) and (2.47):

$$t = t_0 \exp - \frac{Q_t}{f(\gamma_c, M) d^2 \left(\frac{\dot{\gamma}}{\gamma}\right)^2} \qquad (2.48)$$

This relationship suggests that the overall effectiveness of CS process is determined by the relation $(\dot{\gamma}/\gamma)$ (or contact time), characterizes the process of adiabatic shear bands formation at the particle interface. Wright (1987) and Wright and Batra (1985) show that under extreme loading conditions various materials exhibit narrow bands of intense plastic deformation, termed shear bands. The thermomechanical mechanism of shear bands formation is associated with shear localization during high strain rate loading in rate-sensitive materials such as polycrystalline metals or granular materials (Hu and Molinari, 2004). Under this mechanism, thermal softening is thought to overcome hardening effects in the material, and to push forward the shear band formation process. As shown by Wright (1987), Wright and Batra (1985), Walter (1992), and Zerilli and Armstrong (1997), a flow law having an Arrhenius dependence on temperature may be used to model shear localization and shear band phenomena in thermoviscoplastic materials. Flemming et al. (2000) completed the first comprehensive analytical study of shear localization with the Arrhenius flow law, yielding a one-dimensional formulation based on plastic strain-rate function (flow law) analysis. This strain-rate function, $\dot{\gamma}$, depends on both temperature and stress. Physically speaking, the nature of the plastic flow governs shear bands formation, and its mathematical representation is the source of nonlinearity in the problem (Flemming et al., 2000). Theoretical models based on microstructure and the physics of deformation employ the thermal activation of plastic flow, leading to an Arrhenius dependence on temperature in accordance with Kochs et al. (1975), Dodd and Bai (1989), and Zerilli and Armstrong (1997). Such models are used in numerical studies by Wright (1987); however, in many analytical studies the flow laws are approximated so as to simplify calculations. In particular, the following exponential flow law by Flemming et al. (2000) is often used:

$$\dot{\gamma} = \dot{\gamma}_0 \tau^N e^{-Q/RT_D} \qquad (2.49)$$

In this equation, T_D is the deformation temperature, Q is the activation energy, N is the power coefficient, and τ is the shear stress in the area of shear localization. The main feature of (2.49) is the Arrhenius dependence of shear strain on temperature. The form of (2.49) has been chosen by Flemming et al. (2000) to allow two types of stress dependence: $N = 0$ and $N > 0$. In comparing the forms of (2.48) and (2.49), it can be seen that macroscopic characteristics of the CS process such as the average single pass thickness, t, or the deposition efficiency, χ_D, depend on the kinetics of adiabatic shear bands formation that is described by means of the Arrhenius flow law. Indeed, to better understand the role of shear localization as the main coating formation process during particle impact loading in CS, the relationship between Equations 2.48 and 2.49 needs to be further examined.

The detailed analysis of the flow characteristics of gas–particle suspensions in a powder-laden jet seems to be important for the evaluation of particle–particle collisions and particle–turbulence interactions. These processes were studied both for diluted (Senior and Grace, 1998) and concentrated gas–solid suspensions (Eskin and Voropaev, 2004). Despite the relatively high gas velocities (300–900 m/s) in these *fast fluid bed* units (Senior and Grace, 1998), the concentration of suspended solids is neither uniform along the nozzle axis, nor over the cross-section of the flow. High particle concentration generally forms near the nozzle walls where particles are moving with lower velocities. Clusters of particles may also intermittently form and disintegrate in the diluted suspension flows. Thus, the probability of particle cluster formations may be evaluated through the differences in axial particle velocities ΔV_i. The gradient $\Delta V_{(1-50)} = V_1 - V_{50} = \sim600$ m/s, where V_1 and V_{50} are the velocities of particles having sizes of 1 and 50 µm, respectively. In fact, even a small difference in particle size results in a velocity gradient about 100–200 m/s. The scheme of particle clustering for the case of spraying bimodal powder mixtures is shown in Figure 2.18b.

As a first step in attempting to model the CS process, the following two regions of the powder-laden jet must be taken into consideration.

1. The free jet region (whose length is nearly equal to the stand-off distance)
2. The region of the normal shock wave layer

The average concentration of particles rebounding from the substrate is high in the normal shock wave layer. This leads to an increase of particle concentration in clusters moving through the normal shock wave front. Two conclusions may be drawn.

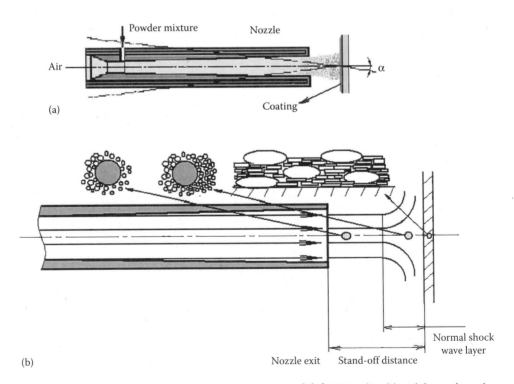

FIGURE 2.18 Composite coating structure formation model for spraying bimodal powder mixture: (a) general scheme of powder jet and coating; (b) particle arrangement in powder jet.

1. The velocity of the cluster depends on the particles rebound process
2. The coating formation in the case of particle clusters impingement with a substrate is similar to powder shock compaction

2.2.2 Particle Deformation during Impact

2.2.2.1 Task Statement

Localized shear deformation and the form of intensive deformation in a narrow band generated during dynamic deformation at high strain rates has been a topic of the great interest for decades. Localized shear is known to be an important mode of deformation that leads to super high strains and development of interparticle bonding between various powder composites. CS technology is known to be the process of metallic particles impact with velocities about 600–1000 m/s that results in particle deformation at strain rates in the range of 10^3–10^9 s^{-1}. That is why it is well recognized that localization of deformation induced by an adiabatic shear band formation in the impacting particles is one of the dominant mechanisms for successful bonding in CS (Assadi et al., 2003). However, an analysis of the localization of deformation during the particle impact is currently lacking in literature (Li et al., 2006; Schmidt et al., 2009). It is well known that the microscopic and macroscopic response of the material under high strain-rate loadings is affected by strain, strain rate, temperature, and microstructure of the material. There are a lot of papers dealing with studying the microscopic features of the localization process at various deformation operations and in dynamic loadings. However, no detailed study of the adiabatic shear band formation and macroscopic mechanical behavior of material during particle deformation at CS has been made yet because of super high strain rates developed at CS.

The bonding mechanisms at CS particle impact are proven to be based on adiabatic shear band formation at the interface (Assadi et al., 2003). Two bonding mechanisms are occurring simultaneously in the process: (1) thin oxide layer is broken and (2) the jetting at the local intensively deformed zones is formed (Li et al., 2006; Yin et al., 2010). The crushing of the oxide film on the surface and bonding in different points of the interface strongly depend on the local strain rate, strain, and temperature. Thus, determination of these parameters at various points at the interface of the particles is very important.

An onset of adiabatic shear instability (strain localization) may be achieved due to optimal thermomechanical conditions (Dykhuizen et al., 1999). The understanding of the real distribution of these parameters along the surface and through the whole particle can be helpful for analyzing the mechanisms of bonding at the interface and structure formation inside the particle. However, the experimental study of adiabatic shear band formation is known to be difficult in many cases because of the need of high-resolution transmission electron microscopy (TEM) and atomic force microscopy (AFM) applications as the main methods of structure parameters determination at the nanoscale. Nevertheless, a search of microstructural methods of process characterization with optical microscopy and scanning electron microscopy (SEM) seems to be very important because they allow us to evaluate the deformation behavior at the micron and submicron scale.

On the other hand, the use of numerical modeling methods to characterize the particle impact process had been proven to be very effective (Assadi et al., 2003). The modeling of the particle impact onto substrate allows defining the stress–strain parameters in each point of the particle. The numerical simulation of the impact process and the investigation of the microstructure changes in various points of particles in the coatings are the main aims of this paper to define the effect of strain localization and evaluate the validity of numerical modeling results.

Therefore, the main aim of this part is to characterize the deformation and strain localization parameters by numerical modeling at various points of particles being consolidated during CS and compare the new numerical results with the real microstructure of stainless steel and ARMCO iron that is known to have different sensitivities to the strain localization process (DiLellio and Olmstead, 2003).

2.2.2.2 Shear Localization Basics

In a fundamental study by Assadi et al. (2003), comprehensive experimental and computational finite element analysis of a copper cold-spraying deposition process was undertaken. The results of this study suggest that the CS bonding mechanism can be attributed to adiabatic shear instability occurring at the particle/substrate or particle/deposited material interfaces for high impact particle velocities. In the papers of Gruijicic et al. (2003, 2004), the analysis of Assadi et al. (2003) is extended to several other metallic material systems. Further, a more detailed analysis of the susceptibility of metallic materials to adiabatic shear instability during CS is undertaken. To better understand the role adiabatic shear instability (or shear localization) plays within the bonding and consolidation processes of GDS, it is beneficial at this time to discuss the basic fundamentals of this phenomenon.

Adiabatic shear instability and the associated formation of shear bands have been considered in detail by Wright (1992, 2002), Bai (1982), Molinari (1997), and others. When metals are subjected to large deformations, shear localization is an important aspect governing material flow. The process of adiabatic shear banding was analyzed by Molinari (1997) from the experimental results of Marchand and Duffy (1988) using thin-walled tubes that were twisted to high strain rates using a Kolsky bar configuration.

As shown in Figure 2.19, there are three primary flow localization processes. In the first (a typical strain-hardening material under nonadiabatic conditions), the shear stress increases due to strain hardening (the curve labeled *Isothermal*). In the second, the plastic strain energy that is dissipated as heat under adiabatic conditions, resulting in a temperature increase, in turn causing material softening whose effect overwhelms that of strain hardening. The primary change in the mechanical behavior of material is indicated by the occurrence a shear stress maximum in the flow curve (the curve labeled *Adiabatic*). It is during this stage that the development of flow heterogeneity may be observed, indicating the onset of a weak instability process (Molinari, 1997). The third type of plastic flow is characterized by the marked development of strain localization, resulting in the formation of an adiabatic shear band that eventually propagates along the circumference of the specimen (Figure 2.19). The fluctuations in stress, strain, temperature or microstructure, and the inherent instability of strain softening can give rise to plastic flow (shear) localization (Grujicic et al., 2004). Thus, shearing and heating (and subsequent softening) becomes highly localized, while the material strain and heating in surrounding regions remains negligible. This, in turn, causes the flow stress to rapidly decrease (the curve denoted *Localization*).

To utilize the adiabatic shear instability modeling techniques, it is essential to understand the flow behavior of the material under processing conditions. Constitutive equations, which relate the flow stress σ in terms of strain ε, strain rate $\dot{\varepsilon}$, and temperature T

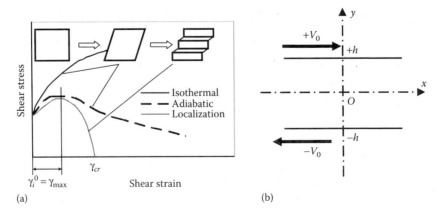

FIGURE 2.19 Flow localization processes. (a) schematics of the stress–strain curves. (After Grujicic et al., Computational analysis of the interfacial bonding between feed-powder particles and the substrate in the cold-gas dynamic-spray process. *Appl. Surf. Sci.*, 219, 211–227, 2003. With permission.) and (b) geometry used for shear analysis.

$$\sigma = f\left(\varepsilon, \dot{\varepsilon}, T\right) \tag{2.50}$$

are often used in deformation problems. To effectively evaluate the deformation and shear localization parameters, the flow stress σ data have to be defined in terms of strain ε, strain rate $\dot{\varepsilon}$, and temperature T. To examine the various deformation models that exist for the regions of flow instabilities during deformation, the following dimensionless parameters must be determined from the measured flow stress data (Murty et al., 2000):

The strain rate sensitivity parameter,

$$m = \frac{\partial \ln\sigma}{\partial \ln\dot{\varepsilon}} \tag{2.51}$$

The flow softening rate,

$$\omega = \frac{1}{\sigma} \cdot \frac{\partial \sigma}{\partial \varepsilon} = \frac{\partial \ln\sigma}{\partial \varepsilon} \tag{2.52}$$

The flow localization parameter for plane strain compression,

$$\alpha = \frac{-\omega}{m} \tag{2.53}$$

The temperature sensitivity of flow stress—a_s,

$$a_s = \frac{1}{T} \cdot \frac{\partial \ln\sigma}{\partial(1/T)} \tag{2.54}$$

In accordance with Chen and Batra (2000), we now consider the simple shearing deformations of an isotropic and homogeneous thermoviscoplastic body that is bounded by the planes $y = \pm h$ and sheared in the x-direction by prescribing velocities $v = \pm V_0$ on the upper and lower bounding surfaces (Figure 2.19b). The bounding surfaces are assumed to be thermally insulated. The equations governing the deformations of the body are

$$\rho\ddot{\gamma} = \tau_{yy}, \dot{\gamma} = v_y \tag{2.55}$$

$$\rho c\dot{T} = kT_{yy} + \beta\tau\dot{\gamma} \tag{2.56}$$

$$\tau = \tau_1 + \tau_2 \tag{2.57}$$

$$\tau_1 = B_1 e^{-(\beta_1 - \beta_2 \ln\dot{\gamma})T} + B_2 \gamma^{1/2} e^{-(\alpha_1 - \alpha_2 \ln\dot{\gamma})T} \tag{2.58}$$

$$\tau_2 = b\alpha_0\mu_0(1 - AT - BT^2)\lambda^{1/2} \tag{2.59}$$

$$\frac{d\lambda}{d\gamma} = f(\lambda, \dot{\gamma}, T) \tag{2.60}$$

Here, ρ is the mass density, γ the shear strain, τ the shear stress, c the specific heat, T the absolute temperature, k is the thermal conductivity, and β is the Taylor–Quinney factor, representing the fraction of plastic work converted into heat. Also, a superimposed dot denotes the material time derivative and a comma followed by y represents the partial derivate with respect to y. Following the work of Zerilli and Armstrong (1997), and Klepaczko (1988), τ is written as the sum of τ_1 and τ_2. The portion of the shear stress given by Equation 1.9 is a result of thermally activated dislocation interactions (Zerilli and Armstrong, 1997). B_1, B_2, β_1, β_2, α_1, and α_2 are constants; B_1 and B_2 are related to the Gibbs free energy, the dislocation activation area at zero temperature, and the magnitude of the Burger vector. The portion of the shear stress given by (2.54) is dependent on the thermomechanical history of the plastic deformation through the internal variable λ, whose evolution is described by (2.55). This internal variable may be identified as the total dislocation density. In (2.59) b, a_0, μ_0, A, and B are constants.

One of the primary goals of an adiabatic shear instability analysis is to determine the critical strain and stress associated with material instability (Figure 2.19a). The instability analysis undertaken by Chen and Batra (2000) showed that, although the strain-rate hardening of the material does not directly influence material instability, it does have an indirect affect. For locally adiabatic deformations, material instability occurs only when thermal softening exceeds the combined effects of the material hardening due to plastic straining and the increase in dislocation density. However, in the presence of heat conduction, higher values of the nominal strain-rate delay the onset of the material instability. The modeling results presented by Chen and Batra (2000) evince the dependence of the instability onset strain γ_i^0, the corresponding shear stress τ_i^0, and the dislocation density λ_i^0 on the nominal strain rate $\dot{\gamma}^0$ within a range of $\dot{\gamma}^0 = 10^2 – 10^5$ s^{-1} (Figure 2.20). These data allow one to clearly characterize the influence of the strain rate and temperature on both the position and value of the initial instability onset. For example, it can be seen that an increase in the strain rate shifts the adiabatic shear instability onset to lower strains and dislocation densities (Figure 2.20a, c). Extrapolation of these functions up to $\dot{\gamma}^0 = 10^7$ s^{-1} suggests an instability onset strain of $\gamma_i^0 = 0.13$; for $\dot{\gamma}^0 = 10^{-2}$ s^{-1}, $\gamma_i^0 = 0.21$. It is interesting to note that an increase in the initial temperature results in a stronger dependence of γ_i^0 on $\dot{\gamma}^0$. At an initial temperature $T = 600$ K, the value of the instability onset strain $\gamma_i^0 = 0.063$ for $\dot{\gamma}^0 = 10^7$ s^{-1}, while $\gamma_i^0 = 0.3$ for $\dot{\gamma}^0 = 10^{-2}$ s^{-1}. Such a strong relationship between the initial temperature and the onset of material instability seems to indicate an intensive material softening at the beginning of the deformation process. This conclusion is confirmed by the fact that the shear stress is dependent on the strain rate at the onset of instability (Figure 2.20b). In fact, the shear stresses remain constant at various temperatures for all strain rates (0.55–0.62 GPa). Also, it should be noted that the dislocation density at the onset of instability determines the level of material hardening, as well as the concentration of defects in the material. The dependence of the dislocation density on strain rate (Figure 2.20c) can be characterized by $\gamma_i^0 = \gamma_i^0(\dot{\gamma}^0)$. Therefore the primary factors that influence the development of shear localization processes are the strain rate and the initial temperature of material.

2.2.2.3 Adiabatic Shear Instability in CS

The general equations described above are able to adequately describe shear instability within the GDS process with the particulars of this theory illustrated by Assadi et al. (2003) and Grujicic et al. (2004). To further this analysis, however, it is beneficial to discuss the model developed by Grujicic et al. (2004), and, in turn, to compare simulated results with experimental data.

To understand the onset of strain softening and adiabatic shear localization, a simple one-dimensional model is utilized by Grujicic et al. (2004). This technique, though relatively basic, is able to reveal the thermomechanical behavior of a small material element at the particle/substrate interface during collision. The model is presented in the terms of the equivalent plastic strain rate $\dot{\varepsilon}^p$ (referred to as shearing by Schoenfeld and Wright, 2003), whose formal definition is given by

$$\gamma^p = \gamma_0^p + \int_0^t \dot{\gamma}^p dt$$

Here, γ^p is the plastic shear strain and $\dot{\gamma}^p$ is an equivalent shear strain rate. Thus, $\dot{\gamma}^p = \dot{\varepsilon}^p$.

This model is based on the following governing equations:

The equivalent plastic strain rate, $\dot{\varepsilon}^p$, where

$$\dot{\varepsilon}^p = \dot{\varepsilon}_0^p \frac{\rho v_p^2}{2} \cdot \frac{1}{s} \qquad (2.61)$$

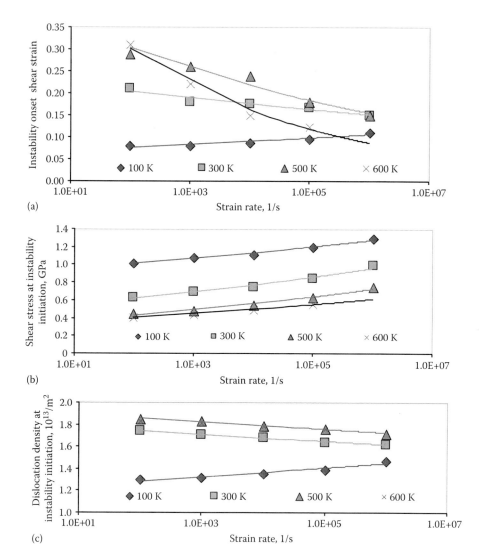

FIGURE 2.20 Shear instability parameters dependences. (After Chen and Batra., Microstructural effects on shear instability and shear band spacing. *Theor. Appl. Frac. Mech.*, 34, 155–166. With permission. 2000). Aluminum (Air) (a) instability shear strain onset, (b) shear stress at instability initiation, and (c) dislocation density at instability initiation.

with the boundary condition

$$\dot{\varepsilon}^p(t=0) = \dot{\varepsilon}_0^p \frac{(\rho v_p^0)^2}{2} \cdot \frac{1}{s_{\text{init}}}$$

The equivalent plastic strain, ε^p,

$$\varepsilon^p = \int_0^t \dot{\varepsilon}^p dt \qquad (2.62)$$

The heating rate, $\dot{T}$,

$$\dot{T} = \frac{s\dot{\varepsilon}^p}{\rho C_p} \qquad (2.63)$$

with the boundary condition

$$\dot{T}(t=0) = \frac{s_{\text{init}} \dot{\varepsilon}^p(t=0)}{\rho C_p}$$

The temperature, T,

$$T = \int_0^t \dot{T} dt \qquad (2.64)$$

with the boundary condition $-T(t=0) = T_{\text{init}}$.

And the particle velocity,

$$v_p = v_p^0 \left(1 - \frac{t}{t_c}\right), v_p(t=0) = v_p^0 \qquad (2.65)$$

Here, $\dot{\varepsilon}_0^p$ is the strain rate proportionally constant, ρ, the (constant) material mass density, C_p, the (constant) specific heat and a raised dot denotes the time derivative of a particular quantity.

A fundamental assumption is made in Equation 2.61, where the equivalent plastic strain rate is taken to be inversely proportional to the equivalent plastic flow strength, s, and, in turn, a function of strain, strain rate and temperature. This quantity is described by the well-known Johnson–Cook model (Johnson and Cook, 1983):

$$s = \left[A + B\left(\varepsilon^p\right)^n \right] \times \left[1 + m \ln\left(\frac{\dot{\varepsilon}^p}{\dot{\varepsilon}_0^p}\right) \right] \times \left[1 - \left(\frac{T - T_{\text{init}}}{T_{\text{melt}} - T_{\text{init}}}\right)^{\vartheta} \right] \qquad (2.66)$$

where A, B, m, n, and ϑ are constants.

The model allows for effective evaluation of the plastic strain rate, heating rate, temperature, and equivalent stress in representative elements at the particle–substrate interface during particle impact. Similar calculations have been made by Assadi et al. (2003). A comparison of the modeling results is shown in Figures 2.21 and 2.22 for equivalent normal stresses and strains. Here the dissimilarities in material deformation can be observed as follows: (i) the peak in the stress is more

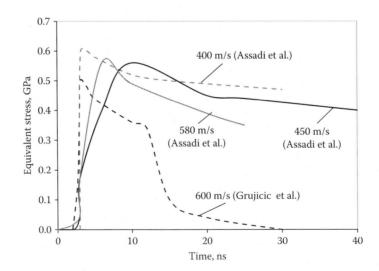

FIGURE 2.21 Comparison of equivalent stresses. (After Assadi et al., Bonding mechanism in cold gas spraying. *Acta Mater.*, 51, 4379–4394, 2003; Grujicic et al., Adiabatic shear instability based mechanism for particles/substrate bonding in the cold-gas dynamic-spray process. *Mater. Design*, 25, 681–688. 2004. With permission.)

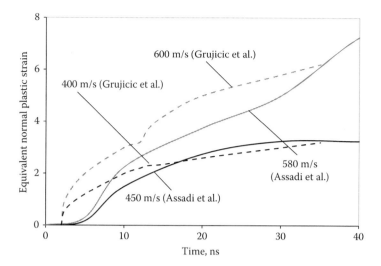

FIGURE 2.22 Comparison of equivalent normal strains. (After Assadi et al., Bonding mechanism in cold gas spraying. *Acta Mater.*, 51, 4379–4394, 2003; Grujicic et al., Adiabatic shear instability based mechanism for particles/substrate bonding in the cold-gas dynamic-spray process. *Mater. Design*, 25, 681–688, 2004. With permission.)

evident, and (ii) the transition to the adiabatic shear band formation process (corresponding to a sharp decline in the equivalent normal stress) is more clearly defined within the model developed by Grujicic et al. (2004). It is shown in Figure 2.23 that, the equivalent normal stress falls to zero. This effect seems to be due to adiabatic softening in the area of localization; the specific duration of this process must be sufficient so as to adequately increase the temperature in this region. Calculated temperature variations with respect to impact times for various impact velocities are presented by Assadi et al. (2003). These are similar to the strain variations that are shown in Figure 2.22 and are due to the strong interdependence of strains and temperatures in Equation 2.66.

It should be noted that the assumptions utilized to determine the contact pressures in the models of Assadi et al. (2003) and Grujicic et al. (2004), though different, do not result in different values. In Grujicic et al. (2004), the contact pressure at the particle/substrate interface is assumed to be

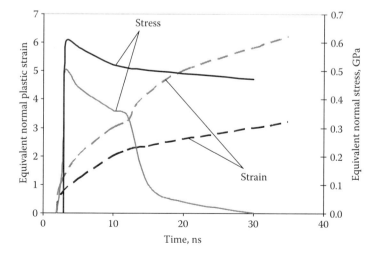

FIGURE 2.23 Equivalent normal stress and strain. (After Grujicic et al., Adiabatic shear instability based mechanism for particles/substrate bonding in the cold-gas dynamic-spray process. *Mater. Design*, 25, 681–688, 2004. With permission.)

proportional to the kinetic energy of the particle per unit volume $\rho v_p^2/2$. This assumption, along with being quite reasonable, is readily utilized in most calculations. There is, however, a fundamental difficulty in determining $\dot{\varepsilon}_0^p$ due to the boundary condition

$$\dot{\varepsilon}^p(t=0) = \dot{\varepsilon}_0^p \frac{(\rho v_p^0)^2}{2} \cdot \frac{1}{s_{\text{init}}}$$

In fact, because of the uncertainty of $s_{\text{init}} = f(v_p^0)$ it does not unambiguously define the value of $\dot{\varepsilon}_0^p$. Thus, the primary requirement for the effective application of either model is the correct determination of the constants and variables that appear in Equations 2.60, 2.63, and 2.66.

The empirical Johnson–Cook formula (Equation 2.66) is used in the computational analysis of one particle impact for determining the equivalent flow strength. In (2.65), A, B, and n represent the work hardening constants, m the strain rate hardening constant, and ϑ the thermal softening parameter. As Johnson and Cook have published tables of these parameters for many common materials, it is relatively simple to apply the flow law (Equation 2.66) in numerous one-dimensional deformation process computations with strain rates up to 10^6 s^{-1}. These constants, however, need to be adequately defined for high impact velocities. For example, the strain rate sensitivity for Al appears to increase dramatically for the strain rates higher that have been achieved in pressure-shear plate impact testing ($>10^6$ s^{-1}, Wright, 2002).

The parameters used in the Johnson and Cook formula are determined from dynamic stress–strain curves for tension and compression at nearly constant strain rates and temperatures. Strain rate sensitivity and thermal softening, however, have a powerful influence on material behavior during strain localization. For this reason, the application of these values in the analysis of particle deformation by impact with strain rates higher than 10^6 s^{-1} need to be revaluated.

Thermal sensitivity, a, and strain rate sensitivity, m, are known to be defined as

$$a = -\frac{1}{\rho C_p} \frac{\partial s}{\partial T} \text{ and } m = \frac{\partial \ln s}{\partial \ln \dot{\gamma}} \tag{2.67}$$

Here, the thermal softening parameter, ϑ, in Equation 2.65 is a function of thermal sensitivity, a.

From an engineering point of view, to analyze the influence of each material parameter on adiabatic shear band formation during high velocity particle impact, it seems to be reasonable to use the constitutive equation in the most simple form—the power law (Daridon et al., 2004; Zhou et al., 2006):

$$\tau(\gamma, \dot{\gamma}, T) = \tau_0 \left(\frac{\gamma}{\gamma_0} \right)^n \left(\frac{\dot{\gamma}}{\dot{\gamma}_0} \right)^m \left(\frac{T}{T_0} \right)^a \tag{2.68}$$

Here, τ_0 is the yield shear stress of the material in a quasistatic simple shear test, n is the work hardening constant, m is the strain rate hardening constant, and $a < 0$ characterizes the thermal softening. Also, γ_0 is the strain at yield point for a shear strain rate of $\dot{\gamma} = 10^{-4}$ s^{-1}, $\dot{\gamma}_0$ is a reference shear rate, T_0 is a reference temperature, and T is the current temperature. A simplified relationship including thermal softening and power-law strain-rate hardening has been presented by Zhou et al. (2006):

$$\tau = \tau_0 g(T)(1 + b\dot{\gamma}^p)^m \tag{2.69}$$

Here, τ_0 is the static yield stress, $g(T)$ is a monotonically decreasing function, b^{-1} is a reference strain rate, and m is the strain-rate sensitivity. For simplicity, strain hardening may be neglected (Nesterenko, 1995). The plastic flow law may be derived from Equation 2.69:

$$\dot{\gamma}^p = \frac{1}{b} \left\{ \left[\frac{|\tau|}{\tau_0 g(T)} \right]^{1/m} - 1 \right\} \tag{2.70}$$

Zhou et al. (2006) consider two kinds of thermal softening:
Linear, as used by Nesterenko (1995),

$$g_1(T) = 1 - \alpha T \tag{2.71}$$

and exponential softening, as used by DiLellio and Olmstead (1997):

$$g_2(T) = \exp(-\alpha T) \tag{2.72}$$

where α^{-1} is a reference temperature. In Equation 2.72, $\alpha^{-1} = T_m - T_r$, where T_m is the melting temperature and T_r the room temperature.

The strain localization during particle impact is governed by strain rate hardening and thermal softening as a consequence of self-heating; it is described by the strain localization parameter (Grujicic et al., 2004) as

$$SL = \left(-\frac{\partial^2 s / (\partial \varepsilon^p \partial \varepsilon^p)}{(\partial s / \partial \dot{\varepsilon}^p) s} \right)_{s = s_{\max}} \tag{2.73}$$

where $s_{\max}$ denotes the maximum flow stress on the stress strain curve at a particular particle velocity. Unfortunately, the direct determination of the partial derivatives of Equation 2.73 is not possible, even in the one-dimensional case. However, adiabatic shear localization evaluation based on the finite element analysis suggests values of SL around 1.6×10^{-4} s/GPa (Grujicic et al., 2004). It is interesting to note that calculations made with the model described in Equations 2.60–2.66 result in comparable tendencies in strain localization for all materials and material combinations. It is the opinion of the authors, however, that due to the absence of real data concerning the thermal and strain rate sensitivity constants this conclusion should not yet be fully accepted. Nevertheless, SL is considered to be a critical condition for the onset of adiabatic strain localization, ultimately allowing authors such as Grujicic et al. (2004) and Papyrin et al. (2002) to calculate the critical particle velocities that are required to attain this critical value of SL, and consequently to achieve successful GDS deposition (see Table 2.2). Critical particle velocities also depend (at least in part) on the fundamental model parameters such as $\dot{\varepsilon}_0^p$, s_{init}, and t_c. This is due to

1. The uncertainty that exists in determining the constants
2. The development of adiabatic shear bands after instability onset
3. Various features specifically related to the shock consolidation of particles

The contact time, t_c, depends on particle velocity at the impact (shown in Figure 2.24).

The numerical modeling of shear localization by DiLellio and Olmstead (1997, 2003) appears to justify this statement. Figure 2.25 depicts a simplified model of this numerical solution (DiLellio and Olmstead, 2003) that has been adapted for the Johnson–Cook equation with brass, nickel, and an iron and aluminum alloy.

All materials exhibit a pronounced drop in shear stress (Figure 2.25a) for distinct shear strains (defined by dimensionless time). Thus, it can be seen that the onset of localization is achieved more readily for nickel than for the iron and aluminum alloy. It is interesting that the sudden increase in temperature is observed to take place at different (nondimensional) times for each metal (Figure 2.25b). Further, the values of nondimensional temperature parameters differ considerably. In fact, the temperature jump for iron is 3 times that of the aluminum alloy. The reason for this is not yet fully understood.

The dependence of the shear strain rate on nondimensional time is shown in Figure 2.25c. These numerical results suggest a strong relationship between the specific material and the onset of localization. The numerical scheme put forth by DiLellio and Olmstead (1997, 2003) provides

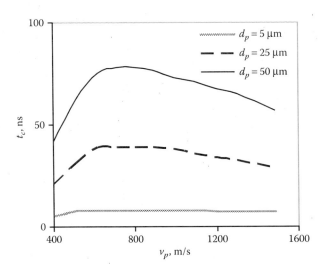

FIGURE 2.24 Contact time versus the impact velocity. (After Papyrin et al., On the interaction of high speed particles with a substrate under cold spraying. In: Lugscheider E.F., Berndt C.C. (Eds.). *Proceeding of the International Thermal Spray Conference 2002*, Dusseldorf (Germany), DVS-Verlag, p. 380, 2002. With permission.)

an efficient approach by which to investigate shear localization in materials that obey the Johnson–Cook flow law (Equation 2.66). Unfortunately, as the simplified Johnson–Cook flow law neglects hardening, it is not without limitation. For this reason, the numerical results that are obtained cannot be quantitatively compared to experiments; qualitative comparison of various materials is, on the other hand, quite appropriate (DiLellio and Olmstead, 2003). This approach, in contrast to works of Assadi et al. (2003) and Grujicic et al. (2004), clearly illustrates the strong influence material type (and consequently strain rate and thermal sensitivity) has on the parameters of localization.

2.2.3 CONSOLIDATION OF THE PARTICULATE ENSEMBLE

2.2.3.1 Identification of the Various Phenomena

The main processing characteristics of densification and bonding structure formation in GDS are considered to be similar to those in processes such as explosive welding or shock wave powder compaction (Assadi et al., 2003; Schmidt et al., 2003, 2006). In all of these processes, bonding is achieved only when a critical impact velocity is exceeded. The critical velocity in these processes is generally perceived to be a function of the material properties (Schmidt et al., 2006).

The most important microstructure formation processes involved in gas dynamic spraying have to be identified and discussed through an energy approach. The primary feature common to both GDS and shock consolidation processes is that energy is dissipated via a shock wave. Interparticle adiabatic shear band formation, vorticity, voids, and particle fracture may also occur due to the plastic deformation. Various energy dissipation processes take place: plastic deformation, interparticle friction, microkinetic energy, and defect generation. Meyers et al. (1999) have introduced an analytical expression that may be used to determine the energy that is necessary to consolidate a powder, as a function of strength, size, porosity, and temperature, based on the materials properties. This formula allows one to predict the pressure that is required to shock consolidate materials, as well to characterize the role of each energy dissipation processes. Based on the utility of this expression, it is therefore beneficial to organize the GDS microstructure formation processes in accordance with the formalism presented Meyers et al. (1999).

The proper identification and quantitative evaluation of the various phenomena that occur during the impingement of a powder ensemble on a substrate is necessary to estimate the overall energy requirements for forming powder coatings. Figure 2.26 provides a schematic representation of

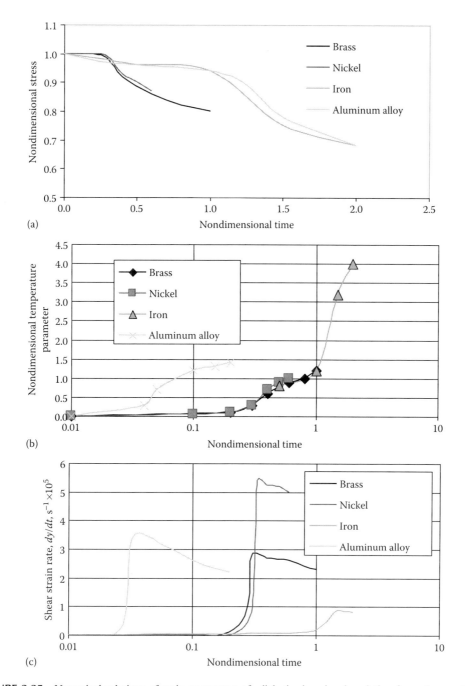

FIGURE 2.25 Numerical solutions of main parameters of adiabatic shear band evolution for various materials. (After DiLellio and Olmstead, Numerical solution of shear localization in Johnson–Cook materials. *Mech. Mater.*, 35, 571–580. 2003. With permission.). (a) nondimensional stress, (b) temperature parameter, and (c) shear strain rate.

these phenomena for GDS. There are in fact two groups of processes that must be characterized: (1) impact processes with energy—these can be computed from phenomenological parameters such as particle velocity, yield stress, material density, and so on; and (2) structure formation processes that define the energy dissipation. The energy for each process may be computed on determining the structural parameters of the system.

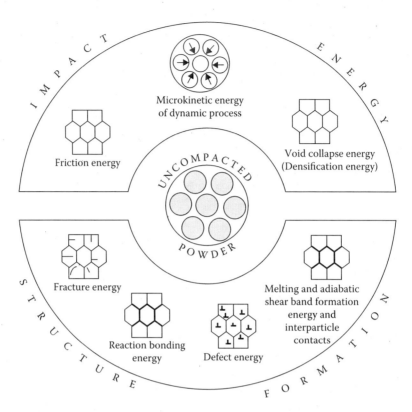

FIGURE 2.26 Various models of energy dissipation in shock compression of powders. (After Meyers et al., Shock consolidation: microstructurally-based analysis and computational modeling. *Acta Mater.,* 47, 2089–2108, 1999. With permission.)

2.2.3.2 Classification of Consolidation Processes

Group 1: Impact Processes with Energy

a. *Plastic deformation energy.* The powder material is being plastically deformed on impact, resulting in the collapse of the voids due to the plastic flow of each particle, as shown in Figure 2.26.

b. *Microkinetic energy.* The localized plastic flow of the material (Figure 2.19b) is achieved by means of the dynamic features of the shock consolidation process. This leads to interparticle impacts, friction, and plastic flow beyond that which is geometrically necessary to collapse the voids, as described in (a). This component is referred to as *microkinetic energy* (Nesterenko, 1995), a process of intense energy dissipation with localized heating. Microkinetic energy results from particle kinetic energy inducing the formation of a jet at the particle–substrate (or particle–particle) interface, in turn bringing about an increased temperature, as observed in CS process. The entire plastic deformation path is replaced by deformation in a localized area. However, the kinetic energy acquired by the material elements (being plastically deformed) eventually dissipates into thermal energy, as in conventional plastic flow.

c. *Friction energy.* The relative movement of particles at the shock front as a result of the applied stress brings about interparticle friction—this plays a significant role in energy deposition. Both stress and deformation within granular materials are known to be nonuniform, particularly at the microscale of particle groups (Molinari, 1997). The various material parameters (or factors such as the coefficient of friction) that yield stress, cohesion loading, and so on influence the characteristics of particle ensemble flow. The high friction forces that can be observed at the particle interface are due to particle flow and high radial velocities.

Group 2: Structure Formation Processes

d. Adiabatic shear band formation and melting at the particle–substrate or particle–particle interface. Although the intensive shear strains that are localized at the interface may eventually lead to material melting, with GDS this is not usually the case.

e. *Defect energy.* Quantitative assessments of defect concentrations in shock-wave deformation reveal this factor to have only a minor influence on total energy losses (Meyers et al., 1999).

f. *Fracture energy.* Brittle particles such as oxides and carbides may fracture due to impact. The comminuted particles can more efficiently fill the voids.

g. *Shock-initiated chemical reactions.* Some chemical reaction at the particle–particle interface may occur as a result of powder mixtures being consolidated during impact. In the case of exothermic reactions, this process contributes additional energy to the powder surface and acts to facilitate bonding.

These GDS impact processes, although they may be conveniently categorized into the distinct groups outlined above, are not entirely independent. Indeed, as presented in Figure 2.26, there exists some degree of overlap. For example, as the transformation from uniform plastic flow (Figure 2.27a) to localized flow (Figure 2.27b) is difficult to define, microkinetic energy and plastic deformation energy are tightly connected. On the other hand, fracture energy and plastic deformation energy are competing processes—densification can occur as a result of either one or a combination of both processes (Meyers et al., 1999). As presented above, however, the simplified explanation of these energy components allows one to more distinctly define the role of each process in GDS structure forming.

2.2.4 CRITICAL VELOCITY

Assadi et al. (2003) and Grujicic et al. (2004) have determined the critical velocities of various metals (Table 2.2). In particular, Assadi et al. (2003) used an experimental method to determine the

TABLE 2.2
Evaluation of the Critical Particle Velocities

		Threshold Particle Velocity (m/s)		
Particle Material	Substrate Material	Assadi et al. (2003)	Finite Element Analysis after Grujicic et al. (2004)	*SL* Factor Analysis after Grujicic et al. (2004)
Copper	Copper	570–580	575–585	571
Aluminum	Aluminum	760–770	760–770	766
Nickel	Nickel	600–610	620–630	634
316L	316L	600–610	620–630	617
Titanium	Titanium	670–680c	650–670	657
Copper	Aluminum	N/A	510–530	507
Aluminum	Copper	N/A	600–630	634
Copper	Nickel	N/A	570–580	571
Nickel	Copper	N/A	570–580	576
Copper	316L	N/A	570–580	574
316L	Copper	N/A	570–580	573
Copper	Titanium	N/A	520–550	514
Titanium	Copper	N/A	570–590	582

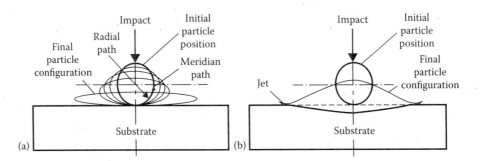

FIGURE 2.27   One particle impact. (a) uniform particle deformation, and (b) localization of deformation.

critical velocity that is detailed by Stoltenhoff et al. (2002). The procedure was based on the combination of calculating the particle velocity distribution using CFD and measuring the corresponding mean velocity by laser Doppler anemometry. As a result, the calculated velocity for the largest and slowest bonded particle was taken as the critical velocity for bonding. Two distinctive features of this method should be noted.

1. The critical velocity for one particle may characterize the GDS process for low particle concentrations in powder-laden jet only when this does not influence the particle consolidation process.
2. The GDS technology parameters of the experiment are not directly related to sprayed material characteristics. Additionally, the properties of the powder materials have not been accounted for.

It is therefore necessary to define the powder material characteristics that ultimately determine the overall effectiveness of the spraying process—this will undoubtedly result in deeper understanding and more effective process modeling. Additionally, the effect of particle concentration on the deposition efficiency should be studied to clarify the nature of powder consolidation in GDS.

In this chapter, the impact parameters of a single particle have been described in a qualitative way. Further, the principal experimental and theoretical modeling methods that are commonly used to examine this process have been briefly reviewed. The main points that must be carried forward to the following chapters are that shear localization and adiabatic shear band formation appear to have a profound effect on the bonding process. Also, it is of great importance to effectively estimate the strain rate and thermal sensitivity of each substrate for various impact velocities. And finally, the role of the initial yield strength and thermal softening processes in adiabatic shear band formation during particle impact has to be further evaluated for both classical and low-pressure GDS. It is only through this analysis that analytical materials models may be used to help interpret results and, in turn, develop an effective approach that permits the accurate calculation of specific GDS parameters.

2.3   MODELING PARTICLE IMPACT AND BONDING

2.3.1   NUMERICAL SIMULATION PROCEDURE

To simulate particle deformation during deposition by the CS process, two sets of simulations were run: (1) stainless steel, and (2) ARMCO iron. Using commercial Lagrangian Finite Element software Abaqus/Explicit Version 6.9 (Hibbitt et al., 2002) a dynamic simulation of particle impact was made. For both particle sizes configurations a 3D FEM analysis was performed. Figure 2.28 shows the scheme of the simulated problem. The substrate is fixed at the bottom, and horizontal displacement is not allowed at the axis of symmetry and at the outer edge (Assadi et al., 2003; Grujicic et al., 2004). Both particles and substrate initial temperatures were defined. Additionally, particles with a diameter of 50 µm hit

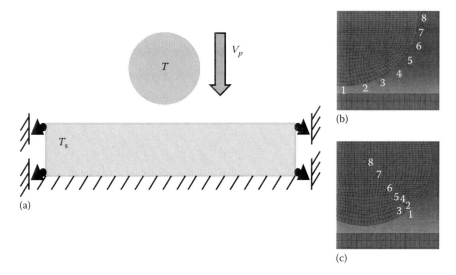

FIGURE 2.28 The schematic diagram of impact model with boundary conditions (a) and series of elements chosen for analysis on the particle surface (b) and series of elements chosen for analysis in the particle core (c).

the substrate with the initial velocity V_p. The particle temperature for both powders was $T_p = 80°C$, substrate temperature $T_S = 20°C$, and impact velocity V_p was of 800 m/s.

Eight node explicit, linear, hexagonal elements for coupled temperature–displacement analysis were taken from Abaqus computer-aided engineering elements library. This element was set to calculate 3D stress with first-order accuracy. Distortion control was set on as well as stiffness type hourglass control. The other parameters were set as default.

Flow stress of both materials was defined using the Johnson–Cook model, for example, (2.66), that is presented by the following form:

$$\sigma = \left(A + B\varepsilon^n\right)\left(1 + C\ln\dot{\varepsilon}^*\right)\left(1 - T^{*m}\right) \tag{2.74}$$

where σ is the (Von Mises) flow stress, A is the yield stress at reference temperature and reference strain rate, B is the coefficient of strain hardening, n is the strain hardening exponent, ε is the plastic strain, $\dot{\varepsilon}^* = \dot{\varepsilon}/\dot{\varepsilon}_0$ is the dimensionless strain rate with $\dot{\varepsilon}$ being the strain rate and $\dot{\varepsilon}_0$ the reference strain rate, and T^* is the homologous temperature and expressed as

$$T^* = \frac{T - T_{\text{ref}}}{T_m - T_{\text{ref}}}$$

Constants A, B, C, m, and n are Johnson–Cook constants listed in Table 2.3.

The Johnson–Cook plasticity model was used both for particle and substrate impact deformation. By using this material model, the evolution of strain rate, flow stress, and temperature with strain was analyzed at various fields of particle impinged with the substrate.

To avoid the influence of meshing element size on the simulation results a mesh convergence study was performed using the *zero elements* method (Assadi et al., 2003). The *zero elements* method is used to obtain mesh size independent results, due to the fact that the numerical simulation output is known to be so sensitive to element size. To find the meshing size that allows us to obtain reasonable simulation results comparable with experimental data a modeling procedure was performed for mesh size of $0.05d_p$, $0.02d_p$, $0.015d_p$. The main criterion of modeling results validity is chosen to be maximal temperature at the particle surface. Additionally, the flow stress dependence on meshing size is discussed as well. The obtained results are compared with the

TABLE 2.3

Material Properties for Stainless Steel and ARMCO Iron Used in Simulations

Property	Stainless Steel	ARMCO Fe
Density (kg/m³)	7900	7890
Young modulus (GPa)	200	207
Poisson ratio	0.3	0.29
Thermal expansion (1/K)	0.0000519	0.000032
Specific heat (mJ/t·K)	440	452
Inelastic heat Fraction	1	1
Johnson–Cook Constants		
A (MPa)	310	175
B (MPa)	1000	380
n	0.65	0.32
m	1	0.55
T_{melt} (°C)	1400	1538
$T*$ (°C)	727	25
C	0.07	0.06

experimental data available in the literature and numerical results obtained by other researchers (Assadi et al., 2003; Li et al., 2009).

2.3.2   MODELING RESULTS

2.3.2.1   Strain Localization Features at the Particle Impact

The main simulation output parameters describing the deformation behavior and effects of the strain localization of a certain node are chosen to be stress–strain and temperature–strain curves. The shear localization is proven to be the main mechanism of deformation at the high velocity particle impact (Assadi et al., 2003; Grujcic et al., 2004; Champagne, 2007). Because the plastic deformation occurs in the particle material, a large portion of the energy of plastic deformation is converted into heat. In the case of the particle impact with velocity of 800 m/s the strain rate is sufficiently high (about 10^9 s^{-1}), and insufficient time survives for this heat to be transferred away from the area of deformation. Thus significant increase of local temperature occurs, that results in strength loss due to local thermal softening. The plastic deformation process becomes unstable and is characterized by formation of adiabatic shear bands (Assadi et al., 2003). So, to define the strain localization effects in the various nodes of the particle being deformed at the impact, the mechanical behavior of the certain nodes needs to be compared with mechanical behavior of the adiabatic shear band. This behavior may be characterized on the base of numerical calculation of the one dimensional model for unidirectional shearing of a slab (DiLellio and Olmstead, 2003) with using the Johnson–Cook simulation parameters (Table 2.3). The temperature–strain curves may be calculated based on (2.75).

$$T = T_0 + \frac{\beta}{\rho c_p} \int \sigma d\bar{\varepsilon}_p \tag{2.75}$$

Here, ρ is density, β is work to heat conversion factor (usually it is taken as 0.9), c_p is the specific heat capacity. The integral $\sigma d\bar{\varepsilon}$ may be calculated numerically as was made by DiLellio and Olmstead (2003). The nondimensional formulation and scaling of the parameters are needed to apply the results of numerical solution (DiLellio and Olmstead, 2003) of one-dimensional problem of shear localization in a finite slab for analysis of the deformation behavior and effects of the strain localization during particle impact. In accordance with DiLellio and Olmstead (2003), the nondimensional variables temperature $\theta = (T - T_0)/T_0$, shear stress $s = S/S^*$, and time $\tau = t/t_0$ are introduced. Here, T_0 is a reference temperature, $T_0 = 300$ K, S^* is a reference shear stress that is defined on the base of Johnson–Cook equation constants, and t_0 is a reference temperature defined as $t_0 = S^*/\mu\dot{\varepsilon}^*$, where $\dot{\varepsilon}^*$ is the reference strain rate.

The simplified numerical solution of (DiLellio and Olmstead, 2003) allows to obtain dependences only for the case of neglecting hardening. So, only part of curves after flow stress maximum may be obtained. The data of DiLellio and Olmstead (2003) demonstrate a dramatic drop of shear stress on the most of shear stress-time curves that is associated with the formation of shear band. An evolution of the dimensionless stress and temperature parameters of ARMCO iron, nickel, 4340 Steel and other materials differs considerably. The bigger times needed to start the shear localization in ARMCO iron indicates that the regime of shear localization of ARMCO iron is impossible to achieve at the little deformation times or high strain rates. To recognize the strain localization process during particle impact deformation at the CS, the modeling results of flow stress and temperature dependences on impact time and strain need to be compared with shear localization calculation results of the finite slab in the nondimensional formulation (Dilellio and Olmstead, 2003).

2.3.2.2 Stress and Temperature Distribution at Particle Impact

The main aim of particle impact modeling is the comparison of the deformation and temperature behavior of particle materials at the various locations that are defined by the nodes as shown in Figure 2.28b, c (mesh size $0.02d_p$). Let us compare a variation of the flow stress of particle and substrate at the particle/substrate interface at the different times of impact. Figure 2.29 depicts the flow stress patterns after 20 and 50 ns of impact.

It can be clearly seen that the flow stress of stainless steel substrate is much higher than that of ARMCO iron particles at the interface (Figure 2.29a, b). The distribution of flow stresses at the surface and in the core of the ARMCO iron particles is more uniform. A similar situation is seen in Figure 2.29c, d after 50 ns of ARMCO iron particle impact. The increase of the impact time by 2.5 times results in the increase of the flow stress up to 600–700 MPa, while the flow stress of the particle at the interface is the same in both cases. For the stainless steel particles/stainless steel substrate the flow stress distribution is similar both for particles and substrate. The flow stress in both cases is in the range of about 800–1000 MPa. However, the thin layer at the interface has much lower flow stress (about 200–300 MPa) that seems to indicate about strain localization (Assadi et al., 2003). Simulation results of SS steel (Figure 2.29b–d) show the strain localization only in impinged particle and not in the substrate. This is because, for simulation purposes, we applied a different mesh size for particle ($\Delta = 0.02d_p$) and substrate ($\Delta = 0.05d_p$), and thus the influence of meshing size on simulation results can be seen.

For a deeper understanding of the modeling results, the dependences of flow stress and temperature on the impact time were calculated in different nodes both on the surface and in the volume of the particle. Figure 2.30 depicts the flow stress variation with the impact time on the surface for stainless steel (Figure 2.30a) and ARMCO iron (Figure 2.30b) at the nodes shown in Figure 2.28b.

The comparison of the flow stress behavior of these two materials shows that the most considerable fall of the flow stress is observed for stainless steel for nodes no. 3, 4, and 5, while the decrease of the flow stress for ARMCO iron is seen to a lesser extent. This conclusion may be supported by the data (Figure 2.31) that show the temperature dependence on time of the same nodes on the surface of the particle. The biggest effect of the temperature rise is observed for the stainless steel for the nodes no. 3–5. The maximal temperature achieved due to the particle impact is about 1300 K at the impact time of 70 ns. Adiabatic heating of the ARMCO particle is observed to a lesser extent. The general conclusion made from the analysis of simulation results shown in Figures 2.30 and 2.31 is that the stress and strain distribution at the particle surface is not uniform, and the favorable area of the considerable fall of flow stress seems to be in the vicinity of nodes no. 3–5. The adjacent areas of the particle surface undergo weak strain localization that results in the development of the

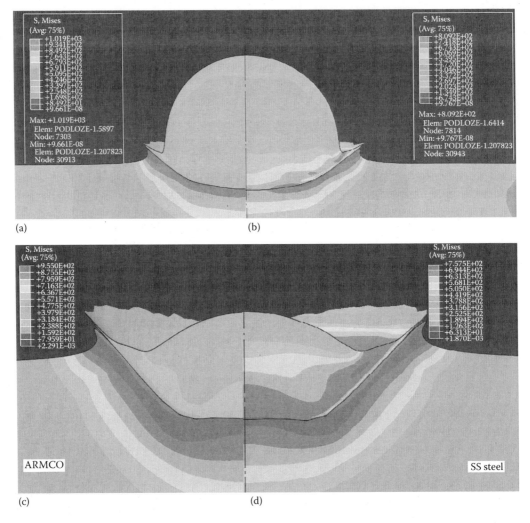

(a)   (b)

(c)   (d)

FIGURE 2.29 Flow stress patterns of the ARMCO Fe and stainless steel at the different times of the impact onto stainless steel substrate. (a) ARMCO Fe, 20 ns, (b) stainless steel, 20 ns, (c) ARMCO Fe 50 ns, and (d) stainless steel 50 ns.

bonding process at the interface to a lesser extent. In these areas, the bonding process seems to be not extensively developed.

Analysis of the impact behavior in various nodes in the particle volume is shown in Figures 2.32 and 2.33.

As in the previous case, the sharp fall of flow stress with the time of impact may be observed in certain nodes of the particle. The dramatic drop of the flow stress can be clearly seen in nodes 1, 2, and 3 near the particle surface, while decrease of the flow stress with time for other nodes is very small. Comparison of the stainless steel and ARMCO iron particle impact behavior shows that the effect of a flow stress fall in certain areas of iron particles is much smaller than that of stainless steel. Additionally, effect of flow stress fall depends on the node coordinate. In this case (simulation with mesh size $\Delta = 0.02 d_p = 1$ μm), the distance between nodes is about 1 μm at the particle subsurface area. Thus, based on these data, the more accurate characterization of the localization area thickness

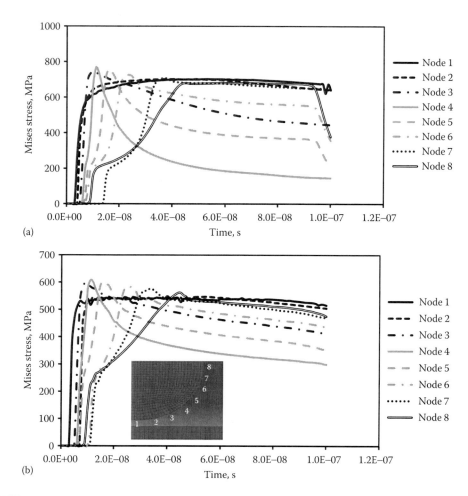

FIGURE 2.30 Dependence of Von Mises stress on the impact time at the surface of particles: (a) stainless steel and (b) ARMCO iron.

for the particle size less than 50 microns seems to be difficult. Most of the nodes shown in Figure 2.28c undergo a hardening and softening processes without the substantial flow stress fall (except nodes no. 1, 2, and 3). This fact has to be taken into account in the analysis of coating structure formation.

The simulation results of temperature-time dependences are shown in Figure 2.33. These data demonstrate difference of the temperature in same nodes of stainless steel and ARMCO iron particles in the range of about 50–300 K.

A mechanical behavior of material during dynamic deformation is believed to be a main indication of shear localization processes (Xu et al., 2008). To differentiate the mechanical behavior during particle impact deformation, the stress and temperature dependences on time of the node no. 4 (Figure 2.30) are compared in nondimensional coordinates with numerical calculation results of shear localization during unidirectional shearing of an SL (DiLellio and Olmstead, 2003) (Figure 2.34). The calculation results (DiLellio and Olmstead, 2003) clearly demonstrate the deformation process may be divided by two periods of time: (1) slow decrease of shear stress and increase of temperature and (2) sharp fall of the shear stress and increase of temperature associated with the formation of a shear

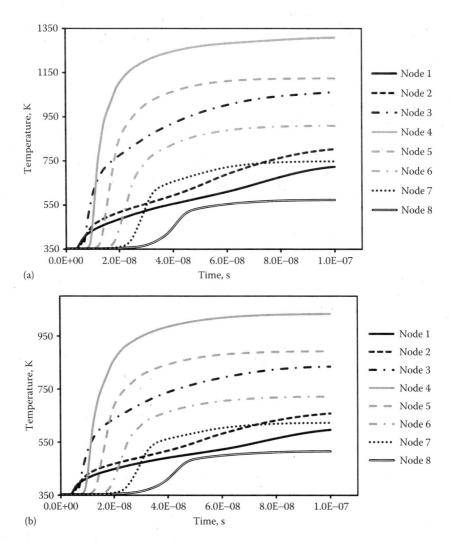

FIGURE 2.31 Dependence of temperature on the impact time at the surface of particles: (a) stainless steel and (b) ARMCO iron.

band. The qualitative comparison of the our modeling curves with results (DiLellio and Olmstead, 2003) show that the drop of flow stress of stainless steel is similar to that of 4340 steel in DiLellio and Olmstead (2003) and occurs at the same nondimensional time ($\tau = 0.02,...,0.07$). Therefore, we can conclude that the dramatic drop of the flow stress of stainless steel is an indication of the formation of shear band. It is interesting to note that the drop of flow stress curve of ARMCO iron is located at the nondimensional time of $\tau = 0.08,...,0.7$ which is much smaller than time of ARMCO iron shear stress fall in (DiLellio and Olmstead, 2003) ($\tau = 0.8,...,2$). It means that the conditions of shear localization of ARMCO iron are believed to be not achieved in the case of CS particle impact.

The qualitative analysis of heat generation during deformation (Figure 2.34b) confirms this assumption. Indeed, the average rate of temperature increase of stainless steel is similar to that of 4340 steel (DiLellio and Olmstead, 2003), while the rate of temperature increase of ARMCO iron is lower than that of (DiLellio and Olmstead, 2003) at localization area ($\tau = 0.8,...,2$). Thus, the shear localization process in ARMCO iron particle is not started at the impact, and the fall of flow stress may be explained by only softening process being developed during the particle impact and being taken into account by Johnson–Cook model. Despite hardening in the model (DiLellio and

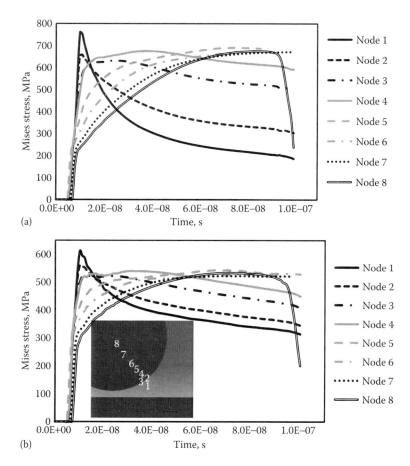

FIGURE 2.32 Flow stress dependence on the impact time for various nodes in the bulk of the particle: (a) stainless steel and (b) ARMCO iron.

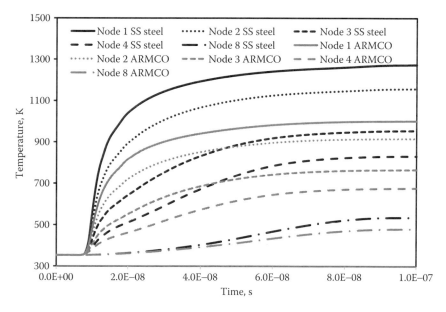

FIGURE 2.33 Comparison of temperature evolution in various nodes of stainless steel and ARMCO Fe particles.

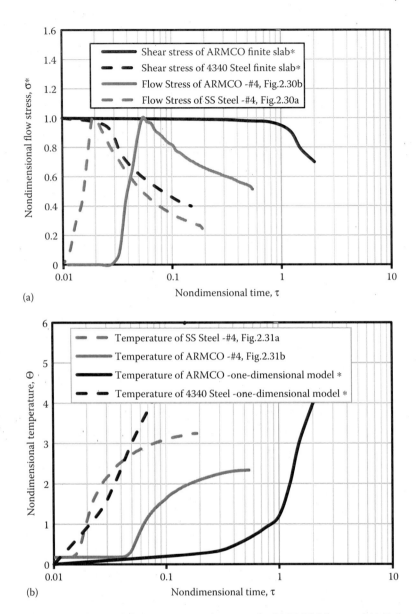

FIGURE 2.34   (a) Stress–time and (b) temperature–time curves of ARMCO iron and 4340 Steel calculated on the base of known constants of Johnson–Cook equation in accordance with DiLellio and Olmstead (*) numerical solution of shear localization in Johnson–Cook materials. For comparison, the similar curves for node no. 4 (Figure 2.30) are shown. (Data from DiLellio J.A. and Olmstead W.E., Numerical solution of shear localization in Johnson–Cook materials. *Mech. Mater.*, 35, 571–580, 2003.)

Olmstead, 2003) has been neglected and only qualitative comparing of various materials is appropriate, the comparison of Figure 2.34 demonstrates that the more conductive ARMCO iron exhibits the effect of shear localization on a bigger timescale.

2.3.2.3   Effect of Meshing Size

As Assadi et al. (2003) notes, the calculated field variables such as temperature, flow stress and strain, only represent the average respective values within particular elements. Thus, for the elements of the interface, where plastic deformation is highly localized, such average values strongly depend on the element size and distance from the interface. To cope with this effect the simulation

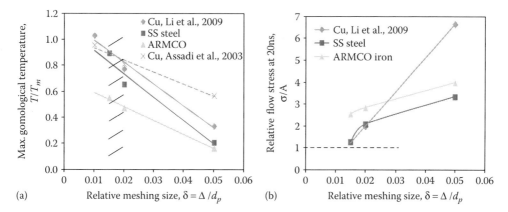

FIGURE 2.35 Effect of meshing size on the particle impact parameters: maximal gomological temperature (a) and nondimensional flow stress (b).

with various meshing size ($0.05d_p$, $0.02d_p$, $0.015d_p$) was made. The main task of this analysis is to define the meshing size that best approaches realistic data.

The simulation results are shown together with Assadi et al. (2003) (Cu particle velocity—600 m/s) and Li et al. (2009) (Cu particle velocity—500 m/s) data in Figure 2.35. To compare the maximal temperature and flow stress at the interface for various materials the nondimensional parameters of temperature and stress were calculated as T/T_m and σ/A, where T_m is the melting temperature of material, A is the Johnson–Cook constant (see Table 2.3). As indicated in Figure 2.35a, the finer meshing size, the higher temperatures are obtained. The real temperatures developed at the interface are known to be below melting temperature (Champagne, 2007). So, the reasonable range of meshing size to obtain the simulation results closest to real parameters is about 0.015–$0.02d_p$ (shaded region in Figure 2.35a). Similar results are obtained by plotting nondimensional flow stress at impact time of 20 ns versus meshing size (Figure 2.35b). It can be clearly seen that materials with high sensitivity to strain localization (Cu, stainless steel) at the chosen particle impact velocities exhibit the flow stress near to the initial flow stress defined by Johnson–Cook equation constant A. It is interesting to note that the calculation results for ARMCO iron at a chosen particle velocity of 800 m/s demonstrate that both the temperature and flow stress at the meshing size 0.015–$0.02d_p$ are far from those of localization. This means that particle velocity is not enough to achieve the effects of adiabatic shear band formation and, consequently, effective particle bonding. Using nondimensional flow stress seems to be more physically based as compared with extrapolation of critical velocities (Li et al., 2009).

2.3.2.4 Particle Deformation Behavior

Analysis of the stress–strain behavior during the particle impact for each node of the particles both on the surface and in the particle volume has to be made based on a detailed description of temperature and strain rate developed in the each node. The numerical simulation allows us to obtain these dependences. An example of such functions for nodes no. 1 and 3 in the volume of the particle (Figure 2.28c) is shown in Figure 2.36.

The comparison of the temperature and strain rate curves of nodes no. 1 and 3 shows that the temperature of node no. 3 is slightly higher than the temperature of node no. 1, while the strain rates of node no. 3 is lower by order than those of node no. 1. Such behavior of the nodes in the particle during its deformation at the impact seems to be the result of variations of thermomechanical parameters during the particle impact. This leads to the specific behavior of material that may be characterized with the flow stress–strain curves. Examples of this curves calculated on the base of modeling results are shown in Figure 2.37.

The deformation behavior of nodes no. 1 and 3 differs considerably for stainless steel. The softening process of node no. 1 is very intensive and that leads to the deformation of the node at

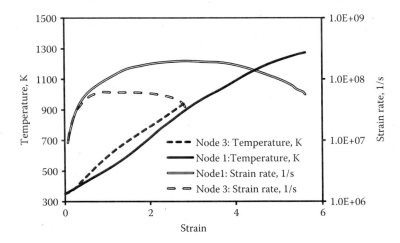

FIGURE 2.36 Temperatures and strain rates of nodes nos. 1 and 3 (Figure 2.28c) versus total strain of stainless steel.

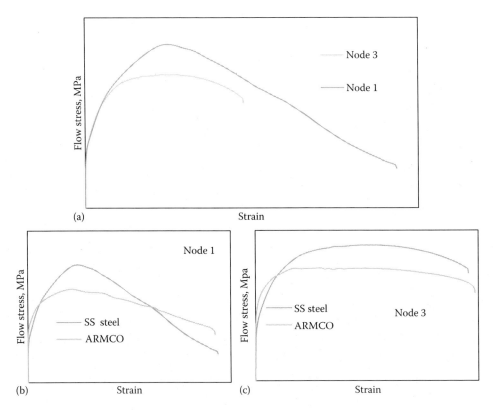

FIGURE 2.37 Strain history of nodes no. 1 and 3 of stainless steel particle (Figure 2.28c) – (a), and comparison of strain behavior of the nodes no. 1 (b) & 3 (c).

low flow stress of 200 Mpa. Intensive plastic deformation of node no. 3 at $\sigma = 500-600$ Mpa and the cupola shape of the stress–strain curve (Figure 2.37a) no. 3 seem to be the evidence of dynamic recrystallization (Wei et al., 2004; Weinert, 2013), because the Johnson–Cook constitutive equation encloses the constants experimentally defined for dynamic recrystallization conditions. Because the distance between nodes no. 1 and 3 is about 2.0–2.5 μm, it may be assumed that the strain localization occurs in the particle subsurface field with a size of about 1 µm. At the same time, a comparison of the stress–strain curves no. 1 for stainless steel and ARMCO iron (Figure 2.37b) demonstrates a slow fall of flow stress of ARMCO iron that says about gradual development of the softening processes without considerable localization effect. Comparison of these data with stress–strain curves no. 3 (Figure 2.37c) demonstrates the similar deformation behavior of ARMCO iron with gradual development of softening processes. Therefore, analysis of the modeled stress–strain curves allows us to uncover the specific features of strain localization during the particle impact.

2.3.2.5 Strain Localization Parameters

The modeling results shown in Figure 2.37 are obtained for all examined nodes at the particle surface and through the core of the particles (Figure 2.28b, c). The main aim of such an analysis is to characterize the strain localization process that results in adiabatic shear bands formation at submicron and nanoscale. For this reason, determination of the strain localization parameters such as width of adiabatic shear bands, temperature distribution in shear bands, and strain localization parameters established by Schoenfield and Wright (Wei et al., 2004) seems to be of great importance. Grujic et al. (2004) suggested to define this parameter as (2.73). Here, σ_{max} denotes the maximum value of the flow stress at a certain node on the particle surface or in the volume of the particle. The numerical solution shown in Figures 2.33 through 2.37 enables the evaluation of the SL parameter taking into account the real dependence of flow stress on the strain rate. We tried to process our results of numerical simulation in accordance with Equation 2.73 to obtain the SL parameter dependence for various nodes of stainless steel and ARMCO Fe. Before this analysis, the general comparison of stress–strain curves for various nodes was made. Numerical simulation results show that the location of the flow stress maximum on the curves σ-ε depends on the coordinate of nodes in the particle (Figure 2.28b, c). The influence of the nodes location on critical strain ε_{cr}, at which the flow stress maximum is reached, is shown in Figure 2.38.

It can be seen that the strain needed to achieve the maximum flow stress is higher for stainless steel as compared with ARMCO iron. The critical strain at the particle subsurface (node no. 2) is similar to that of initial stress–strain curves (Figure 2.28). It is interesting to note that the curves shown in Figure 2.38 have minimum at the distance from the particle surface of about 2–3 µm. The possible reasons of such a behavior of flow maximum at the stress–strain curves is the influence of heat generation and strain hardening at the different strain rates that occur at various nodes. The comparison of strain rates and temperatures of nodes no. 1 and 3 is shown in Figure 2.37. It can be clearly seen that the temperature evolution with the strain is similar for both nodes, while the strain rate of node no. 1 is about 5 times higher than that of node no. 3. Because the plastic deformation of the particles is a thermally activated process, the decrease of strain rate at nodes no. 2 and 3 results in diminishing the heat generation effect that is responsible for softening processes. So,

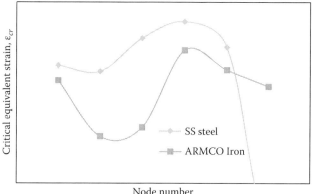

FIGURE 2.38 Critical strains at various nodes of the particle volume (Figure 2.28c).

strain hardening and generation of dislocation at nodes no. 2 and 3 is higher than at node no. 1. For this reason, perhaps, the critical dislocation density at nodes no. 2 and 3 is achieved at lower critical strains.

As previously shown, the *SL* parameter is suggested as a main criterion of strain localization (Wright, 1995; Grujicic et al., 2004). To calculate the *SL* parameter in accordance with (2.73), we defined the derivatives $\partial\sigma/\partial\varepsilon$, $\partial^2\sigma/\partial\varepsilon^2$, and $\partial\sigma/\partial\dot{\varepsilon}$. With this aim, the flow stress–strain curves of various nodes were polynomial approximated. Calculations show that the partial derivative $\partial\sigma/\partial\dot{\varepsilon}$ for our case for node no. 1 is about 7.2×10^{-10} MPa s. The results of these calculations are depicted in Figure 2.39a, b as dependences $\partial\sigma/\partial\varepsilon$ and $\partial^2\sigma/\partial\varepsilon^2$ and *SL* parameter on the strain.

The results show the following differences of deformation behavior of stainless steel and ARMCO iron particles at the area of interface.

- The first derivative curve of stainless steel differs considerably from that of ARMCO iron. Although the critical strain of the flow stress maximum is approximately equal for both materials, there is a minimum on the stainless steel first derivative curve that demonstrates a decrease of softening rate at the low flow stresses due to considerable localization of deformation in the narrow area of adiabatic shear band.

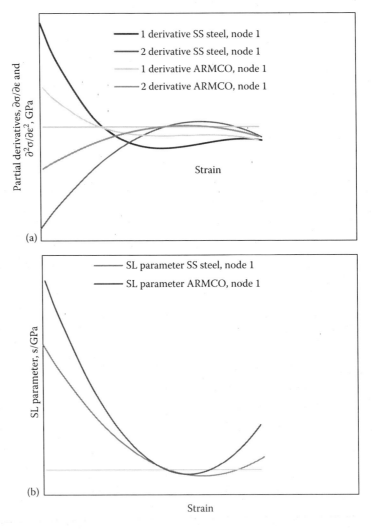

FIGURE 2.39 Dependences of $\partial\sigma/\partial\varepsilon$ and $\partial^2\sigma/\partial\varepsilon^2$ (a) and *SL* parameter and (b) on the strain.

- The more intensive softening processes in the stainless steel proceeds as compared to ARMCO iron. The second derivatives curve allows us to differentiate the behavior of materials at certain strains (Figure 2.39a). $\partial^2\sigma/\partial\varepsilon^2 = 0$ at the strain of about 3.1 at which the flow stress of AISI 904L is about 0.57 GPa. It may be assumed that at these conditions adiabatic shear bands have been created. However, a detailed analysis of adiabatic shear band formation mechanism similar to that made in Dodd and Bai (1984) is needed to prove this assumption.

The general characterization of the shear localization process is made on the basis of the analysis of SL-strain function as shown in Figure 2.39b. The numerical simulation of the particle impact process allows us to obtain this function for all strains achieved during particle impact. It can be seen that minimal values of SL parameter show high sensitivity of material to strain localization.

2.3.3 VALIDATION OF NUMERICAL SIMULATION RESULTS

2.3.3.1 Materials and Deposition Parameters

Stainless steels, carbon steels and other iron-based alloys play an important role among commercially available materials because of their excellent mechanical and exploitation properties such as ductility, corrosion, and wear resistance. It is well known that susceptibility of these materials to adiabatic shear band formation depends strongly on carbon content, structure, and strain rate (Lee et al., 2008). Various materials have their own sensitivity to localization at the CS process, and this results in various critical velocities needed to realize the deposition process (Papyrin, 2007). The experiments of Van Steenkiste et al. (1999) show that the deposition efficiency of ARMCO iron at CS is much lower than that of stainless steel. We obtained similar results in our experiments (Leshchynsky et al., 2011). The main reason for the behavior of these materials during CS seems to be the differences of the deformation mechanisms during particle impact. For this reason, stainless steel and ARMCO iron powders were chosen for this study. Chemical composition of the materials is shown in Table 2.4. The stainless steel grade approved for experiments is AISI 904L because this material belongs to super austenitic stainless steel grade that strikes a good balance between relatively cheap austenitic stainless steel and expensive Ni-based superalloys (Sathiya and Jaleel, 2011). Structurally, 904L is fully austenitic and is less sensitive to precipitation ferrite and σ-phases than conventional austenitic grades. Characteristically, due to the combination of relatively high contents of chromium, nickel, molybdenum, and copper 904L has good resistance to general corrosion, particularly in sulfuric and phosphoric conditions. That is why AISI 904L cold sprayed coatings are planned to be applied for corrosion protection.

ARMCO iron is chosen for the study as a modeling material. ARMCO iron is being widely used in industry applications such as aviation construction, nuclear technology, the production of magnets (pole cores, yokes, and armatures), in automotive construction, as magnetic shielding, as welding rods and fuse wire, as gaskets in the chemical and petrochemical industry, as well as in power station construction.

Figure 2.40 depicts the SEM images of the particles of chosen materials.

TABLE 2.4
Chemical Composition of Powders for Investigation

Material	Chemical Composition (%)									
	C	Cr	Cu	Mn	Mo	Ni	P	S	Si	Fe
AISI 904L	0.018	20.0	1.0–2.0	1.7	4.5	26.0	0.035	0.025	0.9	bal
ARMCO iron	0.019	–	0.06	0.20	–	–	0.015	0.015		bal

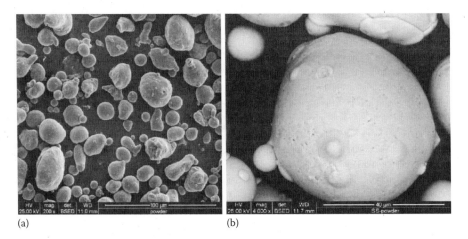

(a)   (b)

FIGURE 2.40 SEM images of the particles. ARMCO iron particles (a) and AISI 904L particles (b).

The structure of ARMCO iron is pure ferrite with a grain size of 10–15 μm. This allows us to characterize the distortion of both particles and grains after particle deformation due to impact in CS. Stainless steel particles have an austenitic structure with a grain size of 2–5 microns (Figure 2.40b). Etching with Glyceregia etchant (3 parts HCl, 2 parts Glycerol, 1 part HNO_3) reveals the equiaxed austenite grain structure. One can note that the similar structure of AISI 904L was shown in paper (Sathiya and Jaleel, 2011).

Evaluation of strains at the certain nodes of the particle deformed due to impact was made by experimental measurement of distortion of the particle grains taking into account an assumption of only equiaxed grain occurrence in the virgin particle.

The atomized powders were used in CS deposition on steel plates. The particles sizes are shown in Table 2.5. It can be seen that the average size of the stainless steel and ARMCO iron particles is about 50 μm.

The Kinetik 4000 CS system was employed to produce one pass coatings. The detailed description of this CS system has been given in Klinkov et al. (2009). The standoff distance from the nozzle exit to the substrate surface was 20 mm. The spray conditions for the powders are also given in Table 2.4. The spray gun was manipulated by a robot and moved at a traverse speed of 300 mm/s across the substrate surface to obtain the coating. The powder mass flow rate was calculated by weighing the powder injected into a closed volume after 10, 30, and 60 s time intervals. Nitrogen was used as a carrier gas with 5 m³/h flow rate. The injected particles are accelerated in the high velocity air stream by the drag effect. To increase the nitrogen velocity and, ultimately, the particle velocity, the compressed gas was preheated to 300°C–600°C. The pressure and temperature of the gas were monitored by a pressure gauge and a thermocouple positioned inside the gun. The gun was installed on an ABB manipulator to scan the gas-powder jet over the substrate surface. The compressed nitrogen pressure was kept constant at 3.5 MPa. The particle velocities at the exit plane of the supersonic nozzle, as measured by a Laser Doppler Velocimeter, were in the range of 600–800 m/s.

TABLE 2.5

The Particle Sieve Analysis and Main Cold Spray Parameters

Powder	Size Distribution (%)			Gas	Pressure (MPa)	Gas Temperature (°C)
	−10 μm	−50 μm	−120 μm			
ARMCO Fe	10	88	2	Nitrogen	40 bar	600
AISI 904L	8	92	–	Nitrogen	40 bar	600

Coated coupons were sectioned perpendicular to the coating surface using a disc cutting machine Delta AbrasiMet and then cold mounted and polished according to conventional metallographic procedures. The powders for cross-sectional examination were also cold mounted and polished (up to 1 μm diamond) using conventional procedures. The microstructure of the precursor powders and sprayed coatings was examined using an optical microscope and SEM using the SEM microscope FEI.

The experimental procedure (Li et al., 2006) was approved for grain strain localization evaluation. The cross-sectional morphologies of the coatings were observed by using a SEM and the values of flattening ratio of the deformed grains were estimated from the cross sections.

The flattening ratio (R_f) is defined according to the definition in (Li et al., 2006) as $R_f = D_s/D_p^o$, where D_s is the spreading diameter of the flattened grain, D_p^o is the original grain diameter. The spreading diameter is defined as the maximum diameter of the deformed grain perpendicular to impact direction. The maximum width and height of the grains in cross section were measured first. Then, the equivalent grain diameter was estimated from the area of the particle cross section. The flattening ratio was calculated using the relationship $R_f = D_s/D_p^o$ by taking the maximum width as D_s and the equivalent diameter as D_p^o. These parameters were applied to characterize the grain strains at various distances from the interface.

2.3.3.2 Comparison of the Results

Because of the very short duration of the deformation process during particle impact there are some difficulties to record stress and strain parameters. For this reason, only deformed particles were observed. The results of the metallographic examination of AISI 904L and ARMCO iron coating structure are shown in Figures 2.41 and 2.42. The particle shape after deposition is seen in Figure 2.41. The results demonstrate the visible difference of particle shape. ARMCO iron particles are of ellipsoid shape with a bigger flattening ratio, while AISI 904L particles are distorted to a lesser extent. This seems to indicate a higher hardening coefficient of AISI 904L as compared with ARMCO iron (see Figure 2.29 and Table 2.4). For this reason, the multiple impact of the ARMCO iron particles during CS coating formation results in higher total particle strains. However, a detailed analysis of the grain shape before (Figure 2.40) and after CS (Figure 2.42) show the effect of strain localization of AISI 904L.

The SEM images of AISI 904 particles after coating formation show considerable deformation of its grains at the areas of interface (Figure 2.42a). We defined the flattening ratio of the grains at the different distances from interface (y-direction on Figure 2.42). If the initial equiaxed grain size of AISI 904L is about 3–4 μm (Figure 2.40), the grains after deformation at the interface in the area adjacent to the interface are distorted and have a lens shape with the size about 10–12 μm. As can be seen in Figures 2.41a and 2.42a, the severe deformed interface areas of AISI 904L are intensively

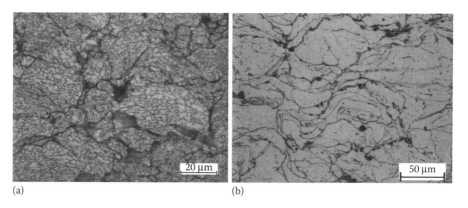

(a)   (b)

FIGURE 2.41 Optical images of deposited coating microstructure: (a) AISI 904L and (b) ARMCO iron.

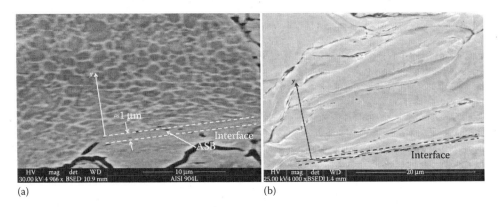

(a)   (b)

FIGURE 2.42   SEM images of AISI 904L (a) and ARMCO iron and (b) grains in the particles.

etched and some amount of material from the interface has been dissolved. The areas of dissolution seem to be the ASB. The opposite is the case for ARMCO iron particles (Figure 2.41b, 2.42b); the severe but relatively uniform deformation of the whole big sized grains can be seen. While the initial equaxed grain size of ARMCO iron particles is about 10 μm, most of the ARMCO iron particle grains after deformation have an ellipsoid shape (Figure 2.42b), and there is no evidence of severe deformation of grains adjacent to the interface. Thus, numerical simulation and experimental investigation of the particle deformation process at CS shows similar results.

To compare the numerical simulation and metallographic study results, the influence of distance from the interface on the flow stress at the time of impact of 80 ns and on grain flattening ratio of the coating was determined (Figure 2.43). In spite of high scatter of experimental results of flattening ratio determination the effect of strain localization of AISI 904L can be clearly seen. It is interesting to note that there is no correlation between grain flattening ratio and distance from interface for ARMCO iron that shows evidence of the absence of strain localization of this material.

The numerical simulation shows the strong dependence of the flow stress on the distance from interface. The numerical procedure allows us to accurately define the width of strain localization area due to an inflection of flow stress–distance curve. This analysis will be made in the following work.

A thorough analysis of deformational behavior and numerical modeling of cold spraying of the two particulate materials (stainless steel and ARMCO iron) result in the following specific features of this behavior.

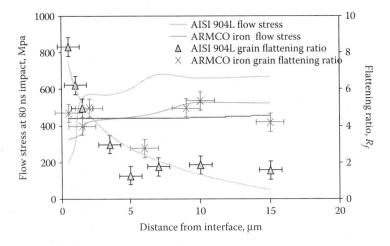

FIGURE 2.43   Comparison of numerical simulation and experimental data.

- The dependences of both stress and temperature on the strain of the two materials at high velocities were calculated. The graphs indicate the presence of maximum on both stress–strain curves, which is an evidence of the occurrence of recrystallization processes.
- The flow stress pattern images of the two materials at the different impact tiles indicate lower sensitivity to localization of ARMCO iron. The localization of deformation is observed in the stainless steel particle at 50 ns of impact, but not in the substrate of the same material. Different mesh size used for particles and substrate may have influenced this result.
- Several nodes on the surface and through the core of the particle were introduced to conduct a more detailed investigation of stress and temperature distribution in the particle. It can be seen that some nodes undergo an excessive deformation while others are deformed to a lesser extent. The difference of the maximum attained temperatures and flow stresses demonstrates lower sensitivity of ARMCO iron to localization.
- The sharp fall of flow stress at several nodes in stainless steel particle suggests the occurrence of deformation localization at these pointes, whereas others mostly undergo the dynamic recrystallization process.
- Based on the analysis of dimensionless flow stress—meshing size dependences, the optimal meshing size is found to be 0.015–0.02 *dp*.
- Strain rate is found to vary significantly in different nodes of the particle; that needs to be taken into account during evaluation of nodes' deformation behavior.
- Based on modeling results critical strains of different nodes of the particle were defined. Although critical strain achieved through the body of stainless steel particles is higher than that of ARMCO iron, the general behavior of the curves is the same for both materials. This indicates that the intensity of occurring processes of softening due to heat generation and strain hardening varies considerably from node to node.
- Strain localization parameter was calculated. Experimental method of strain localization parameter definition by using grain flattening ratio defined on the base of metallographic examination was compared with modeling results (flow stress data). It was found that the modeling results allow us to define the area of strain localization more accurately than the experimental data.

3 Cold Spray Powders and Equipment

Paolo Matteazzi, Alberto Colella, Volf Leshchynsky,
Kazuhiko Sakaki, Hirotaka Fukanuma, and Roman Gr. Maev

CONTENTS

3.1 POWDER MATERIALS FOR COLD SPRAY

Paolo Matteazzi

Alberto Colella

Volf Leshchynsky

3.1.1 INTRODUCTION

A particle shock consolidation encountered in the coldspray (CS) process is defined by a high-velocity impact of powder particles onto the substrate. The particle impact results in the generation of high stresses and strains both in the particles and the substrate. As shown in the basic monographs (Champagne, 2007; Papyrin et al., 2007), the powder material to be sprayed must feature sufficient ductility to ensure particle strains and cold welding without its failure. In some cases (at very high particle velocities) Ti alloy particle melting occurs (Vlcek et al., 2002). However, impact of other metallic powder materials is characterized by high stresses and particle shear strains. Because a lot of information is available on various powders and powder mixtures developed and applied for cold spraying (Jeandin et al., 2014; Moridi et al., 2014, and others), the goal of this chapter is to describe and discuss the concept of CS material selection, basic criteria for evaluation of its suitability for cold spraying, and particle behavior during the deposition process (acceleration and formation of interfaces).

As we have shown in Chapter 2, the concept of evaluating materials by the critical velocity necessary for their deposition is well accepted to explain bonding and jetting phenomena (Champagne, 2007). According to the theory of impact loading (Meyers, 1994; Edwards, 2006), a pressure pulse created by impact and the resulting plastic shock wave lead to intensive particle strain, providing the strain distribution through the particle is quite nonuniform. By means of the equations of state, the maximum impact pressures can be estimated, showing that for the particle velocities well above 10^3 m/s, peak shock pressures of 40–50 GPa are possible for iron- and copper-base materials (Vlcek et al., 2005). A key variable of the powder shock compaction process is the contact stress pulse time needed for good bonding between powder particles. If an elastic load relief would be more than bonding strength, debonding would also be possible. The deformation kinetics is determined by the material properties, mainly the crystal and grain structure, as well as by the type of bonding within the material. The Grüneisen parameter is key in the calculation of shock pressure and dynamic yielding and combines the mechanical and thermodynamic properties of a material. The calculation results made by Vlcek et al. (2005) demonstrate that feedstock properties are the key variables in the CS process. Vlcek et al. (2005) clearly show that a coating buildup by the CS process will mainly be governed by the material sprayed. Let us analyze these data in detail.

3.1.2 IMPACT DYNAMIC CRITERIA OF MATERIALS EVALUATION

Vlcek et al. (2005) described the main equations of impact loading based on high-velocity impact dynamics developed by Zukas (1990). The behavior of shock waves in the material is described by the laws of the conservation of mass, the conservation of impulse, and the conservation of energy versus the shock front. For the general case of a body traversed by a shock wave and having the state t_0 before the shock-wave front and the state t_1 just behind the shock, the general balance equations are analyzed. Based on general balance equations, two types of equations can be calculated: (1) Hugoniot curve (the p–V relationship) and (2) simple representation of mass velocity u (u_1) versus shock-wave velocity U. For most metals, a linear mass velocity dependence on shock-wave velocity may be observed at mean pressure levels. So, the u–U curve can be described by the following straight line (Zukas, 1990):

$$U = C_0 + S \cdot u \quad \text{with} \quad C_0 = \sqrt{\frac{K}{\rho_0}} \qquad (3.1)$$

where S depends on the material properties, which are combined in the Grüneisen parameter Γ (Zukas, 1990):

$$S = \frac{1}{2} \cdot (1 + \Gamma) \quad \text{with} \quad \Gamma = \frac{3 \cdot \alpha K}{\rho_0 \cdot c_v} \tag{3.2}$$

The Grüneisen parameter Γ describes the relationship between internal energy and pressure at a constant volume. The linear coefficient of expansion α, the thermal capacity c_v, the density ρ_0, and the modulus of compression K are included. The flow of the material occurs at the flow limit σ_{so}, defined by the criterion established by Mises and Tresca $\sigma_{so} = \sigma_x - \sigma_y$ at

$$\sigma_{HEL} = \left(\frac{K}{2G} + \frac{2}{3} \right) \cdot \sigma_{so} \tag{3.3}$$

and is termed the hugoniot elastic limit (HEL) or dynamic flow limit. At this limiting stress, the elasticity of the material is exceeded and a transition occurs from the elastic to the plastic shock wave, with plastic deformation of the material. The results of σ_{HEL} calculations made by Vlcek et al. (2005) are presented in graph $\sigma_{HEL} = f(\sigma_{so})$ (Figure 3.1) for Al and Al alloys, copper, stainless steel, and Ti and Ti alloys. Aluminum and copper are characterized by the low dynamic flow limit $\sigma_{HEL} < 0.5$ GPa, while Ti and Ti-6Al-4V alloy exhibit $\sigma_{HEL} \approx 3$ GPa. These results reveal that the metal powders with high dynamic flow limit such as Ti alloys have poor compressibility during shock compaction due to substantial stresses needed to achieve the transition from an elastic to plastic shock wave.

In accordance with Zukas (1990), for mean shock-wave pressures of $t_o \approx 10$ GPa, bulk modulus, K, and shear modulus, G, are pressure-independent constants. These constants for different metals are extracted from the work by Vlcek et al. (2005) and are shown in Table 3.1. One has to note that K and G were measured at ambient temperature T_o. However, the melting temperature T_m of metals differs considerably. So, the thermodynamic state at which these constants are measured is different too. For this reason, it is reasonable to define the normalized or homological ambient temperature for each material as $T_{oh} = T_o/T_m$ in K.

A combined influence of K and G on material properties, according to Equations 3.1 and 3.2, seems to better evaluate as a dependence $K \times G = f(T_{oh})$ (Figure 3.2). Three groups of evaluated

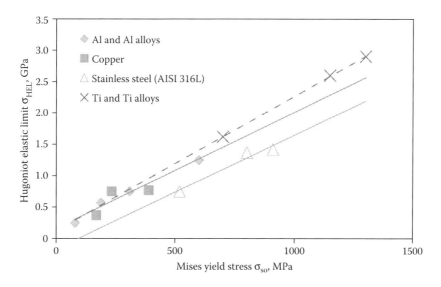

FIGURE 3.1 Dependence of the Hugoniot elastic limit on the yield stress. (After Vlcek et al., A systematic approach to material eligibility for the cold spray process. *J. Therm. Spray Technol.,* 14, 125–133, 2005. With permission.)

TABLE 3.1

Impact Loading Characteristics

Metal	Homological Temperature, $T_h = T_oK/T_mK$	$K \times G/1000$ (GPa²)	Grüneisen Parameter (Γ)
Al	0.28	2	2.2
Cu	0.225	2.5	–
Au	0.2	4	2.85
Cu	0.2	5.5	1.8
AISI 316L	0.17	14	–
Pt	0.13	16	2.6
Cd	0.52	1	2.35
Zn	0.38	3	2.25
Mg	0.3	1.2	1.5
Co	0.15	13	1.8
Ti	0.14	5	1.25
Co	0.15	13	1.75
V	0.13	7.5	1.3
Cr	0.13	17.5	0.9
Nb	0.1	6	3
Mo	0.08	32.5	1.6
Ta	0.05	13.5	1.6
W	0.02	50	1.6
Fe	0.12	14	1.5

Source: Vlcek et al., A systematic approach to material eligibility for the cold spray process. *J. Therm. Spray Technol.,* 14, 125–133, 2005. With permission.

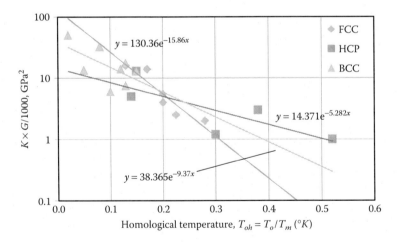

FIGURE 3.2   The dependence $K \times G = f(T_{oh})$.

materials are compared: face-centered cubic (FCC), body-centered cubic (BCC), and hexagonal closed packed (HCP). The results demonstrate a strong dependence of $K \times G$ on the homological temperature. One has to note that the intersection of three exponential approximations of each group of metals occurs at one point with coordinates $T_{oh} = 0.2$ and $K \times G/1000 = 5.5$ GPa (Cu) (Table 3.1). This means that the graphs, Figure 3.2, may be divided into two areas of temperatures, and the ability for deformation and shock consolidation at high-velocity impact is perfect for metals with homological ambient temperature $T_{oh} \geq 0.2$. So, from the viewpoint of their CS ability the

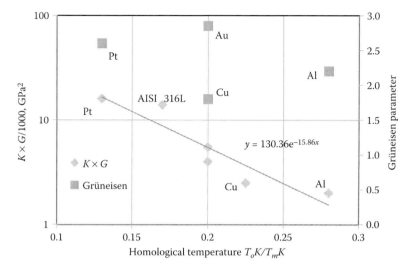

FIGURE 3.3 Grüneisen parameter dependence on homological temperature.

metals and alloys may be shared by two groups: (1) with melting temperature lower than 1465 K—good CS ability and (2) with melting temperature higher than 1465 K—low CS ability. Materials of the second group have $K \times G > 7 \times 10^3$ GPa.

The data of Figure 3.2 reveal that the elasticity parameter $K \times G$ correlates well with homological temperature of metal or alloy with certain crystal symmetry (FCC, BCC, and HCP) contrary to Grüneisen parameter (Figure 3.3). A simulation veracity of the exponential approximation of the dependence $K \times G = f(T_{oh})$ for FFC metals is about $R^2 = 0.858$.

3.1.3 POWDER MATERIAL SYSTEMS FOR CS

The approach developed in a review by Moridi et al. (2014) is employed as the base for description of the CS coating materials.

Development of new industries and new commercial product applications based on reduced material consumption need the advanced coating technologies and, in particular, CS technology with its unique capabilities as previously described. While at the beginning, CS technology was developed for the deposition of the soft metallic materials, such as Al, Cu, and Zn, but at the present time a wide range of materials, including hard ones, have been successfully sprayed to form coatings with various properties. For instance, one could mention successful cold spraying of hard and brittle compounds on a ductile matrix by the utilization of the particle deep penetration effect. Currently, studies in this field are expanding and the main purpose of this section is to shortly describe the various powder materials being applied and being developed for cold spraying. From this viewpoint according to Moridi et al. (2014), material systems involved in CS technology are divided into the following important categories: metals, metal-matrix composites, ceramics, polymers, and nanostructured powders.

In the first instance, during CS the particle impact results in a particle and substrate plastic deformation process. One has to note that the type of crystal lattice defines the mechanisms of plastic deformation during the particle impact of the CS process. Based on modeling results, the specific features of strain localization during the BCC and FCC metal particle impact are discussed in Chapter 2, and stress–strain curves are calculated.

The deformation strain rate on particle impact is extremely high ($d\varepsilon/dt > 1 \times 10^5 \text{s}^{-1}$), which is why additional strain-hardening effects due to the accumulation of dislocations, interactions of the atoms among one another, and, if appropriate, a temperature-induced decrease in the yield criterion value should be taken into consideration. For example, some observations made by Vlcek et al. (2005)

show that for the strain-hardening alloy 316L it is difficult to generate dense coatings. Ti-6Al-4V alloy exhibits a decrease in strain hardening at temperatures below the β-transition and can only be cold-worked with difficulty, which explains the high porosity and cracks in the corresponding coatings. In the CS process, the response of the materials to high-velocity deformation and the associated strain rates of $>1 \times 10^5$ s^{-1} is decisive. Usually the strain rates during particle deformations are in the range of 1×10^5 to 1×10^7 s^{-1}. The deformation process is determined by the mobility of the dislocations and their interactions in the deformation process. Consequently, in particular, the crystal structure and the type of bonding as well as further parameters, such as structure, grain size, and foreign atoms or phases, determine the mechanisms of deformation. Comparing the deformation properties of different metals clearly show that those metals of one crystal structure and type of bonding feature similar deformation mechanisms (Frost and Ashby, 1982). Hence, the polycrystalline solids can be classified into the following three groups (Table 3.1):

1. Al, Cu, Ag, Au, Pt, Ni, and γFe—FCC
2. W, Ta, Mo, Nb, V, αFe, and βTi—BCC
3. Cd, Zn, Co, Mg, and Ti—HCP

The FCC metal structure features the greatest packing density (PD = 12), coordination number (KZ = 12), and a large number of sliding planes, which explains the good deformability (Kittel, 2004). The HCP crystal structure features the same PD, but due to the spatial arrangement in a tight stacking sequence the number of sliding planes is large resulting in less deformability. In addition, the stacking faults greatly influence the strain mechanisms. The BCC metal structure has a significantly lower PD (0.68) and KZ (8), which is why it should be assigned the lowest deformability of the BCC structures. We did not classify the groups of amorphous, tetragonal, or trigonal crystal systems including oxides, which due to their low plasticity are not suitable for the CS process at the present time. A correlation between deformation properties and melting temperature for the BCC, FCC, and HCP groups is shown in Figures 3.2 and 3.3. The product of shear modulus and modulus of compression takes into consideration two important material parameters and clarifies the distribution of the data points (Vlcek et al., 2005). Figure 3.2 illustrates the specific distribution of the BCC, FCC, and HCP groups, whereby it is generally more difficult to process BCC metals in the CS process because, at deformation under high strain rates, screw dislocation mobility is strongly hindered by the Peierls stress.

Simple estimates of the CS ability of γ Fe-based (stainless steel) and Cu-based powder materials on the base of stress–strain curves modeling feature a relatively low yield stress, and high impact pressures and temperatures are reached, which is why the localized severe plastic deformation can be assumed. Dense aluminum material can be sprayed easily in the CS process, which is reasonable to be understood by the effect of localization at the high shock temperatures, low dynamic flow limit, and low yield strength value. In contrast, the poor deformation and strain localization ability of BCC-structured metals reproduces the poor suitability of this group for the CS process. Classifying the metals by isomechanical groups (i.e., the groups that combine the similar mechanical properties), defined according to the classification of Frost and Ashby (1982), reproduces the general suitability for the CS process. As previously shown, the mechanisms of deformation of the materials are correlated with their crystal structure. BCC materials usually prove to be difficult in the CS process (i.e., processing with helium is the alternative to obtain pore-free coatings). At moderate strain rates, some BCC metals may deform in a manner that is similar to that of FCC metals, but at the strain rates that are typical for the CS process the necessity to activate screw dislocation movement (which needs more energy than edge dislocations in an FCC matrix) makes this type of metal significantly more difficult to deform plastically. Low-melting-point HCP metals have been found to be easy to process. This proves that, besides the mechanical characteristics, other features such as the bonding energy must be taken into consideration.

The parameterization of the materials versus the Grüneisen parameter Γ (Equation 3.2, Figure 3.3) does not provide a comparable classification because of the great influence that K and c_v

have on Γ. Attempts to classify the ability of the powder material or the CS process based on the Peierls stress appear to be appropriate from the view of dislocation theory because the Peierls stress together with stacking fault energy is a well-known measure for the energy necessary to move dislocations or enable twinning in a lattice (Hirth and Lothe, 1992).

Based on the above analysis, one could conclude that CS suitability of powder materials depends on their deformation properties including the ability for strain localization and adiabatic shear band formation. Materials with relatively low melting point and low mechanical strength such as Zn, Al, and Cu are ideal materials, as they have a low yield strength and exhibit significant softening at elevated temperatures. No gas prewarming or only low process temperatures are required to produce dense coatings using these materials. It is worth mentioning that the deposition of Al is somewhat more difficult than other soft materials such as Zn and Cu. This is attributed to its high heat capacity, which makes it more difficult to achieve shear instability conditions during impact, in spite of its low melting point and low yield strength. As shown above, for the majority of BCC-based alloys with high strength such as carbon steels and Ni base materials, high particle velocities may provide enough energy for successful deposition, and CS technology is being developed to deposit stainless steels, titanium and its alloys, nickel and its alloys, and tantalum (Champagne, 2007).

Amorphous metals may also be deposited using CS. An amorphous metal (also known as metallic glass or glassy metal) is a solid metallic material, usually characterized by its lack of crystallographic defects such as grain boundaries and dislocations typically found in crystalline material (Schuh et al., 2007). The absence of grain boundaries, the weak spots of crystalline materials, leads to better resistance to wear and corrosion in amorphous metals. Therefore, they are potential candidates to form a strong coating. Amorphous metals, while technically glasses, are also tougher and less brittle than oxide glasses and ceramics. In fact, they exhibit unique softening behavior above their glass transition temperatures. A number of alloys with low critical cooling rates have been produced; these are known as bulk metallic glasses (Peker and Johnson, 1993). The most useful property of bulk amorphous alloys is that they are true glasses, which means that they soften and flow on heating during the particle impact.

3.1.4 Particle Behavior at Acceleration

The particle size, shape, and structure influence on coating structure formation during cold spraying, and these factors have been analyzed in Chapter 2 for atomized ARMCO iron, stainless steel, and copper. The most important parameter characterizing the deposition process is known to be particle velocity, which depends on the particle shape, size, and powder-laden gas jet characteristics. The powder sieve analysis provides the particle size distribution of certain powder material (an example is shown in Figure 3.4). The results reveal that atomized stainless steel powder contains about 10% of particles less than 10 μm and about 90% of particles of the sizes of 10–50 μm. The acceleration of various particles depends on its size. For this reason, particles impingement and formation of clusters is an inherent feature of the acceleration process. While the particle impingement with the substrate has been discussed in Chapter 2, impingement of the particles and formation of the particle clusters during particle acceleration in de Laval nozzle are not analyzed yet.

The scanning electron microscopy (SEM) images (Figure 3.5) of Ti and Al powders illustrate the presence of the relatively small particles in the feedstock. The detailed analysis of the flow characteristics of gas-particle suspensions in a powder-laden jet seems to be important for the evaluation of particle–particle collisions and particle–turbulence interactions (Maev and Leshchynsky 2007). These processes were studied both for diluted (Senior and Grace, 1998) and concentrated gas–solid suspensions (Eskin and Voropaev, 2004). Despite the relatively high gas velocities (300–900 m/s) in the CS nozzles similar to *fast fluid bed* units of Senior and Grace (1998), the concentration of suspended solids is neither uniform along the nozzle axis nor over the cross section of the flow. High particle concentrations generally form near the nozzle walls where particles are moving with lower

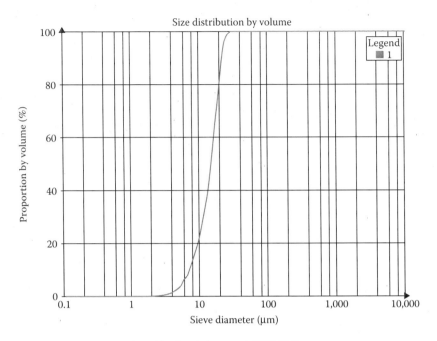

FIGURE 3.4 Sieve analysis (by volume%) of stainless steel AISI 904L.

(a)   (b)

FIGURE 3.5 SEM images of Ti (a) and Al powder (b).

velocities. Clusters of particles may also intermittently form and disintegrate in the diluted suspension flows. Thus, the probability of the formations of particle clusters may be evaluated through the differences in axial particle velocities ΔV_i. The gradient $\Delta V_{(1-50)} = V_1 - V_{50} = \sim600$ m/s, where V_1 and V_{50} are the velocities of particles having sizes of 1 and 50 µm, respectively. In fact, even a small difference in particle size results in a velocity gradient of about 100–200 m/s. The Al powder and scheme of particle clustering for the case of spraying bimodal powder mixtures are shown in Figure 3.6. In this case, particles having low velocities and high weights—which are the cores of the clusters—play the role of shot balls in the powder shock consolidation process. These particles are distributed within the matrix of small particles (Figure 3.6b). The ratio of the particle sizes,

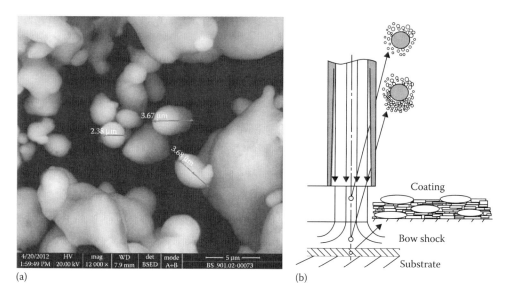

FIGURE 3.6 The closed view of Al particles at magnification 12,000× (a) and cluster formation scheme (b).

weights, and concentration governs the efficiency of the powder shock compaction process, resulting in bonding and the densification of powder layers.

Impingement between particles during its acceleration in the de Laval nozzle is possible to characterize by SEM and energy-dispersive X-ray spectroscopy (EDS, EDX) examination of the particles after blowing out the nozzle. The dendrite Cu and polyhedron tungsten particles are chosen for cold spraying experiments. The Cu+10 wt% W powder mixture was used for acceleration in low-pressure cold spraying machine.

The powder was collected after spraying into free space and examined by SEM and EDS. The Cu particles surface topography examination results before (Figure 3.7a) and after (Figure 3.7b) mixing with tungsten particles demonstrate the slight surface deformation of the dendrite Cu particles. A surface densification of Cu particles may be observed (Figure 3.7b).

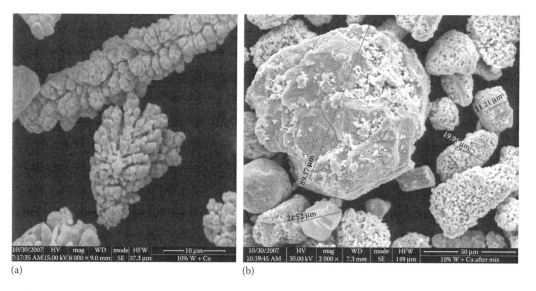

FIGURE 3.7 SEM image of surface topography of Cu particles (a) and Cu + 10 wt% W powder mixture (b) after mixing in Turbula mixer during 3 hours.

Some interactions between hard tungsten and soft copper particles occur during mixing that leads to mass transfer between the W and Cu particles. So it seems to be reasonable to characterize this process by EDX examination of the particle surface. The EDX spectra of the W particle surface reveal that Cu content on the W surface is about 3 wt%, which demonstrates a low extent of W and Cu particles interaction during mixing. The analogous SEM and EDX examination of W and Cu particles after acceleration in the CS nozzle demonstrate that the Cu particles are deformed by W particles. The indentation imprints are seen in Figure 3.8b. Hence, relatively high contact stresses may arise due to W and Cu particles collision during powder acceleration in the powder-laden jet that results in the plastic deformation of Cu particles. The particle interaction results in remarkable mass transfer, which is easily detected by EDX. The EDX analysis made at the Cu and W particles surfaces (points #1, #2, Figure 3.8b). While the Cu content on the W particle surface (#2) is about 11.54 wt%, the W content on the Cu particle imprint area is about 7.46 wt%. The great particle collision effect seems to be the result of a big difference between W and Cu particle velocities. It increases the probability of particle clusters formation.

3.1.5 PARTICLE STRUCTURE

3.1.5.1 Grain Size

As shown in Chapter 2, particle microstructure considerations that influence CS are important in the CS process analysis, and powder microstructure characteristics need to be defined. Several researchers investigated some specific features of powders and CS coating microstructures using electron backscattered diffraction (EBSD) and other methods, which were observed in the review of Jeandin et al. (2014). The first parameter to be taken into account is grain size, which depends on atomizing parameters and type of alloy. The second, characteristics of the particles to be sprayed, is dislocation structure, which is being considerably changed due to deformation processes under the particle impact. The third parameter is the presence of oxide films on the particle surface, which greatly influences interparticle bonding and deposition efficiency.

The grain size determination is performed by the conventional SEM and OM methods using etching. As an example, Figure 3.9 depicts the SEM images of the powder and particle structure of AISI 904L and ARMCO iron powder materials for grain size definition. The structure of ARMCO iron is pure ferrite (Figure 3.9a and c) with the grain size of 10–15 μm. Stainless steel

(a)   (b)

FIGURE 3.8 SEM images of Cu–10 wt% W particles after acceleration in the CS nozzle. (a) Cluster of Cu–W particles and (b) imprints of W particles at Cu particles.

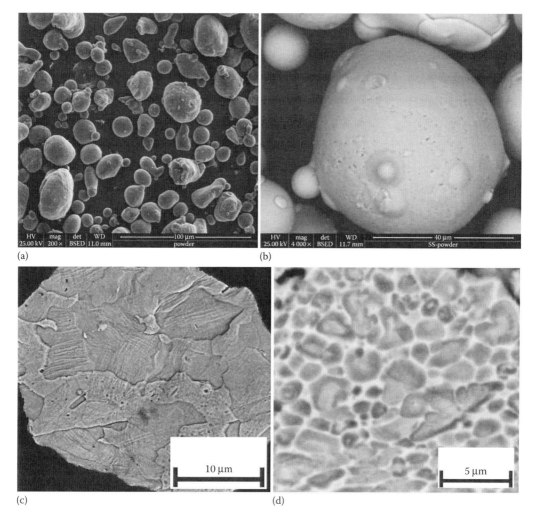

FIGURE 3.9 SEM images of the particles and particle structure. ARMCO iron particles (a) and microstructure (c) after etching with Nital (3% HNO_3 in alcohol). AISI 904L particles (b) and microstructure (d) after etching with Glyceregia.

particles have an austenitic structure with the grain size of 2–5 μm (Figure 3.9b and d). Etching with Glyceregia etchant (three parts HCl, two parts glycerol, and one part HNO_3) results in vision of the austenite grain structure (Figure 3.9d). Based on the initial microstructure examination results, the determination of a grain distortion by characterization of the grain shape after impact may be performed. The results of EBSD examination of iron powder shown in Figure 3.10 reveal that grains depicted in Figure 3.9c consist of subgrains size of 0.2–1 μm with low-angle and high-angle boundaries.

3.1.5.2 Particle Oxidation

Particle velocity is very important for the deposition of dense coatings. It has been well established that impacting particles must exceed a critical value of particle velocity, called critical velocity, to deposit on the substrate instead of bouncing off. The magnitude of the critical velocity generally depends on several factors, including particle and substrate material properties, particle temperature and size, particle and substrate surface conditions, and so on. Of special interest is the presence of an oxide layer on the particle surface and its ability to considerably affect the critical

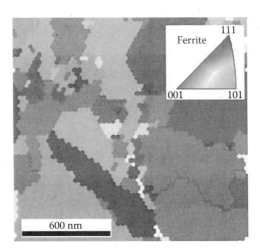

FIGURE 3.10 The EBSD examination of iron powder.

velocity. Most metals and their alloys can be deposited by cold spraying (Champagne et al., 2007; Papyrin et al., 2007). It has been widely accepted that, for an accelerated particle of a given material, there is a critical velocity resulting in a transition from erosion of the substrate due to impact of the particle to the deposition of the particle. Only the particles achieving a velocity higher than the critical one can be deposited to form a coating. The critical velocity is associated with the properties of spray material (Li et al., 2010), substrate state (Assadi et al., 2003), and particle temperature (Li et al., 2006). However, the critical velocity for a given powder material often may vary. For example, as far as Cu particles are concerned, different critical velocities have been reported by different research groups: from approximately 500 m/s to 550–570 m/s to 640 m/s. Furthermore, when Cu powder with low oxygen content has been used, investigators obtained quite a low critical velocity of only about 320 m/s (Li et al., 2010). In the mentioned study, the large discrepancy of the critical velocity data was attributed to the difference in oxygen content of the used copper powders. Indeed, as it was clearly demonstrated in Li et al. (2010), the oxidation conditions of the starting feedstock had a significant effect on particle critical velocity. The same holds true for aluminum powder. Recently, Kang et al. (2008) have examined the oxidation dependence of critical velocity for Al powder by analyzing different values of the critical velocities reported in Van Steenkiste et al. (2002). The effect of oxygen content in the sprayed powder material is especially noticeable at relatively low carrier gas temperatures. Thus, for a carrier gas temperature of about 500°C, the deposition efficiency for Cu powder with high oxygen content was greatly reduced (Champagne, 2007). Therefore, diminishing the oxygen content in the sprayed powder as well as the thickness of the oxide layer on the powder surface is very important to make the CS process more efficient. Plasma processing technologies (Anders, 2004) applied for the pretreatment of feedstock metal powders are of definite interest from the point of view of oxygen content reduction in metal powders. Figure 3.11 show an example of such a plasma-based metal powder pretreatment system.

The plasma system with a downward plasma torch is adapted to treat solid powders. In this process, the coarse-grained powders are injected into the plasma at a predetermined position along the reactor axis by a powder feeder with the feed rate of about 50–200 g/min. The diameter of the plasma arc is about 10 mm, and the flow rate of hydrogen plasma gas is about 20 L/min. Another gas line is used to deliver hydrogen as an additional reduction agent as well as a gas blanket to prevent both the precursor and produced powders from depositing onto the inner wall of the cooling chamber. Argon separately passing through the powder feeder is used as the carrier gas. The described system may be applied for reducing various metal powders, in particular, Cu, Ni, Fe, and V. This specific feature of the installation enables one to collect treated powder in a protective atmosphere to prevent its further oxidation during storage.

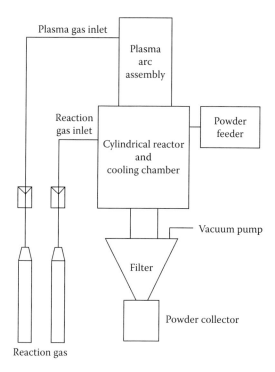

Plasma gas inlet

Plasma arc assembly

Reaction gas inlet

Powder feeder

Cylindrical reactor and cooling chamber

Vacuum pump

Filter

Powder collector

Reaction gas

FIGURE 3.11 Schematic diagram of the plasma arc system for the processing metal powders (Kong et al., Synthesis and characterization of V$_2$O$_3$ nanocrystals by plasma hydrogen reduction. *J. Cryst. Growth.*, 346, 22–26, 2012).

As noted by Moridi et al. (2014) and shown in Chapter 2, there is no established criterion to ensure the success of the CS process during the deposition of composite materials. All numerical studies in the literature have focused on powder deposition rate and the critical velocity as a criterion for bonding. Because there are no criteria for successful deposition of metal and composite powders, different studies in this field have used experimental tools such as optical, transmission, and SEM, as well as microhardness and bond strength tests to evaluate the coatings characteristics, which are controlled by particle–particle and particle–substrate interface structure formation. This topic is discussed in the following section.

3.1.6 INTERFACES AND METAL-MATRIX COMPOSITES

3.1.6.1 MMC Definition

Metal-matrix composite (MMC) coatings are being used for various applications such as corrosion protection, wear resistance, high conductance, thermal barrier, and other applications. The feedstock powders are being applied normally by MMC coating formation. Powder mixtures are designed based on achieving both technological (good spray ability) and exploitation properties. Usually, the powder mixtures consist of soft metallic and hard (carbide, oxide, and intermetallic) particles to get metallic particle deformation during impact at high strain rate and, consequently, to form localized shear deformation at the particle–substrate interface that results in bonding and coating buildup. On the other hand, by increasing demand for coating tribological properties, hard particles with enhanced hardness are needed. Because of lack of ductility, they cannot be deposited directly. Hence, the deposition of the soft and hard powder mixture feedstock is one of the main directions of cold spraying MMC coatings (Table 3.2).

TABLE 3.2

Mechanical Properties of Al-Al$_2$O$_3$, Al-SiC, Al-B$_4$C, and Al-TiC Composite Coatings Made with Cold Spray Technology and Composites Made with PM Technology

Materials (PM—Sintered at 600°C—Rolled 500°C; CS—as Sprayed)	Density ρ (g/cc)	Relative Density ρ (%)	Elastic Modulus E (GPa)	Specific Modulus E/ρ (GPa/[g/cc])	Yield Strength σ$_s$ (MPa)	Ultimate Tensile Strength σ$_b$ (MPa)	Specific UTS σ$_b$/ρ (MPa/[g/cc])	Elongation (φ4 mm, Gauge length 25 mm) δ (%)
Al—PM	2.72	1.0	70	25.9	64	90	33	21
Al CS coating	2.65	0.974	62	23.4	76	80	30.2	2
Al CS coating—annealed 600°C	2.656	0.976	65	24.47	62	84	31.7	15.3
Al+SiC (20 vol%)—PM	2.77	1.0	102	36.8	117	200	70.3	10
Al+SiC (20 vol%) CS coating	2.68	0.967	95	35.44	102.5	187	69.8	1.5
Al+SiC (20 vol%) CS coating—anneal	2.7	0.975	93	34.4	98	192	71.1	8.5
Al+TiC (20 vol%)—PM	3.14	1.0	116	36.9	148	233	74.1	9
Al+TiC (20 vol%) CS coating	3.03	0.965	102	33.7	153	228	75.2	1.5
Al+TiC (20 vol%) CS coating—anneal	3.07	0.978	98	31.9	145	237	77.2	7.1
Al+B$_4$C (20 vol%)—PM	2.75	1.0	105	38	143	208	75.4	9.2
Al+B$_4$C (20 vol%) CS coating	2.62	0.953	96.3	36.7	135	192	73.3	1.5
Al+B$_4$C (20 vol%) CS coating—anneal	2.65	0.964	94.2	35.5	132	195	73.5	5.3

Source: Kuruvilla A.K., et al., Effect of different reinforcements on composite-strengthening in aluminium. *Bull. Mater. Sci.*, 12, 495–505, 1989.

According to Raabe et al. (2010), corresponding material systems of MMC coatings can be grouped according to a microstructural or chemical classification scheme. From a microstructural perspective, multiphase systems can be classified as either particle-like alloys after primary synthesis or as lamellar or filament-type micro- or nanostructured materials. From a chemical perspective, these alloy systems can be classified as immiscible pure metal–metal-matrix compounds, intermetallic–metal-matrix compounds, or carbide–MMCs. In pure metal–MMCs, it is possible to observe the formation of supersaturated solid solutions. In the case of composites consisting of intermetallics or carbides dispersed in a metallic matrix, one can additionally observe phase changes and interfacial chemical reactions.

The interface between the matrix and reinforcement phase plays a crucial role in MMC structure formation that results in certain MMC properties. The processes influencing the interface structure formation during cold spraying and postspraying define the important techniques by which the interfacial properties can be controlled. The aim of this section is to analyze the main factors influencing the particle–substrate and particle–particle interfaces in MMC coatings during cold spraying. Let us make such an analysis on the base of Al-based MMC coatings.

Among the various matrix materials available, copper and aluminum and its alloys are widely used in the fabrication of MMC coatings. This is because of the fact that they are light in weight, economically viable, amenable for production by various processing techniques and possesses high strength and good corrosion resistance. Some of the important reinforcement materials used in the copper and aluminum MMCs are carbon/graphite, silicon carbide, alumina, zirconia, and zircon in particulate, whisker or fiber form. The major fabrication methods used for aluminum MMCs are stir casting, squeeze casting, infiltration, spray deposition, powder metallurgy (Rajan et al., 1998).

The interface between the matrix and the reinforcement particles defines the properties of the composite coating. The main processes controlling the interface structure formation are the interfacial chemical reaction, change of powder particles surface structure at the CS deposition, wetting with the matrix at the melting operations, and so on.

3.1.6.2 Interface Structure

The nature of interface has a strong influence over the properties of the MMCs. The mechanical properties of MMC coatings are defined by the matrix and reinforcement phases, which depend on the strength of the interfacial bond between them. A strong interfacial bonding permits transfer and distribution of the load from the matrix to the reinforcement particles. The properties such as stiffness, fracture toughness, fatigue, coefficient of thermal expansion, thermal conductivity, and creep are also affected by the interface. It seems to be possible to divide the bonding nature of MMC interfaces into the following two groups (Rajan et al., 1998).

1. A mechanical bonding that arises from mechanical interlocking between the matrix and substrate in the absence of all chemical sources of bonding.
2. A chemical bonding that occurs when the atoms of matrix and reinforcement phase or substrate surface are in direct contact, and bonding is accomplished by the exchange of electrons. This type of bonding can be metallic (metallurgical), ionic, or covalent. An interface with a metallic bond is more ductile than other bonds, and is desirable in MMCs.

Mechanical interlocking. An example of the first type of bonding is shown in Figure 3.12 for Cu coating made by high-pressure cold spraying on a carbon steel substrate. The severe deformation of Cu particles results in vortex formation and intensive dispersion of Cu oxide films at the particle–particle interface. Moreover, the vortex formation due to severe straining is achieved for the carbon steel surface layers that is possible only at specific conditions (high temperature, contact stresses, and strains).

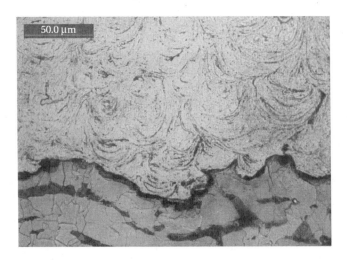

FIGURE 3.12   Cu coating on carbon steel substrate with particle–particle and particle–substrate interlocking due to vortex formation and severe plastic straining both Cu particles and steel substrate.

Chemical bonding. In some systems, during processing of MMCs, a chemical reaction occurs at the interface between the matrix and the reinforcement particles. In such cases, it leads to the formation of an interface reaction product layer with properties differing from those of either the matrix or the reinforcement particles. Kinetics of the chemical reaction and the type of reaction products depend on the particle composition, cold spraying parameters such as temperature, pressure, and atmosphere, particle velocity, and surface chemistry of the particles. Interfacial reaction can decrease the interfacial energy of the metal/reinforcement interface and improve adhesion through chemical bonding (Rajan et al., 1998). The reaction products formed during processing may continue to form during service as well, thereby resulting in progressive improvement or degradation of the properties.

As an example, based on the work of Rajan et al. (1998) it seems to be reasonable to mention that the following interfacial reactions may be observed during the cold spraying aluminum MMC coatings because the temperature in the areas of adiabatic shear band formation is near to melting:

$$4Al_{(l)} + 3SiC_{(s)} \rightarrow Al_4C_{3(s)} + 3Si_{(s)}$$

$$4Al_{(l)} + 3SiO_{2(s)} \rightarrow 2Al_2O_{3(s)} + 3Si_{(s)}$$

$$3Mg_{(l)} + Al_2O_{3(s)} \rightarrow MgO_{(s)} + 2Al_{(l)}$$

$$3Mg_{(l)} + 4Al_2O_{3(s)} \rightarrow 3MgAl_2O_{4(s)} + 2Al_{(l)}$$

Here (l) is the index of liquid phase, and (s) is the index of solid phase. These interfacial reactions are clearly defined with liquid Al and Mg during MMC synthesis. We believe that during the particle impact with supersonic velocity these reactions do proceed due to considerable temperature, stress, and strain increase during adiabatic shear band formation in localization areas at the interface. The interfacial solid-state diffusion reactions during particle impingement seem to be enhanced as well by deformation processes and considerable increase of temperatures and dislocation and other defect density. The dynamic recrystallization processes proceed too.

3.1.6.3   Interfaces and Composite Properties

Most of the mechanical and physical properties of the MMCs such as strength, stiffness, ductility, toughness, fatigue, creep, coefficient of thermal expansion, thermal conductivity, and damping characteristics are dependent on the interfacial behavior. The interface plays a crucial role in transferring the load efficiently from the matrix to the reinforcement. The strengthening and stiffening of composites depend on the load transfer across the interface. The high bond strength is required at the interface for effective load transfer. A strong bond is usually formed with the reaction between the matrix and the reinforcing phases and the reaction products, which define the nature of the bond. For example, presence of intermetallic precipitates at the interface, as in Al-Cu-Mg-SiC composites, is also detrimental to mechanical properties (Rajan et al., 1998).

An examination of mechanical properties of Al-Al_2O_3, Al-SiC, Al-B_4C, and Al-TiC composite coatings made by the cold spray technology (authors unpublished results) and the same composites made with the PM technology (Kuruvilla et al., 1989) demonstrates that both Al-TiC PM composites and coatings have the yield and ultimate tensile strengths higher than those of Al-SiC and Al-B_4C. The main reason for such an effect is the better bond integrity at the Al-TiC interface with as-sprayed and annealed conditions (Table 3.2).

The ductility of the composites is also largely influenced by the interfaces. Al-TiC PM composites are more ductile compared to Al-Al_2O_3 (Earvolino et al., 1992). The similar effect is observed for Al-Al_2O_3 and Al-TiC composite coatings made by cold spray. It is a difficult exercise to design optimized interfaces common and suitable for all systems. Some of the methods to obtain desired interfaces with better properties are optimization of the MMC composition, modification of the particle surface by mechanical alloying, and other specific treatments. Among these, the most important technique is to improve interfacial properties of the particles before the cold spray procedure (for example, to diminish the influence of oxide films on the particles).

3.1.7 Nanostructured Powders

The fundamental monograph of Gleiter (1990) shows that the physical and, in particular, mechanical, thermal, and electrical properties of metals and alloys made of nanocrystals are superior to those of conventional polycrystalline metals and alloys. The nanocrystalline materials have been the subjects of widespread research during the past 40 years. Of their outstanding mechanical properties, one could mention high strength, increased resistance to tribological and environmentally assisted damage, and potential for enhanced superplastic deformation at lower temperatures and faster strain rates (Kumar et al., 2003). Many techniques are being applied to nanocrystals manufacturing such as sonochemistry (precipitation from solution), gas phase synthesis, ball milling, rapid solidification, and crystallisation from amorphous phases. The separate nanocrystals or micrometer-sized powders (ball milling) containing nanocrystals are being produced by these techniques.

From the application viewpoint of nanocrystalline powders for cold spraying, one could note the difficulties to accelerate the nanoparticles because of small particle size, whereas the acceleration of microsize particles permits the particle kinetic energy needed for coating formation during impact. CS appears to be an appropriate spraying technique for depositing nanocrystallne powders because of its low deposition temperatures. Therefore, the combination of CS and the ball milling process provides the opportunity to produce a complete nanostructured coating. Deposition of different nanostructured powders including metallic and composite powder mixtures is presently a very hot topic (Moridi et al., 2014).

3.1.7.1 Mechanically Alloyed Powders for Cold Spraying

Mechanical alloying (MA) is a powder solid-state processing method involving repeated welding, fracturing, and rewelding of powder particles by grinding means in a container. This method allows production of advanced and nanostructured materials starting from blended elemental powder mixtures with unique advantages in terms of material chemistry and microstructure.

A wide range of ball mills of different types have been used for powder production of alloys. These range from very high energy bench top units such as shaker/vibratory ball mills capable of producing batch quantities of up to 10s of grams of powder alloys (in hours to 10s hours), through to planetary ball mills that are capable of processing a few kg of powder (in 10s hours to days). Larger quantities of powder for industrial scale manufacture are possible, 10s to 100s kg, using either vertical or horizontal attritors (Soni, 1999). Each type of the mills has advantages and disadvantages. A shaker mill rapidly produces very small quantities of MA powders that avoids excessive contamination from milling and is useful for microstructural or alloy concept studies; planetary mills take longer (lower collision energy) to achieve steady state (MA) but process powder quantities that enables consolidation, thermomechanical processing, and property assessment of test quantities of powder alloys. Attritors are capable of producing 100s kg quantities but are lower energy than either vibratory or planetary mills so take longer to achieve MA (e.g., 80 h), which often leads to increased contamination of the powder charge from ball and chamber materials (particularly pick-up of C from the hardened steel balls), causing up to 95% of observable contamination and even higher for the case of hard materials. Moreover, each mill type tends to produce a different MA "signature" that is evident in the processed powders, affecting quality control across powder batches. These traits reflect factors such as the ball–ball collision energy, the relative proportion of normal incidence or shear collisions, and the ball mill layout.

MA is best performed in conditions in which pure mechanical impacts are transferred to small volumes of particles that are entrapped during the impact in order to trig mechanochemical reactions that such impacts release shock waves to the material as well combine effects of "aggregation" and breaking. This is why it was developed and installed by MBN Nanomaterialia (Italy), an industrial high-energy ball mill (HEBM) Mechanomade® plant in which the grinding means are accelerated to the impact working surfaces at high velocities (several m/s) in an ordered and more controlled manner. This configuration allows the kinetic energy to be purely transferred during the impact to the powders and partly to the container, whereas ball/(powder)/ball interactions are minimized. When comparing HEBM to attritors, which operate through disordered impact upon grinding and abrasion leading to many more uncontrolled ball/powder/ball interactions, the levels of contamination are greatly reduced.

Thanks to the complete control of atmosphere during processing and the lining design of the milling vials of itanium-based alloy and Ni-based superalloy, powders can be produced by MA for biomedical purposes according to the quality standards required in terms of interstitial level of elements and composition homogeneity. Furthermore the processing time of a fully MA product can indeed be reduced by increasing the productivity of the machine. The concept of the Mechanomade® ball mill, besides the above advantages, has the fundamental benefit of already being scaled up in an industrial platform capable of delivering up to hundreds of kilogram of powder per day depending on the type of product (powder density and process time).

Powder particles produced by mechanical alloying do not usually meet the morphology and flowability targets for spraying deposition processes. Ball-milled powders are usually angular in shape because of flattening during high energy impacts between particles and mill balls. CS process requires a good flowing powders that are both roundish in shape and fall within a specific particle size distribution generally in the range 15–40 μm. Powders that are irregular in shape and/or too small will reduce the ability of the powder to flow, causing either clogging problems or scarce efficiency in deposition with greater coating porosity and poor surface roughness. In addition, low energy milling steps to improve the morphological characteristics of MA ball-milled powders have been purposely adopted for CS powder, which helps improve yield in the particle size window without affecting the material properties and composition. This method, which is integrated into the powder classification system, allows us to eliminate fine agglomerated particles by removing the sharp edges for brittle materials and finally to deliver more spherical particles with high flowability. The output of this combined process is constituted by mechanically alloyed powders in the micron-size range (i.e., 15–45 μm), roundish in shape, and constituted of stable aggregates that can be effectively used by CS.

The HEBM technique is able to generate numerous variants of alloys and nanocomposites materials and structures to promote different synthesis effects depending on materials and process conditions used. Cold Spraying can be seen as a very complementary technology of HEBM because it is able to retain and even to refine the initial powders microstructure limiting the grain growth effect in the material by exploiting purely kinetic effects during the deposition. The combination of the two techniques allows indeed to obtain nanostructure coatings that are able to deliver unique physical and tribological properties. The distinctive powder fine structure that can go down to the nanostructure level obtained by HEBM allows ceramic-metallic (CerMet) materials to be deposited by CS with a relatively high deposition efficiency overcoming the intrinsic limits that are usually experienced in using this technique for hard materials (with hardness more than 800 HV) even for depositing thick coatings up to some millimeters. Moreover the HEBM powders can develop self-sustaining reactions during CS triggered by the kinetic impact to complete the formation of alloys and intermetallic phases into the coating. This principle can be exploited thanks to the very fine material structure and short (nanometric) diffusion paths of elements that enhance the chemical activity and reaction kinetic among reactive elements in the composition.

Oxidation is one of the main limiting factor for some of the CS powder and the main contamination in MA powders. Suitable processing atmosphere together with dedicated post MA treatment are necessary to limit the oxygen, nitrogen, and hydrogen content for materials that require, for example, high conductivity (for example, WCu) or that need to fall within specific standards (for example, Ti biomedical powder).

The following set of material class has been successfully developed by mechanical alloying for cold spraying:

- Cermet and reactive materials: WC-Ti, Ti-SiC, Ti nano composite, Fe-Cu-Alumina
- Metal composites: immiscible alloys as W-Cu
- Intermetallic alloys: MA of two or more materials is aimed to form a fine dispersion of intermetallic phases in a metallic matrix as Ni-Sn

The powder materials of the first class are described next.

3.1.7.2 CerMets and Reactive Deposition

Reactive nanostructured materials represent an innovative way to employ hard metals and metal composites by the CS technology. The reactive deposition allows the formation of hard reinforcement improving the mechanical properties of the coating not accessible with standard powders and the CS process. Reactive powders are conceived in such a manner as to have a residual chemical energy stored due to incomplete chemical reactions. The impact energy of the CS process is exploited to release the stored energy by triggering the incomplete reactions. As a result of this process, it is possible to deposit hard coatings via the CS technology.

When the kinetic energy of the CS process is not sufficient, the incomplete reactions can be finalized by a post-, localized, heat treatment. Several studies have been done to achieve the expected reactions during the deposition process and/or the following heat treatment. The innovative concept related to the design of the reactive powders has been patented by MBN Nanomaterialia and can be applied starting from different self-propagating high-temperature synthesis (SHS) reactions in order to form a fine dispersion of carbides and/or oxides in a metallic matrix that ensures the necessary toughness for an efficient deposition.

Possible reactions are used in the Fe-Cu-Al-alumina (FAC-Al) composite that exploits, first during HEBM and then in the following CS deposition, the termite reaction between hematite and aluminum to form alumina.

$$2Al + Fe_2O_3 \xrightarrow{\text{HEBM+CS}} 2Fe + Al_2O_3$$

The harder alumina phase results finely dispersed and well embedded in the metal matrix with Cu to deliver a trade-off between the lubricating effect of soft areas and the resistance to abrasion of the harder areas. The addition of Cu and Al as alloying elements in the metallic matrix decreases the coefficient of friction, whereas the particles of alumina improve the tribology of the final coating.

This kind of materials can be deposited as a coating for those applications requiring fretting resistance properties with a certain level of hardness (up to 700–750 HV), as for example for cylinder bores in special series of automotive engines (Georgiou et al., 2013). This kind of material is not achievable by other manufacturing processes since the Cu and the Al can form phases characterized by a low melting point. In addition, aluminum forms intermetallic phases with iron and, moreover, the aluminum oxide is difficult to be dispersed in the metallic matrix. FAC-Al Coatings sprayed by CS exhibit outstanding behavior in dry sliding conditions and similar to WC-Co based CerMet in fretting conditions.

As in the Fe-Cu-Al-alumina composite, the SHS reaction is exploited to obtain an oxides dispersion in other material options. The formation of carbides as strengthening phase has been achieved. An example is represented by the formation of titanium carbide in a metal matrix by exploiting the SHS reaction at cold spraying and following heat treatment (HT):

$$Ti + C \xrightarrow{\text{CS+HT}} TiC$$

By combining HEBM and CS, it is possible to achieve a percentage of stoichiometric titanium carbide, near to the 100%. The SHS reaction starts during the CS deposition and can be completed during a following heat treatment. In this case, the thermal treatment is needed to guarantee the complete stoichiometric CerMet formation.

Exploiting the high affinity of titanium with carbon, it is possible to manage and to design other useful SHS reactions. The adoption of this concept has made possible the design of interesting materials such as the Ti-WC and the Ti-SiC. The main aim of the Ti-WC material design is the substitution of cobalt with titanium in WC-based CerMet. This has a potential of introducing a new generation of coatings in the market. This kind of material exploits the following SHS reaction:

$$2Ti + 2WC \xrightarrow{\text{CS+HT}} TiC + WC + W + Ti$$

Titanium acts both as a metallic matrix and as carbide former. Because of the strong chemical bonds established between the different phases, the final material has a high density and mechanical properties. The phase changes, initiated by the CS process, can be completed in the following heat treatment. The reactions between titanium and tungsten carbide confer an enhanced compactness and hardness to the material. The coatings so obtained exhibit an improved wear resistance and a hardness comparable to the standard WC-Co coating sprayed by HVOF with the advantage of the possibility to raise the thickness of the coating without an increase in cobalt content in the composition. The hardness of starting powder is about 500 HV, thus allowing a good deposition of efficiency (up to 80%–85% with a CS high pressure gun) and consequently the spraying of larger thicknesses, in the order of 1–3 mm.

Thanks to the outstanding physical and mechanical properties, and to the advantages derived by the adoption of the CS technology, such materials find applications in various areas: in the automotive area for the deposition on Mg parts used in racing competitions, in the extrusion dies, and in the devices for the rolling mills. When substituting WC with a SiC in a titanium matrix, the reaction that takes place during mechanical alloying brings to the formation of TiC a set of Ti silicide intermetallics (mainly TiSi2) and some residual metallic Ti. As for the Ti-WC, titanium acts both as a metallic matrix and as a carbide former. Also in this case, the phase change, initiated during the CS process, is completed in a following heat treatment.

$$2Ti + 2SiC \xrightarrow{\text{CGS+HT}} TiC + SiC + Si + Ti$$

This kind of agglomerated powder represents a new experimental line of materials for wear resistant surface applications. Coatings are fully dense and a good adhesion both to steel and aluminum substrate. The lower initial powder hardness allows us to obtain a good deposition efficiency, and the coating thickness up to 1 mm is achievable by depositing this material via CS. The chemistry of the composite can be conceived to obtain a low absolute density and powder coatings. The coating density after heat treatment is up to four times lower than that of the conventional WCo hard coatings with the related advantages in terms of the weight of the material deposited. The hardness after heat treatment achieves values up to 1300 HV by starting from a 400 to 700 HV in the virgin powders according to the initial composition. This kind of material is particularly suitable for those applications where Cr hard coating could be substituted with environmental friendly Ti-SiC materials solutions.

So, the hardness and thickness of reactive CerMet coatings obtained by CS + HT are as follows:

- FAC-Al (thickness up to 500 μm)—HV = 650–750 MPa according to the composition;
- WTi (thickness up to 3 mm)—HV = 600–750 MPa for as-sprayed coating according to the equipment used, and HV = 1250 MPa for coating after heat treatment;
- Ti-SiC (thickness up to 1 mm)—HV = 400–700 MPa for as-sprayed coating according to the equipment used, and HV = 1300 MPa for coating after heat treatment;

3.2 COLD SPRAY EQUIPMENT

3.2.1 BASICS OF COLD SPRAY SYSTEM
Kazuhiko Sakaki

3.2.1.1 Principles

Based on the results of study conducted at many researchers over the world, several type of cold spray (CS) equipment[1] have been developed by different companies including ITAM SB RAS (Novosibirsk, Russia), Kteck Corporation (Albuquerque, USA), Oerlikon Metco (Truebbach, Switzerland), Impact Innovations (Rattenkirchen, Germany), Plasma Giken Co. Ltd (Saitama, Japan), OCPS (Obninsk, Russia), Centerline (Windsor, Canada), and others. This section presents a description of the CS equipment, the nozzle design, and geometry of CS gun. To illustrate the main aspects of CS system parameters determination, the numerical simulation and experimental data are presented in Section 3.2.2.

A schematic diagram of a CS system is shown in Figure 3.13. CS system consists of the following major components.

1. High-pressure gas supply (nitrogen, helium, air, or those mixed gases)
2. Gas control module

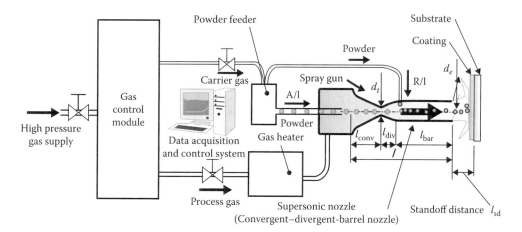

FIGURE 3.13 Schematic diagram of a cold spray system.

3. Resistive coil gas heater and power supply
4. Prechamber and supersonic nozzle assembly
5. Powder feeder (high-pressure type powder feeder for an axially powder injection [A/I], low-pressure type powder feeder for radially powder injection [R/I])
6. Process control and data acquisition system
7. Robot and its controller

The process gas is introduced through a gas control module to a manifold system containing a gas heater and powder feeder. The high-pressure gas is heated to a preset temperature, often using a coil of an electrical resistance-heated tube. The gas is heated *not* to heat or soften spray particles but instead to achieve higher sonic flow velocities, which ultimately result in higher particle impact velocities. The high-pressure gas is introduced into the converging section of a de Laval-type nozzle (i.e., a convergent–divergent [CD] nozzle, but convergent–divergent-barrel [CDB] nozzle in Figure 3.13), the gas is accelerated to sonic in the throat region of the nozzle, and the flow then became supersonic (Mach numbers ranging from 2 to 4) as it expands in the divergent section of the nozzle (Tucker 2013). The powder is delivered by a precision metering device called *powder feeder* and typically is injected axially into the center of the high-pressure side of the nozzle intake (axial injection of powder: A/I) (in most case). The powder is accelerated in the divergent and the barrel section of the nozzle, and powder and the gas gush out in the atmosphere and collide with a substrate. And the powder which collided is higher than the critical velocity deposits and forms a coating.

One of the major elements of the CS process is the high-speed gas jet, which is governed by gas dynamics. In gas dynamics, supersonic flows are obtained with CD (or de Laval) nozzles. Thus, in CS, the nozzle is one of the most important components.

3.2.1.2 CS Process Classification

Table 3.3 shows CS process classification. As for the current CS system, a pressure of the process gas is classified roughly into *high-pressure types* more than 1 MPa and two of the following *low-pressure types*. Furthermore, the case of installation of CS device in a decompression chamber is called as *the vacuum cold spray*. The aerosol deposition method (Akedo 2008) is out of explanation.

High-pressure/low-temperature gas type shown at (a) in Table 3.3 is the first-generation CS system with less than approximately 600°C of process gas temperature at nozzle intake. High-pressure/high-temperature gas type shown at (b) is the second-generation CS system with high temperature (1000°C) of gas more than (a), and make fireproof metal, such as tantalum and niobium, deposit and improve quality of coating to have improved an adhesion and deposing rate of its coating.

The possibility to obtain coating from the WC/Co composition with use of additional preheater of the powder (i.e., increasing its initial temperature at nozzle intake) in CS system of Taekwang Tech (Kim 2005).

In (a) the high-pressure/low-temperature type CS system, powder size to use was approximately 5–50 μm, and marketing spray powder was not often available because it was smaller than a commercial powder size distribution for conventional thermal spray. There is kinetic spray® (KS) (only as for the plot type) that improved this fault by nozzle entrance part extension (Van Steenkiste et al. 2000). The gas flows in the entrance convergent section of the nozzle exhibit a relatively higher temperature and are subsonic; thus, this region is most suitable for heating spray particles, mentioning it later in Section 3.2.1.4. In the second-generation of commercial high-pressure/high-temperature-type CS system (Table 3.3, b), the extension of the convergent section of the nozzle and a higher gas heating temperature allowed to deposit the particles of relatively bigger size. In addition, the high-pressure portable system (c) in Table 3.3 was developed for repair in 2009 too.

New CS device (SISP: shock-wave-induced spray process) such as an explosion spray device (f) in Table 3.3 intermittent continuation shaped a detonation that researchers of Ottawa University developed in 2005, and they continue this work now (Yandouzi et al. 2007).

TABLE 3.3

Classification of Cold Spray Process and Device

No.		Process Gas Type Gas Type	Stagnation Pressure (MPa)	Stagnation Temperature (°C)	Main Characteristic	Commercial Device < >
	Cold Spray					
(a)	High press./low temp.	a	1–(4)	Below 600	First-generation CS device	Developed by researchers <SM Kinetiks 3000>[4] <Taekwang Tech (Korea)>[5]
(b)	High press./high temp.	a	~4 (max. 5)	Up to 1000	Second-generation CS device Application to fireproof metal	<SM Kinetiks 4000, 8000>[4] <II 5/8, 5/11>[6] < PG PCS-800, 1000>[7]
(c)	High press. portable	a	1–2	Below 400	Al, Cu, Zn, Ag, and so on	<SM Kinetiks 2000>
(d)	Low press. portable	Air	Below 1	(Below 600)	Low-melting-point metal, repair use	<OCPS (Russia) DYMET 412, 403>[8] <Centerline (Canada) SST>[9]
(e)	Low press. sonic	Helium	Below 1	(Below 400)	Little gas consumptions, controlled by lower than sound speed	<Inovati (USA) KM-CDS, PCS,MCS>[10]
(f)	Low press./high temp.	a		Below 900		<Medicoat (Switzerland) ACGS>[11]
(g)	Shock-wave-induced spray process (SISP)	Helium Nitrogen	2	550–900	Al, Cu, SUS	Developed by Ottawa Univ. in 2005[12]
	Low Temp. HVOF					
(h)	Warm spray	Combustion gases + nitrogen (air)	(Below 1)	(600–2000)	Improvement of HVOF, superior coating of Ti, WC-Co	Developed by researchers NIMS/Kagoshima Univ. No commercial device
(i)	High-pressure warm spray	Combustion gases + nitrogen	Below 4	1430–2350	Making warm spray high pressurized, superior coating of Ti	<PG/NIMS/Kagoshima Univ.>[13]

SM: Sulzer Meteco (Truebbach, Switzerland), II: Impact Innovations GmBH (Rattenkirchen, Germany), PG: Plasma Giken Co. Ltd (Saitama, Japan), NIMS: National Institute for Materials Science (Tsukuba, Japan).

a Nitrogen, helium, air, or those mixed gases.

In low-pressure types of (d)–(f) in Table 3.3, Kashirin et al. have developed a downstream injection of the powder (R/I) (Kashirin et al. 2002); therefore, the powder is supplied to the nozzle by low pressure, and a high-pressure powder feeder vessel became needless. However, it should be noted that this technique results in a subsonic particle beam, hence requires a peening agent to hammer the particles deposited onto the substrate, and to form coating. The low-pressure CS process yields coatings of selected ductile materials with acceptable coating characteristics, as compared to the high-pressure process that yields coating of almost any materials with highest coating quality (Tucker 2013).

On the other hand, a low-temperature high-speed flame spray (HVOF) device (h) (Table 3.3) for warm-spray process was developed by researchers (Kuroda et al. 2011) in 2000. The temperature of the supersonic gas flow is generated by the combustion of kerosene and oxygen with HVOF flame spray gun that is controlled by diluting the combustion gas with an inert gas such as nitrogen[19]. As for this warm spray, the new model that raised combustion pressure for a further technological advance to 4 MPa was developed recently (Molak et al. 2014).

At all events, determination of a shape and the dimensions of the nozzle is very important rather than elevating temperature and pressure of the process gas to accelerate and heat particles efficiently. This nozzle design and geometry will be explained in the next section.

3.2.1.3   Nozzle Design and Geometry

3.2.1.3.1   Introduction to Nozzle Design for CS Process

As shown in Chapter 2.1, one of the major characteristics of the CS process is the high-speed gas jet, which is governed by gas dynamics. In gas dynamics, supersonic flows are obtained with CD nozzles (or de Laval nozzle), which are used for rocket motors. Rocket motors, including their nozzle design, have been studied and analyzed in detail. While the principal purpose of the design of a rocket motor nozzle is to maximize the thrust, in thermal spraying (including CS) the main purpose is to obtain better coating quality (Champagne 2007).

CS systems employ various kinds of gun nozzle contours, such as CD (or de Laval nozzle) after Alkhimov et al. 1994, CDB after Sakaki et al. 2002, 2006) and convergent-barrel (CB) after Li and Li 2005.

Previous studies on thermal spraying show that the coating properties are principally determined by the thermal and kinetic energy states of particles on impact with the substrate. To have a balance between these two states, various changes in the design of the high-velocity oxygen fuel (HVOF) gun nozzle have been attempted. However, works concerning the influence of nozzle geometry on the thermal spray process are sparse (Hackette and Settles 1995; Kopiola et al. 1997; Sakaki and Shimizu 2001). Previously, the effects of throat diameter and exit divergence of the gun nozzle on the HVOF process have been considered (Sakaki and Shimizu 2001). The combustion gas flow (such as pressure, velocity, temperature, and expansion state of gas jet from the nozzle exit), the particle behavior and, therefore, the nature of coatings were found to be significantly influenced by these nozzle parameters. In addition, the effect of the expansion state of the combustion gas jet on the HVOF process was investigated using a diverging nozzle exit. The particle velocity reached a maximum with the correct expansion state of the gas jet due to an increased gas jet velocity. This resulted in an increase in the bonding strength of the NiCrAlY coating (Sakaki et al. 2006).

The nozzle geometry is also important with regard to the CS method. In the CS method, a coating is formed by exposing a substrate to high-velocity solid-phase particles, which have been accelerated by supersonic gas flow at a temperature much lower than the melting temperature of the feedstock. In this section, the influence of contours (shape), expansion ratio (exit diameter/throat diameter), nozzle length, and cross-sectional shape of nozzle on the CS process (i.e., the behavior of the process gas and spray particles) is investigated by experiments and a numerical simulation prior to designing the CS equipment and producing coatings. The governing equations of a numerical simulation were shown in Chapter 2.1 in detail, and the results of experiments and a numerical simulation are described below.

3.2.1.3.2 Nozzle Shape Influence (Modeling Results)

Figure 3.14 shows schematic diagrams of the CS nozzles used for numerical simulation of gas flow within the nozzles: (a) CB nozzle, (b) CD (or de Laval) nozzle, and (c) CDB nozzle. The cross sections of these nozzles are circular. Nozzle total length l, entrance converging length l_i, entrance diameter d_i, and throat diameter d_t are shown in Table 3.4. The results provided by this simulation could be a little larger than the real values.

The numerical simulation results for the effect of nozzle geometry on gas pressure, velocity of gas, and particles, and temperature of gas and particles are given in Figure 3.15. The CS condition data used and the initial conditions are shown in Figure 3.15. In this section, the modeling results of the CD nozzle, which is used by typical CS devices (CD nozzle in Figure 3.15) are explained.

The acceleration of the gas takes place predominately in the area of the nozzle throat and in the first third of the diverging section. Here the gas velocity U_g has already reached 85% of its exit velocity. Namely, gas velocity reaches sonic velocity at the nozzle throat and reaches supersonic velocity with the increasing cross-sectional area of the nozzle in the diverging section (see Figure 3.15c). At the same time the gas temperature drops to values far below room temperature as

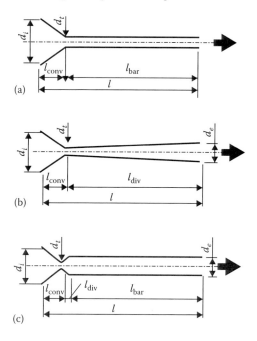

FIGURE 3.14 Schematic cross section diagram of the cold spray nozzles: (a) convergent-barrel nozzle, (b) convergent–divergent nozzle, and (c) convergent–divergent-barrel nozzle.

TABLE 3.4
Nozzle Configuration

Nozzle No.: Nozzle Geometry	Powder Injection Position	d_i (mm)	d_t (mm)	d_e (mm)	Expansion Ratio A_e/A_t	l (mm)	l_{conv} (mm)	l_{div} (mm)	l_{bar} (mm)
CB: convergent-barrel nozzle[a]	Axial	40	2	4	4	300	50	–	250
CD: convergent–divergent nozzle[a]	Axial	40	2	4	4	300	50	250	–
CDB: convergent–divergent-barrel nozzle[a]	Axial	40	2	4	4	300	50	10	240

[a] Use in numerical simulations in Figure 3.14.

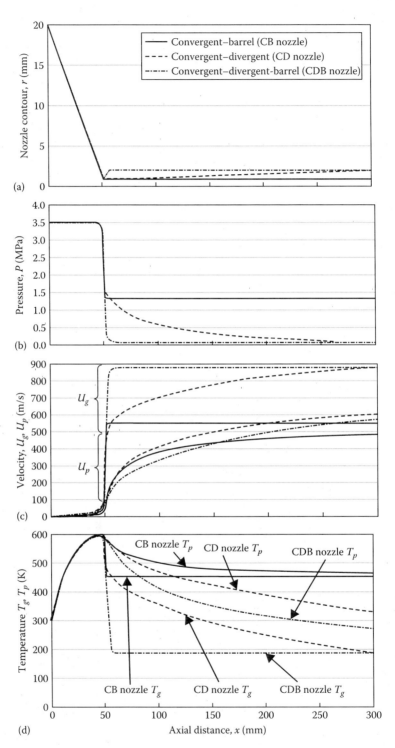

FIGURE 3.15 The effect of nozzle geometry on the calculated results of (a) nozzle contour; (b) gas pressure; (c) velocity of gas and particle; and (d) temperature of gas and particles for nitrogen as the process gas, copper particle (15 μm). Gas initial (stagnation) pressure P_i = 3.5 MPa, gas initial temperature T_{gi} = 600 K, gas initial velocity U_{gi} = 10 m/s, T_{pi} = 300 K, and particle initial velocity U_{pi} = 10 m/s (Sakaki et al. 2002).

the gas expands in the diverging section of the nozzle. The particle is introduced into the gas flow immediately upstream of the converging section of the nozzle and is accelerated by the rapidly expanding gas. The dwell time of the particle in contact with hot gas is brief, and the temperature of the solid particle at impact remains substantially below the initial gas preheat temperature.

The numerical simulation results of the effect of nozzle geometry on gas pressure, velocity of gas and particles, and temperature of gas and particles are shown in Figure 3.15b–d, respectively. It is clearly seen that for the CD nozzle and the CDB nozzle the gas pressure (or density) decreases significantly to a very low value on leaving the nozzle throat owing to the fast gas expansion (see Figure 3.15b). On the other hand, for the CB nozzle, the gas pressure is not significantly decreased along the nozzle up to the exit. Moreover, the oscillation of the gas variables outside the nozzle exit indicates that expansion waves are generated, and those are more obvious for the CB nozzle (Sakaki et al. 2002; Li and Li 2005). This is due to the fact that at the nozzle exit the high-pressure gas expands sharply to ambient for the CB nozzle, while for the CD and CDB nozzles, the gas has expanded gradually from the throat.

Figure 3.15c shows that the gas velocity increases remarkably to a high value after leaving the nozzle throat with the CD and CDB nozzle, while the velocity is relatively low (sonic velocity) with the CB nozzle. Although the particle velocity with the CB nozzle is lower than that with CD and CDB nozzles, the ratio of the particle velocity to the gas velocity is higher for the CB nozzle. This shows that particles can achieve more effective acceleration with a CB nozzle, because the particle velocity is not only influenced by the driving gas velocity but also by the gas density, as shown in Chapter 2.1. The results demonstrate that the particle velocity at the nozzle exit for the CD nozzle is higher than that for the CDB and CB nozzles. On the other hand, the particle temperature at the nozzle exit for the CB nozzle is higher. The divergent section with the CD nozzle (Figure 3.14b) is conical. The numerical codes of fluid dynamics, in particular the method of characteristics (MOC), were also used to develop new nozzle designs that allow a more uniform particle acceleration (Heinrich et al. 2005; Gartner et al. 2006). According to the MOC, the bell-shaped diverging section nozzle can produce better and more uniform particle acceleration over the diameter of the nozzle cross section. Using the standard trumpet-shaped nozzle, a copper particle may be accelerated to a velocity of 500 m/s, which is not high enough to allow the bonding of copper. At the same parameter setting using the bell-shaped MOC-designed nozzle, the velocity of a 20 μm copper particle is increased to 580 m/s, which is well above the critical velocity of copper of 550 m/s (Gartner et al. 2006).

3.2.1.3.3 Influence of the Nozzle Expansion Ratio

The spray gun is fitted with a CD nozzle (or a conical de Laval nozzle) designed to produce perfectly an expansion gas jet (Sakaki et al. 1998), which is supersonic at its exit, and free shock diamonds. Namely, the nozzle exit pressure P_e of the gas fed at varied nozzle intake pressures P_i matches the ambient pressure by changing the nozzle exit diameter d_e. Schlieren photographs illustrating the influence of the expansion ratio of cross-sectional areas at the throat A_t and exit of the nozzle A_e on the expansion state of the gas jet are presented in Figure 3.16. Gas expanded within the diverging section of the nozzle and the expansion state of the gas jet changed from significant underexpansion (Figure 3.16a) to overexpansion (Figure 3.16c) with an increasing expansion ratio.

Figure 3.17 shows the change in the deposition efficiency η (that is experimental value) of cold-sprayed copper coatings with a change in the expansion ratio under different nozzle intake gas pressures P_i at $T_{gi} = 400°C$. In the same way, Figure 3.18 shows the change in the deposition efficiency η with a change in the nozzle intake gas pressure for different A_e/A_t ratios at $T_{gi} = 300°C$. It was clearly found that the deposition efficiency η increased with an increasing expansion ratio and reached a peak at a particular expansion ratio, and then decreased with a further increase in the expansion ratio under different nozzle intake gas pressures P_i. For example in Figure 3.16, η under $P_i = 1.5$ MPa reached a peak of 60% when the expansion ratio was 9, and η under $P_i = 4.0$ MPa reached a peak of 90% when the expansion ratio was around 16. In Figure 3.18 at gas temperature T_{gi} of 300°C, η were lower than those at 400°C in Figure 3.17. Thus, these results show the existence of

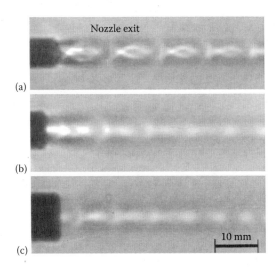

FIGURE 3.16 Schlieren photograph illustrating the influence of the expansion ratio of the nozzle on the expansion state of nitrogen gas jet without substrate. The expansion ratio A_t/A_e of the gun nozzle: (a) 4; (b) 9; and (c) 16 (T_{gi} = room temperature, P_i = 3.0 MPa).

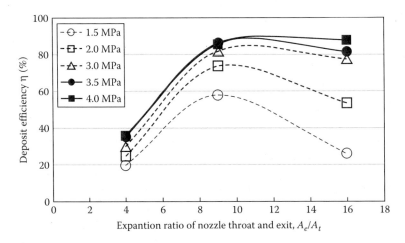

FIGURE 3.17 Influence of the expansion ratio of the gun nozzle and gas pressure on the deposition efficiency of cold sprayed copper (10 μm) coatings at T_{gi} = 400°C, with powder R/I (Sakaki et al. 2006).

an optimal expansion ratio, for which the particle velocity and deposition efficiency are a maximum under a given intake pressure P_i estimated. In addition, the optimal expansion ratio increases with increasing gas pressure P_i. The nozzle intake gas pressure and ratio A_e/A_t of cross-sectional areas at the throat and exit of the gun nozzle affect the deposition efficiency of cold-sprayed copper coatings.

From these experimental results, the expansion ratio of the nozzle maximizing a deposition efficacy each nozzle intake gas pressure P_i in MPa is expressed as

$$\frac{A_e}{A_t} \geq 1.5P_i + 6.9 \tag{3.4}$$

Now in the low-pressure CS patent (Kashirin 2002), the cross-sectional areas of the supersonic nozzle at the juncture of the nozzle and the powder feeder conduit should be related to the throat area per the following relation, too

$$\frac{S_i}{S_k} \geq 1.3P_0 + 0.8 \tag{3.5}$$

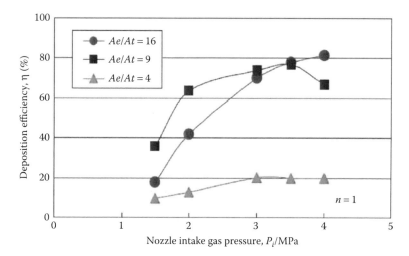

FIGURE 3.18 Influence of the gas pressure and expansion ratio of the gun nozzle on the deposition efficiency of cold-sprayed copper (10 μm) coatings at $T_{gi} = 300°C$, with powder R/I (Sakaki et al. 2006).

where S_i is the cross-sectional area of the supersonic nozzle at the juncture of the nozzle and the powder feeder conduit, S_k is the supersonic nozzle throat area, and P_0 is the full gas pressure at the supersonic nozzle inlet, expressed in MPa.

3.2.1.3.4 Influence of Nozzle Length

Assadi et al. (2003) can express a relation of the critical velocity V_{cr} and effect of these parameters in the following simple formula:

$$V_{cr} = 667 - 14\rho + 0.08T_m + 0.1\sigma_u - 0.4T_i \qquad (3.6)$$

where ρ is the particle density in g/cm³, T_m is the melting temperature in °C, σ_u is the ultimate strength of particle in MPa, and T_{pi} is the initial particle temperature before impact in °C.

In this formula, particle temperature has significant effects on the critical velocity in cold gas spraying. In other words, the critical velocity V_{cr} decreases with an increase in the particle impact temperature T_{pi}, and the particle becomes easy to form a coating (Sakai and Shimizu 2001; Sakai et al. 2006).

The gas flows in the entrance convergent section of the nozzle exhibit a relatively higher temperature and are subsonic; thus, this region is most suitable for heating spray particles. The effect of the nozzle entrance convergent section length l_{conv} on the CS deposition efficiency is shown in Figure 3.19, which depicts the dependence of the deposition efficiency and titanium particle velocity (measured and calculated) on l_{conv}. The deposition efficiency reached a peak of 75% with the nozzle with $l_{conv} = 100$ mm.

The change of the entrance convergent section length (rather than barrel part length or total length) of the gun nozzle had a significant effect on the deposition efficiency. The calculated particle temperature at nozzle exit (not shown in Figure 3.19), particle velocity measured by PIV, and deposition efficiency increase with the convergent section length l_{conv}. However, the particle velocity and deposition efficiency decrease when $l_{conv} > 100$ mm (Figure 3.19). Thus, the parameter of length l_{conv} has to be optimized during the CS nozzle design development.

3.2.1.3.5 Influence of Nozzle Cross-Sectional Shape

In the CS process, cross-sectional shape of the nozzle has a significant effect on spray pattern of coatings (Sakaki et al., 2014). There are a rectangular and a circular cross-sectional shape on the CS nozzles. It is known that spray pattern of the rectangular spray nozzle is wider than that of the circular one. Accordingly, the rectangular nozzle has better capacity than the circular one at cold

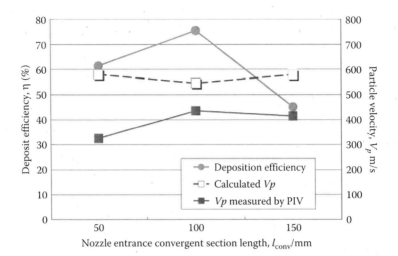

FIGURE 3.19 Effects of the entrance convergent section length of the de Laval (convergent–divergent) nozzle on the deposition efficiency and the measured and calculated particle velocity with titanium powder (mean particle size: 25 μm) with nitrogen gas initial (stagnation) pressure $P_i = 3.0$ MPa and gas initial temperature $T_{gi} = 350°C$ (Sakaki et al. 2006).

spraying passes of a constant width. However, commercially CS nozzle is mostly circular. The goal of this section is to establish a design for the CS gun nozzle to gain a flatter spray pattern of coatings. Sakaki et al. (2014) performed the computational fluid dynamics (CFD) numerical simulation and experiments to study the effect of cross-sectional shape of the CS nozzle on spray pattern of copper coatings. The CS nozzle shapes are shown in Figure 3.20.

The two different rectangular nozzles, such as a CDB rectangular nozzle with an expansion ratio of 6.1 and CD rectangular nozzle with expansion ratios of 6.1, 11.2, and 15.0, were made to compare their spraying parameters with those of the circular nozzle. It is found that the spray pattern fabricated by the CD rectangular nozzle has become more flat, uniform, and wider than that of the circular nozzle. However, a spray pattern of coating with the CDB rectangular nozzle has a concave shape. The numerical simulation results reveal that at divergent section, the gas axial velocity and the velocity along a nozzle width axis y (V_{Gy}) are different for the CDB rectangular nozzle and the CD rectangular nozzle. The velocity along a nozzle width axis V_{Gy} is about 150 m/s for the

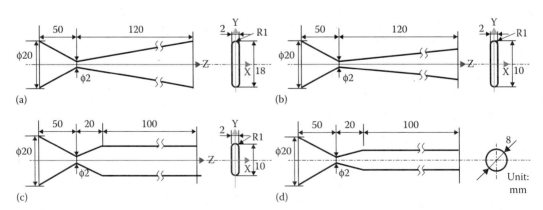

FIGURE 3.20 Schematic diagrams of rectangular and circle cold spray nozzles: (a) CD Rectangular nozzle (18); (b) CD Rectangular nozzle (10); (c) CDB Rectangular nozzle (10); and (d) CDB Circular nozzle. (After Sakaki et al., Influence of cross sectional shape of cold spray nozzle. *Proceedings of the JSME/ASME 2014 International Conference on Materials and Processing, ICMP2014*, June 9–13, Detroit, Michiga, USA, ICMP2014-4961, 2014. With permission.)

CDB rectangular nozzle, while $V_{Gy} = 30$ m/s for the CD rectangular nozzle. It is defined that the optimized expansion ratio of CD rectangular nozzle is about 11.2.

3.2.2 NUMERICAL SIMULATION AND EXPERIMENTAL VALIDATION OF SUPERSONIC SYSTEMS
K. Sakaki

3.2.2.1 Introduction

Calculation and optimization of thermal spraying nozzles geometry is known to be a key issue of spraying equipment development. The advanced supersonic spraying equipment such as HVOF thermal spray and CS systems being developed over the last decade are the main representatives using de Laval nozzles. The fields of application of the HVOF and CS spraying technologies are expanding rapidly. That is why calculation and optimization of thermal spraying nozzles geometry is of great importance.

One of the major characteristics of the HVOF process is the high-speed gas jet, which is governed by gas dynamics. In gas dynamics, supersonic flows are obtained with convergent–divergent nozzle, which are used for rocket motors. Rocket motors, including their nozzle design, have been studied and analyzed in detail. While the principal purpose of the design of a rocket motor nozzle is to maximize the thrust, in thermal spraying the main purpose is to obtain better coating quality.

The HVOF systems employ various kinds of gun nozzle contours, such as convergent barrel (Mason and Kris, 1984; Kowaisky et al., 1990) convergent–divergent (or de Laval nozzle) (Diamond Jet 1995), convergent–multistage divergent (Carbide Jet System 1995) and convergent–divergent-barrel (Thope and Richer 1992; Heath and Dumola 1998). The HVOF gun systems without any nozzle are also in use (Diamond Jet 1995).

Previous studies on thermal spraying show that the coating properties are principally determined by the thermal and kinetic energy states of particles on impact with the substrate. To have a balance between these two states, various changes in the design of the HVOF gun nozzle have been attempted. However, the works concerning the influence of nozzle geometry on the thermal spray process are sparse (Hackette and Settles 1995; Korpiola et al., 1997; Sakaki et al., 1998a).

Previously, we have considered the effect of throat diameter and exit divergence of the gun nozzle on the HVOF process (Sakaki et al., 1997, 1998a). The combustion gas flow (such as pressure, velocity, temperature, and expansion state of gas jet from the nozzle exit), the particle behavior, and, therefore, the nature of the coatings were found to be significantly influenced by these nozzle parameters. In addition, the effect of the expansion state of the combustion gas jet on the HVOF process was investigated using a diverging nozzle exit. The particle velocity reached a maximum with the correct expansion state of the gas jet due to an increased gas jet velocity. This resulted in an increase in the bonding strength of the NiCrAlY coating (Sakaki et al., 1999).

In this chapter, the effect of increasing the length of the entrance convergent section of the nozzle on HVOF thermal spraying process has been investigated by a numerical simulation and experiments with a Jet Kote® system (Stellite Coatings, Goshen, IN). The gas flow in the entrance convergent section of the nozzle is of relatively higher temperature and is subsonic; therefore, this region will be convenient for heating of the high-melting-point spray materials such as ceramics and refractory metals. The goal of this analysis is to establish a design for the HVOF gun nozzle to gain better coating quality of any material powder.

The nozzle geometry is also important with regard to the CS method (Papyrin et al. 2007). In the CS method, a coating is formed by exposing a substrate to high-velocity solid-phase particles, which have been accelerated by supersonic gas flow at a temperature much lower than the melting or softening temperature of the feedstock. The effect in HVOF process will also be applied to the nozzle design for the CS method. The influence of nozzle geometry and gas initial conditions on the CS process (i.e., the behavior of the carrier gas and spray particles) within the nozzle is investigated by a numerical simulation prior to designing the CS equipment and producing coatings.

3.2.2.2 Equipment and Methods

3.2.2.2.1 HVOF Thermal Spray Equipment

The Jet Kote HVOF thermal spray system with a mass-flow-controller attachment was used for this study. A schematic diagram of the Jet Kote gun and nozzle is shown in Figure 3.21. Propylene, C_3H_6, gas was used as the fuel and the spray conditions used are given in Table 3.5. Some technical characteristics of this system are as follows: (1) it employs an internal combustion chamber to generate the hot, extreme velocity exhaust jet to spray, and (2) it injects powder axially into the center of the exhaust gas jet at the nozzle intake. In this system, combustion gas flows from the combustion chamber through four holes in the combustion head (changing its direction at a right angle) to the nozzle intake with an initial velocity of U_{gi}.

To study the effects of the gun nozzle geometry on the combustion gas behavior, spray particles, and coating properties, two different nozzle shapes, namely a *straight nozzle* (equal to factory standard-made barrel one) and a *convergent nozzle*, were used for this study. Further, the dimensions such as total length l of the straight nozzle and convergent length l_{conv} at the entrance of the convergent nozzle were varied,

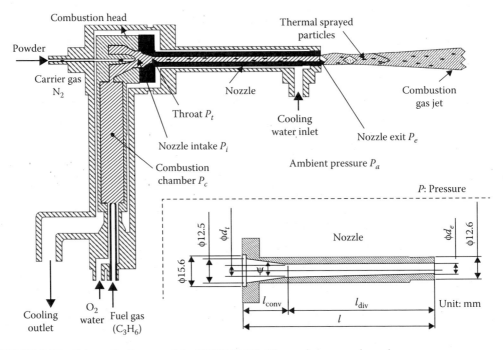

FIGURE 3.21 Schematic diagram of the HVOF (Jet Kote) spraying gun and nozzle.

TABLE 3.5
HVOF Spraying Parameters and Initial Conditions

Fuel/oxygen flow rate (C_3H_6/O_2): 90/486 L/min (normal)

Carrier gas (N_2): 35 L/min (normal)

Powder feed rate: 7.7 cm³/min

(NiCrAlY alloy: 55 g/min, Al_2O_3–40 mass% TiO_2: 12 g/min)

Gun traverse speed: 100 mm/s, gun traverse pitch: 5 mm

Spray distance: 200 mm

Substrate: SUS304, SS400

<Initial conditions for numerical simulation>

Particle velocity U_{pi}: 10 m/s

Particle temperature T_{pi}: 300 K

TABLE 3.6

Shape and Size of Gun Nozzle Used and State of Combustion Gas Stream

Nozzle	d_t (mm)	ψ (°)	d_e (mm)	l (mm) {in}	l_i (mm)	l_s (mm)	P_i^a (MPa) (abs)	P_e^b (MPa) (abs)	U_{ge}^b (m/s)	T_{ge}^b (K)	Gas Jet[c]
12S7.8[d]	7.8	(60)	7.8	304.8{12}	8.9	295.9	0.42	0.245	1055	2832	Weak underexpansion
6S7.8[d]		(60)		156.2{6}	8.9	143.7	0.42	0.245	1055	2832	Weak underexpansion
3S7.8[d]		(60)		76.2{3}	8.9	67.3	0.42	0.245	1055	2832	Underexpansion
3-26 Conv7.8		12		76.2{3}	25.5	50.7	0.42	0.245	1055	2832	Underexpansion
3-48 Conv7.8		6		76.2{3}	48.0	28.2	0.42	0.245	1055	2832	Underexpansion T_1

d: nozzle diameter, ψ: nozzle intake angle, l: nozzle length, P: pressure, U: velocity, T: temperature, g: combustion gas, i: nozzle intake, t: nozzle throat, e: nozzle exit, s: nozzle straight part.

[a] Measured mean value.

[b] Calculated value.

[c] Results of observation of HVOF free jet by photography.

[d] On the market.

as shown in Table 3.6. The length l of a straight nozzle (in the following, straight nozzle is indicated by *total nozzle length* l–S.) was varied to three different levels namely 76.5 mm (3 in.), 156.2 mm (6 in.) and 304.8 mm (12 in.). A convergent nozzle (in the following, the converging nozzle is indicated by l–*length of entrance convergent part* l_{conv} Conv) was made by increasing the l_{conv} of straight nozzle as shown in Figure 3.22. The throat diameter d_t and exit diameter d_e were fixed at 7.8 mm.

3.2.2.2.2 Thermal Spray Powder

NiCrAlY alloy powder of Ni-13Cr-5Al-0.5 mass% Y composition (Shocoat® MA-90, gas atomized powder (Showa Denko, Tokyo) was used for this study. An increase in the entrance part of the nozzle can cause an increase in the heat input of particles and oxidize NiCrAlY particles. Therefore, Al_2O_3–40 mass% TiO_2 powder (Shocoat® K-40F, crushed powder) was also used. The properties of these powders are shown in Table 3.7.

3.2.2.2.3 Evaluation Approach

The expansion state of HVOF jets from the gun nozzle exit with nozzles used without spray powder was evaluated by means of visual observation and by photographic methods. The nozzle intake pressure was measured through a powder feed port by a pressure indicator (PF-30KF, Kyowa Electronic

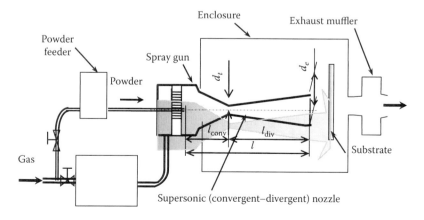

FIGURE 3.22 Conceptual drawing of the cold spray equipment.

TABLE 3.7
HVOF Spraying Powder Properties

Property	NiCrAlY	Al_2O_3–40 mass% TiO_2
Diameter (μm)	10–45 (30)[a]	5–25 (12)[a]
Melting point (K)	1727	2133
Density (kg/m³)	8300	3710
Specific heat J/(kg K)	444	1183
Latent heat of fusion (×10⁶ J/kg)	0.3	1.0

[a] Mean diameter of the powder.

Instrument, Tokyo). (Sakaki et al., 1997). The velocity, surface temperature, and diameter of the sprayed particles during HVOF spraying were measured using an in-flight particle measurement system, namely the DPV-2000 (Tecnar Automation Ltd, St-Bruno, QC, Canada), which detects the thermal radiation emitted by hot in-flight-particles (Gougeon et al., 1994).

The nature of sprayed coatings was characterized by means of electron probe microanalysis, microhardness, and X-ray diffraction analysis.

The deposition efficiency was ascertained by measuring the weight gains on mild steel substrates of 75 × 105 × 6 mm dimensions for a spray time of 30 s taking into account the known feedstock powder flow rate.

3.2.2.3 Numerical Simulation of Thermal Spray Process

3.2.2.3.1 HVOF Process

There are several techniques that can be used to calculate the gas flow of HVOF systems. Recent analyses have used CFD methods to simulate this complex phenomenon in two dimensions. In this paper, the internal nozzle flow was treated as quasi-one-dimensional isentropic flow. This one-dimensional approximation is simple and sufficiently correct for the present purpose of expressing an effect of nozzle geometry on the internal nozzle flow (Sakaki et al., 1997). The detailed derivation of the modeling can be found in Sakaki et al. (1997, 1998). A brief description of the simulation is given below.

Modeling of combustion gas flow within the nozzle. The following assumptions were made to model the gas flow in the HVOF gun nozzle (Sakaki et al., 1997).

- Combustion gas flow within the nozzle is the quasi-one-dimensional isentropic flow of semiperfect gas.
- The chemical reaction of combustion in the combustion chamber follows Equation 3.4 so that combustion gas is composed only of CO_2, H_2O (gas), excessive O_2, and carrier gas N_2.

$$C_3H_6 + (9/2 + X)O_2 + YN_2 = 3CO_2 + 3H_2O + XO_2 + YN_2 + 1926 [kJ/mol] \qquad (3.7)$$

- The combustion gas flows from the combustion chamber to the nozzle intake with an initial velocity U_{gi}, temperature T_{gi}, and pressure P_i.
- The principles of heat transfer only apply to heat exchange between the hot combustion gas and combustion chamber wall/cooling water.

In this manner, the pressure P, density ρ_g, temperature T_g, and velocity U_g of the gas flow can be calculated from the ratio of the nozzle cross-sectional area at a given point to the nozzle throat area (Matuo, 1994).

Modeling of particle behavior within the nozzle. Particle acceleration and heating in a gas flow within the nozzle are given by solving the equations of motion and heat transfer as described below. These equations in this paper are based on the following four assumptions.

1. The spray particle is spherical with negligible internal temperature gradients.
2. The particle specific heat is independent of its temperature and constant.
3. The gravitational effect and the interaction between particles are ignored.
4. The influences of particles on gas flow are neglected. This is equivalent to stating that the gas energy decrease along the nozzle due to acceleration and heating of the particle is neglected.

Under the above assumptions, the equations of motion of a particle in the HVOF process can be written as

$$\frac{dU_p}{dt} = \frac{3}{4}\frac{C_d}{D_p}\frac{\rho_g}{\rho_p}(U_g - U_p)|U_g - U_p| \tag{3.8}$$

where U_p is the particle velocity, t is time, C_d is the drag coefficient, D_p is the particle diameter, ρ_g is the combustion gas density, ρ_p is the particle density, and U_g is the gas velocity. C_d for sphere is a function of particle Reynolds number (Clift et al., 1987).

Heating of a particle in a gas flow can be expressed as follows:

$$\frac{dT_p}{dt} = (T_g - T_p)\frac{6h}{c_p\rho_p D_p} \tag{3.9}$$

where T_p is the particle temperature, T_g is gas temperature, h is the heat transfer coefficient, and c_p is the specific heat of particle. The heat transfer coefficient h in Equation 3.6 can be found by means of the semi-empirical Ranz–Marshall equation, in which h is a function of the Reynolds number and the Prandtl number. The influence of radiant heat between the combustion gas and particle was neglected.

For the calculation of the Reynolds number and the Prandtl number, the values of the specific heat of gas, the gas viscosity, and the gas thermal conductivity are used in the film temperature T_f, which is defined by Bejan and Kraus (2003):

$$T_f = \frac{(T_g + T_p)}{2} \tag{3.10}$$

When T_p reaches the melting point of the particle T_{mp}, the heat from the gas to particle, Q, will be the heat of fusion of the particle. The particle state is represented by the degree of melting of the particle as follows:

$$\frac{\sum Q}{Q_f} = \frac{h\int(T_g - T_{mp})dt}{\rho_p D_p L} \tag{3.11}$$

where Q_f is the heat of fusion per particle and L is the latent heat of fusion of particle material.

Numerical approximation of the HVOF process. An outline of a numerical approximation of the HVOF process is as follows: (1) initial conditions were given; (2) the pressure P, the temperature T_g, and the velocity U_g of the gas flow were calculated from the ratio of the cross-sectional area of the nozzle at the intake point to the nozzle throat area; (3) the above differential equations (3.8) and (3.9) concerning particle behavior were solved numerically by the Euler method; and (4) the process of (2) and (3) were repeated from the nozzle intake to the nozzle exit.

The thermal spray condition data and the initial conditions shown in Table 3.5 (except the powder feed rate and the spray distance) were used. The values of diameter, density, melting point, specific heat, and latent heat of fusion of spray power used in calculation are given in Table 3.7.

The results provided by this simulation could be a little larger than the real values because of the above assumptions.

3.2.2.3.2 CS Process

A conceptual drawing of the CS equipment for the present study is shown in Figure 3.22. Compressed nitrogen gas is introduced to a heater and a powder feeder. The pressure gas is heated in an electric furnace. The feedstock powder is injected axially and centrally into the gas flow at the gun nozzle intake. The spray gun is fitted with a convergent–divergent nozzle (or a conical de Laval nozzle) designed to produce a correct expansion gas jet (Sakaki et al., 1998), which is supersonic at its exit, and free shock diamonds. Namely, the nozzle exit pressure P_e of the gas fed at varied nozzle intake pressures P_i matches the ambient pressure by changing exit diameter d_e. The total nozzle length l and the throat diameter d_t are fixed at 300 and 5 mm, respectively.

The spray powder used in this study is Ni–Al bronze because prior studies (Bhagat et al., 1997) have created a coating with this powder by the CS method. Alkhimov et al. (1990) have reported that there exists a critical velocity U_{pcr} for each coating and substrate material combination, above which the particles have sufficient kinetic energy to build a coating. The value of U_{pcr} for a Ni–Al bronze particle in this study is 600 m/s, because the typical values of U_{pc} for copper, zinc, nickel, and iron range from 550 to 650 m/s for a copper substrate (Alkhimov et al., 1990).

The spray parameters used in this simulation are shown in Table 3.8. The basic treatment for the CS simulation is the same as that for HVOF simulation. The same assumptions as those applied to the HVOF process were made to model the CS process with an additional one that the gas flow within the nozzle is the quasi-one-dimensional isentropic flow of semiperfect gas.

The equations of motion and heating of a particle in the CS process can be written as Equations 3.5 and 3.6 under the above assumptions. An important distinction between the modeling of the two processes is that the influence of radiant heat of the gas on the particle might be neglected, because the gas temperature was lower. The numerical approximation of the CS process and that of the HVOF process are basically the same.

3.2.2.4 Validation of the Numerical Simulation

3.2.2.4.1 *Effect of Convergent Section Length on the HVOF Spraying Process*

The numerical simulation results on the effect of increasing the nozzle entrance convergent section length l_{conv} on gas velocity U_g, temperature T_g, particle velocity U_p, and other properties are given in Figure 3.23 and as part of Table 3.6. The results show that the gas flow in the entrance convergent section of the nozzle is of a higher relative temperature and subsonic. Therefore, the degree of particle melting Q/Q_L increases and U_p decreases slightly with an increase in l_{conv}. The calculated U_{pe} agreed approximately with the values measured by the in-flight particle measurement system for NiCrAlY HVOF processes in change with l_{conv}. This tendency was confirmed by observing the

---

TABLE 3.8

Initial Conditions of Cold Spray Process Simulation

Gas: N_2
  Gas initial pressure P_i: 0.5–5.0 (2.0) MPa (abs)
  Gas initial velocity U_{gi}: 0–100 (0) m/s
  Gas initial temperature T_{gi}: 300–2000 (750) K
Powder: Ni–Al bronze
  Melting point: 1340 K
  Density: 7600 kg/m³
  Specific heat: 440 J/(kg K)
  Latent heat of fusion: 0.205×10^6 J/kg
  Diameter: 1–50 (20) μm
  Particle velocity U_{pi}: 0–100 (10) m/s
  Particle temperature T_{pi}: 300 K

( ): baseline condition.

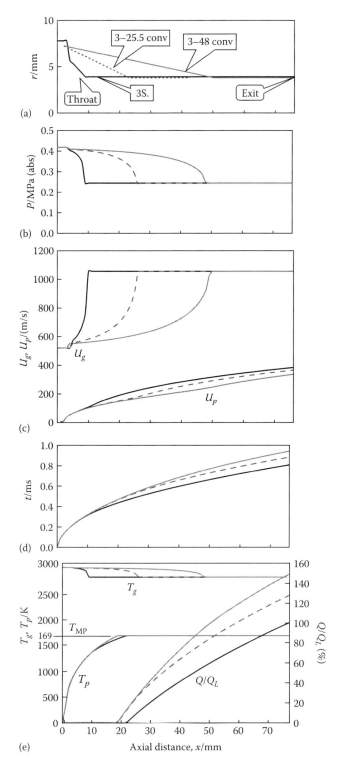

FIGURE 3.23 Effect of increasing in nozzle entrance convergent section length (for three-nozzle shape) on the calculated results of the (a) nozzle contour, (b) gas pressure, (c) velocity of gas and particle, (d) particle resident time, and (e) temperature of gas and particles and the degree of melting particle.

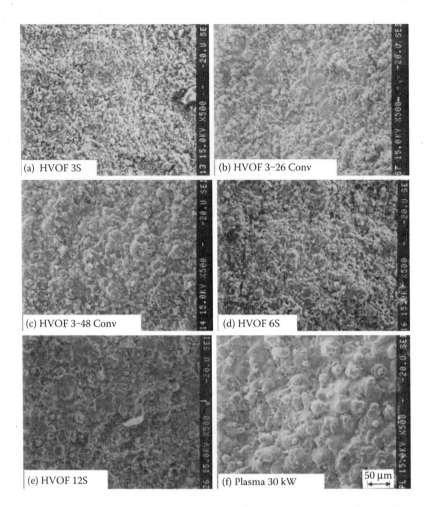

FIGURE 3.24 Surface SEM structures of Al$_2$O$_3$–40 mass% TiO$_2$ coatings sprayed by HVOF (with several nozzle shape) and plasma spraying. 3 S. is a 3 in. straight nozzle, 3–26 Conv is a 3 in. converging nozzle with 26 mm converging part length, 3–48 Conv is a 3 in. converging nozzle with 48 mm converging part length, 6 S. is a 6 in. straight nozzle, 12 S. is a 12 in. straight nozzle (the similar nozzle shapes are shown on Fig. 3.14).

nature of the sprayed coatings, namely, cross-sectional hardness and oxygen content of NiCrAlY coatings increase and deposition efficiency decreases with increasing l_{conv}. As an example of the effect of increasing l_{conv} on the coating property, surface structures of Al$_2$O$_3$–40 mass% TiO$_2$ coatings are shown in Figure 3.24, in which coatings sprayed with a longer nozzle by HVOF and by plasma spray are compared.

It was found that splat morphologies vary with l_{conv} and nozzle length l, because input heat of particles from the gas or degree of melting increases with increasing l_{conv} and l.

Figure 3.25 shows some typical structure of Al$_2$O$_3$–40 mass% TiO$_2$ splats sprayed by HVOF and collected on 304 stainless steel substrates at room temperature. To obtain isolated splats, a shielding plate on which several holes of 1 mm were distributed was placed parallel to the substrate at a distance of about 5 mm. The splat patterns are roughly divided into the following three categories: (a) nonmolten particles, (b) semimolten particles, and (c) molten particles with splash. The different morphologies arise due to variations in the input heat of particles, substrate temperature, and impact velocity of particle to substrate. From (a) to (c), the input heat of particles increases.

The proportion of the various splat patterns of Al$_2$O$_3$–40 mass% TiO$_2$ powder with respect to a change in the nozzle entrance convergent section length l_{conv} is shown in Figure 3.26. In this figure,

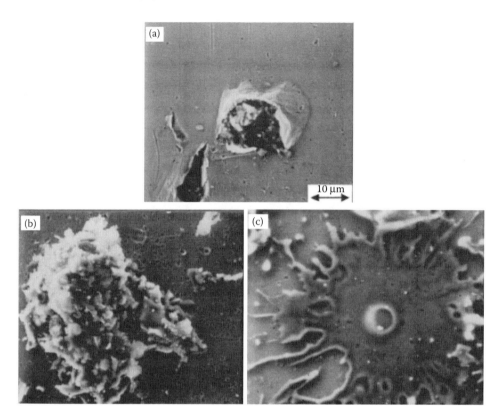

FIGURE 3.25 Typical patterns of the structure of Al_2O_3–40 mass% TiO_2 splats sprayed by HVOF collected on a 304 stainless steel substrate of (a) nonmolten particle with trusting, (b) semimolten particle, and (c) molten particle with splash.

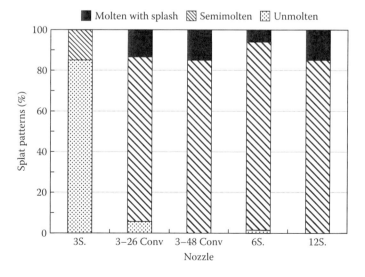

FIGURE 3.26 Parentage of patterns of the structure of Al_2O_3–40 mass% TiO_2 splats sprayed by HVOF with change in nozzle shape. 3S. is a 3 in. straight nozzle, 3–26 Conv is a 3 in. converging nozzle with 26 mm converging part length, 3–48 Conv is a 3 in. converging nozzle with 48 mm converging part length, 6S. is a 6 in. straight nozzle, 12S. is a 12 in. straight nozzle.

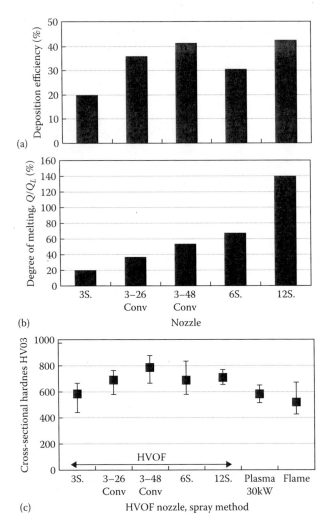

FIGURE 3.27   Effect of the nozzle shape on (a) deposition efficiency, (b) degree of melting of particle, and (c) cross-sectional hardness of sprayed Al_2O_3–40 mass% TiO_2 coatings.

the percentage of splat morphology (b) semimolten particles and (c) molten particles with splash increases slightly with l_{conv} at three 76.2 mm (3in.) nozzles (from 3S., 3–26 Conv to 3–48 Conv nozzle). This tendency shows that an increase in l_{conv} causes an increase in the gas input heat of particle.

Figure 3.27 shows the results of the deposition efficiency, the calculated degree of melting of particles, and the cross-sectional hardness of Al_2O_3–40 mass% TiO_2 coatings with respect to a change in l_{conv} and total nozzle length l. The figure shows an increase in the deposition efficiency and coating hardness with an increase in l_{conv}. Moreover, the deposition efficiency and the cross-sectional hardness of the coating with 3–48 Conv nozzles are higher than those with longer nozzles such as 12S and 6S. This result can be explained from heat losses by nozzle cooling and pipe fraction loss, which were observed in terms of the expansion state of the gas jet as shown in Table 3.8. Therefore, these losses increase with the nozzle length, so that velocity and temperature of the gas and particles decrease with an increase in the nozzle length.

In summary, the effect of increasing the length of the entrance convergent section of the particle-heating nozzle is larger than that of increasing the length of the barrel part or total nozzle length in the HVOF thermal spraying process. Therefore, a combination of increasing the entrance convergent section length and the total length of nozzle is more effective.

3.2.2.4.2 *Influence of Gas Parameters and Nozzle Geometry on the CS Process*

3.2.2.4.2.1 *Influence of Particle Diameter*
The numerical simulation results (gas velocity U_g, temperature T_g, particle velocity U_p, and temperature T_p) with a change in the particle diameter of Ni–Al–bronze powder are given in Figure 3.28. The entrance convergent length l_{conv} of the conical de Laval nozzle used is 9 mm, and the following initial conditions are used as baseline conditions: gas pressure of 2.0 MPa (abs), temperature T_{gi} of 750 K, and velocity U_{gi} of 0 m/s, particle temperature T_{pi} of 300 K, and velocity U_{pi} of 10 m/s.

Using the conical de Laval (convergent–longer divergent) nozzle, the gas velocity U_g increases along the axial distance within the nozzle to reach 950 m/s (Mach number M of 2.7) at the nozzle exit, and the gas temperature decreases to reach 290 K (which is equal to room temperature).

Figure 3.28 shows that the particles, even the larger ones, are accelerated and heated very quickly. The results indicated that 20 μm and smaller particles reach the critical velocity of 600 m/s before arriving at the nozzle exit and attain the gas temperature within the entrance convergent part of the nozzle.

The value of T_g becomes lower than T_p in the nozzle, because the heat capacity of the gas is much lower than that of particle and the gas initial temperature was much lower. In conventional thermal spray processes such as HVOF, plasma spray, and so on, using higher temperature gas, this tendency is not observed.

3.2.2.4.2.2 *Influence of Gas Initial Conditions*
Figure 3.29 shows the effect of initial gas pressure P_i on the CS process (velocity of gas U_{ge} and particle U_{pe}, temperature of gas T_{ge}, and particle T_{pe} at nozzle exit) with nozzles designed to produce a perfect expansion gas jet according to P_i. In this figure, U_{ge} and U_{pe} increase with an increase in P_i. Particles 20 μm and smaller reach the critical velocity U_{pcr} at P_i above 2 MPa. However, particles 50 μm and larger cannot attain this velocity at

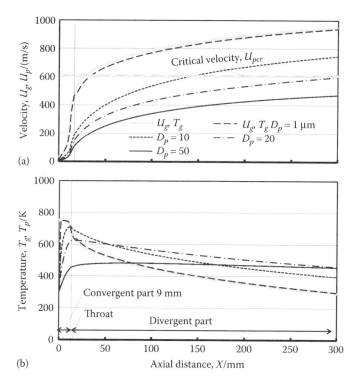

(a)

(b)

FIGURE 3.28 Numerical simulation results with a change in the particle diameter of Ni–Al–bronze powder by cold spray: (a) velocity of gas and particle and (b) temperature of gas and particle (baseline: $P_i = 2$ MPa, $U_{gi} = 0$ m/s, $T_{pi} = 750$ K, $U_{pi} = 10$ m/s, and $T_{pi} = 300$ K).

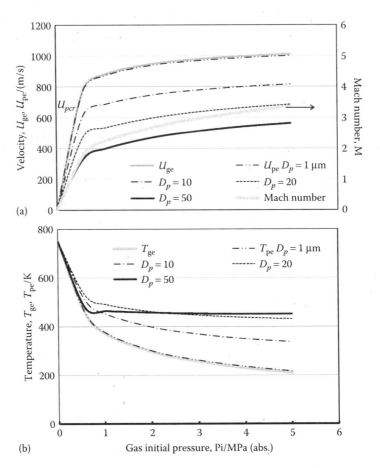

FIGURE 3.29 Effect of the gas initial pressure on the calculated results of the cold spray process: (a) velocity of gas and Ni–Al bronze particle and (b) temperature of gas and particle (U_{gi} = 0 m/s, T_{gi} = 750 K, U_{pi} = 10 m/s, and T_{pi} = 300 K).

P_i up to 5 MPa. The values of T_{ge} and T_{pe} decrease with an increase in P_i. For example, T_{ge} at P_i of 5 MPa drops to 200 K (or –73°C).

The effect of gas initial temperature T_{gi} on the CS process under the baseline conditions, except for T_{gi}, is provided (Figure 3.30). At most, the gas for CS might be used at a temperature up to 1000 K. In this figure, the results up to 2500 K are shown to compare with the results of the HVOF process. The values of U_{ge} and U_{pe} increase with T_{gi}. The value of U_{pe} is higher than U_{pcr} at T_{gi} above 700 K. It is clear that the minimum T_{gi} exists for these initial gas pressures, which allow particles to reach U_{pcr}. For the HVOF process, the pressure of the combustion gas is lower than that of the CS gas because the temperature of the combustion gas is much higher.

The effect of the initial gas velocity, U_{gi}, on the CS process under the baseline conditions except, for U_{gi}, is indicated in Figure 3.31. With an increase in U_{gi}, U_{ge} and U_{pe} increase slightly, while T_{ge} and T_{pe} decrease slightly. To heat powder effectively, gas might be supplied to the nozzle at lower velocity. However, particle behavior is independent of the gas initial velocity up to 100 m/s.

3.2.2.4.2.3 Influence of the Nozzle Entrance Convergent Section Length The effect of increasing the nozzle entrance convergent section length, l_{conv}, on the CS process under the baseline conditions is given in Figure 3.32. The value of T_{pe} increases steadily and T_{ge} increases slightly, while U_{pe} decreases steadily and U_{ge} decreases slightly with increasing l_{conv}. Therefore, increasing l_{conv} affects particle heating but does not accelerate it in the same fashion as for HVOF. Thus, l_{conv}

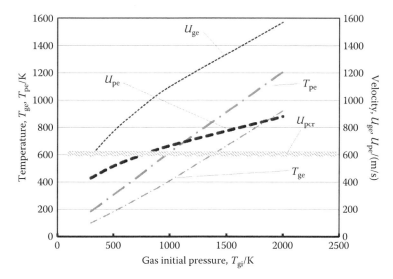

FIGURE 3.30 Effect of the gas initial temperature on the calculated results of the cold spray process: velocity of gas and Ni–Al bronze particle (20 μm), temperature of gas and particle (P_i = 2 MPa, U_{gi} = 0 m/s, U_{pi} = 10 m/s, and T_{pi} = 300 K).

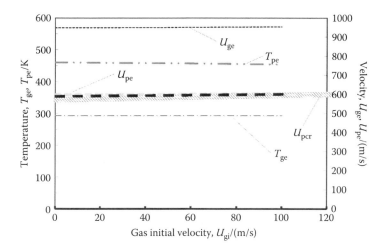

FIGURE 3.31 Effect of the gas initial velocity on the calculated results of the cold spray process: velocity of gas and Ni–Al bronze particle (20 μm), temperature of gas and particle (P_i = 2 MPa, T_{gi} = 750 K, U_{pi} = 10 m/s, and T_{pi} = 300 K).

must be up to 100 mm under these spray conditions to obtain a critical velocity of 600 m/s. However, now the influence of particle and substrate temperatures on the critical velocity is not revealed. Increasing l_{conv} to lower the critical velocity can raise the particle temperature.

3.2.2.5 Conclusion

Numerical simulation and experiments have investigated the effect of increasing the nozzle entrance converging section on the HVOF process. The numerical simulations also investigate this effect with regard to the spray parameters of the CS process. The results are summarized as follows.

- The particle temperature or the degree of melting of particles increases, but the particle velocity decreases slightly with an increase in the entrance convergent section length of the nozzle of HVOF and CS equipment. Therefore, increasing this length has an effect on particle heating.

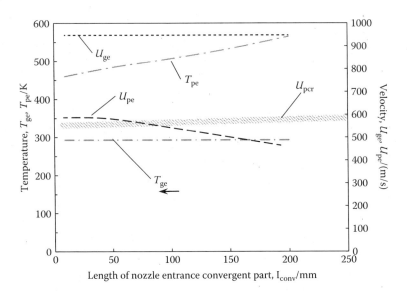

FIGURE 3.32   Effect of the length of the nozzle entrance convergent part on the calculated results of the cold spray process: velocity of gas and Ni–Al bronze particle (20 μm), temperature of gas and particle (baseline: $P_i = 2$ MPa, $U_{gi} = 0$ m/s, $T_{pi} = 750$ K, $U_{pi} = 10$ m/s, and $T_{pi} = 300$ K).

- The surface structure and morphology of the splat pattern of Al_2O_3–40 mass% TiO_2 coatings sprayed by HVOF vary with an increase in the nozzle entrance convergent section length.
- The deposition efficiency and cross-sectional hardness of Al_2O_3–40 mass% TiO_2 coatings sprayed by HVOF significantly increase with the nozzle entrance convergent section length.
- In CS, gas velocity increases along the axial distance within the nozzle to reach 950 m/s (Mach number of 2.7) at the de Laval nozzle exit and gas temperature decreases to 290 K under the baseline initial conditions: gas pressure of 2.0 MPa (abs), temperature of 750 K, velocity of 0 m/s, and particle temperature of 300 K and velocity of 10 m/s. Ni–Al bronze particles 20 μm and smaller reach the critical velocity.
- Initial gas pressure and temperature affect accelerating particles in the CS process. The initial gas velocity has a slight influence.

3.2.3 PCS-1000 COLD SPRAY EQUIPMENT

Hirotaka Fukanuma

PCS-1000 cold spray (CS) System has been developed not only for research at universities and institutes, but mainly for industrial uses. The equipment is designed to endure hard uses in various production environments. By strictly controlling the operating parameters, it can stably and continuously run for hours with nitrogen or helium gas. For research and development purposes, it is suitable to study the processes, spray materials, characteristics of the deposits, and industrial applications.

3.2.3.1 Features of PCS-1000 Equipment

3.2.3.1.1 General Description

Two types of CS equipment currently available are low- and high-pressure systems. These systems are divided by operating gas chamber pressure (at around 1 MPa) (Maev and Leshchynsky 2008). PCS-1000 is a high-pressure and high-temperature CS system. Unlike the portable low-pressure systems, PCS-1000 gun should be mounted on to a robot.

Distinctive features of the equipment are as follows.

(a) PCS-1000 can be operated at high temperature and pressure of 1000°C and 5 MPa, respectively. For this, the particle velocity can reach more than 900 m/s even when operated with nitrogen. High impact velocity of particles leads to dense deposits and good bonding strength.
(b) The system uses a water-cooled convergent–divergent nozzle. By water cooling, it is possible to keep the nozzle at a lower temperature. This gives more range for the selection of gas sealing materials and can prevent deterioration due to oxidation.
(c) The equipment can spray at high deposition rates of more than 30 kg/h due to high powder feed rates.

PCS-1000 consists of gun and heater integrated unit, gas control unit, powder feeder unit, heater power sources, touch panel controller, and water cooling unit. These units are shown in Figure 3.33 (except for the water cooling unit). Figure 3.34 shows a schematic diagram of the system.

The gun unit is composed of gas chamber, convergent–divergent nozzle, powder tube, and water jacket. The chamber is directly connected with the gas heating coil. By making the gas transfer tube length short as possible, it minimizes heat loss. Also, it allows the operators to handle the gun without restraints by getting rid of a gas transfer pipe from the preheater. The heater-integrated gun weighs around 45 kg. The recommended load capacity of a robot is to be more than 60 kg. Figure 3.35 is an image of PCS-1000 gun mounted on a six-axis robot.

3.2.3.2 Gas and Electric Power Consumption

Nitrogen and helium gas consumptions of PCS-1000 with 3 mm throat diameter are shown in Figures 3.36 and 3.37. The gas flow rate is determined by gas chamber temperature and pressure if the throat diameter is constant. The gas flow rate increases with pressure and decreases with temperature. The following equation was used to calculate the gas flow rates. The equation is valid when the flow is isentropic (Anderson 2004):

$$\dot{Q} = \frac{\pi d^2 P_0}{4} \sqrt{\frac{\gamma}{RT_0} \left(\frac{1}{\gamma = 1} \right)^{\gamma+1/\gamma-1}} \tag{3.12}$$

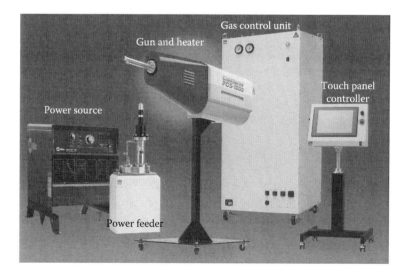

FIGURE 3.33 PCS-1000 cold spray system and units.

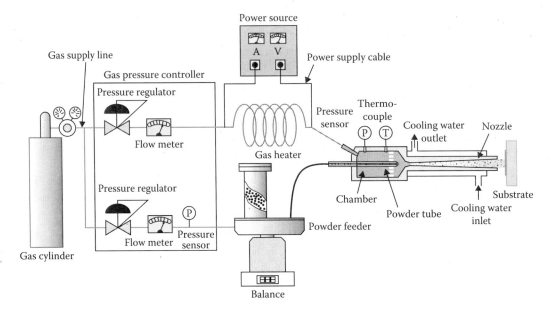

FIGURE 3.34   Schematic illustration of cold spray equipment.

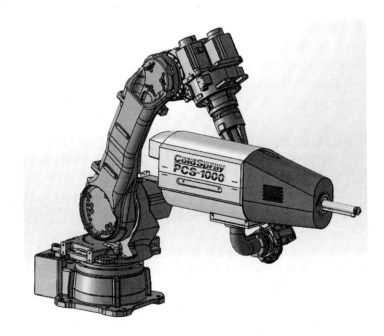

FIGURE 3.35   Schematic illustration of a PCS-1000 heater integrated gun mounted on a robot.

where $\dot{Q}$, P_0, T_0, γ, R, and d are gas flow rate, gas pressure and temperature (in the gun chamber), ratio of specific heats, gas constant, and throat diameter, respectively. When helium is used, the gas flow rate becomes too much using a 3 mm throat nozzle. Therefore, nozzles with smaller throats should be used. Nozzles with 2.0 and 2.5 mm throat can be supplied from Plasma Giken Co., Ltd.

The relations between electric power consumption and gas flow rates using nitrogen and helium gases are shown in Figures 3.38 and 3.39.

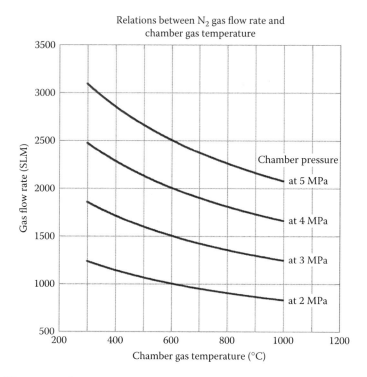

FIGURE 3.36 Nitrogen gas flow rates operated at certain temperatures and pressures with PCS-1000.

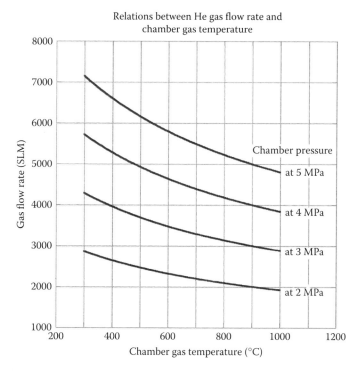

FIGURE 3.37 Helium gas flow rates operated at certain temperatures and pressures with PCS-1000.

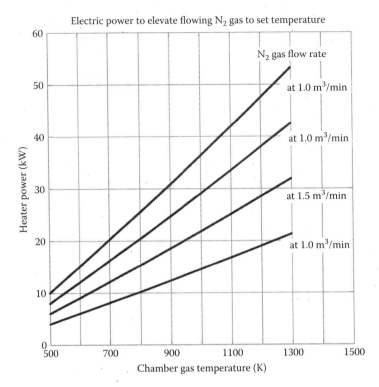

FIGURE 3.38 PCS-1000 electric power consumption at certain temperatures and pressures with nitrogen gas.

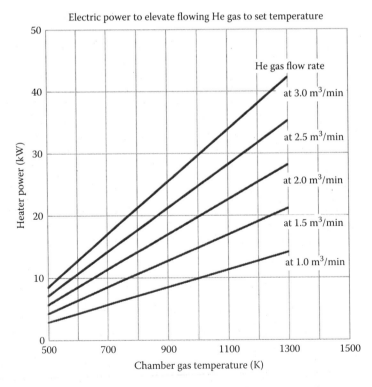

FIGURE 3.39 PCS-1000 electric power consumption at certain temperatures and pressures with helium gas.

3.2.3.3 Deposition Efficiency and Particle Velocity

Various kinds of material with low- and high-melting temperature can be deposited with the PCS-1000 system. Figure 3.40 shows deposition efficiencies of several metals and alloys most of which exceed 90% in efficiency. The deposition efficiency greatly depends on particle velocity when impacted onto the substrate. The gas velocity in the nozzle increases as gas temperature rises along with particle velocity.

The velocity distribution of Inconel 718 powder is shown in Figure 3.41 in connection with gas temperature from 600°C to 980°C. The particle morphology and size distributions used for

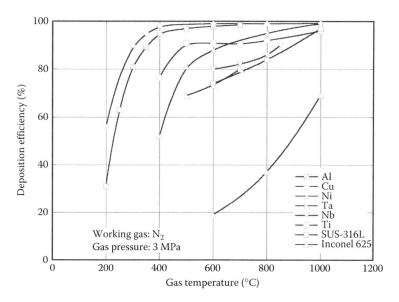

FIGURE 3.40 Deposition efficiencies of various metals and alloys at different chamber gas temperatures sprayed with PCS-1000.

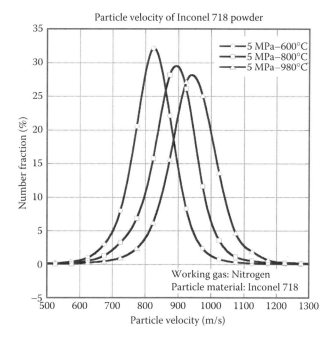

FIGURE 3.41 Particle velocity distributions with different chamber gas temperatures.

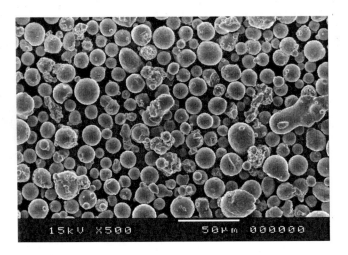

FIGURE 3.42   Particle morphology of Inconel 718 powder.

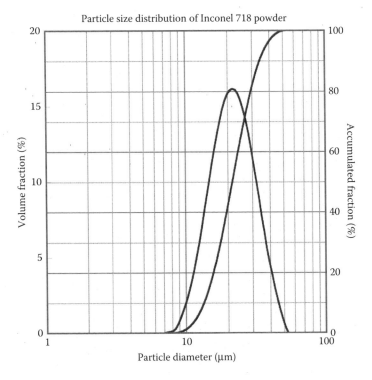

FIGURE 3.43   Particle size distribution of Inconel 718 powder.

measuring the velocity are shown in Figures 3.42 and 3.43. Most of the particles are spherical and the average size is around 25 μm. The velocity measurement was carried out at the distance of 30 mm from the nozzle outlet. Figure 3.41 shows that average velocity of nearly 1000 m/s can be obtained with PCS-1000 equipment even with nitrogen.

   The particle velocity when gas pressure was at 3 and 5 MPa is shown in Figure 3.44. The average velocity at 5 MPa is about 100 m/s faster than that of 3 MPa. It is known that gas pressure affects particle velocities as well as temperature. The deposition efficiency of Inconel 718 at 3 MPa and 1000°C is about 70% as shown in Figure 3.40. The velocity data at 3 MPa in Figure 3.44 suggests that the critical velocity is more than 800 m/s because the velocity faster than 800 m/s occupies

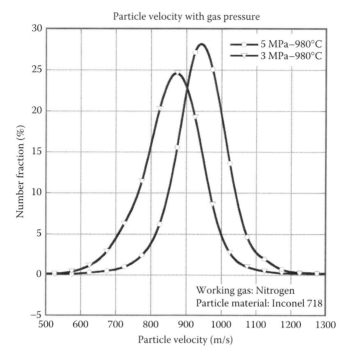

FIGURE 3.44 Particle velocity distributions with different chamber gas pressures.

about 70% of the curve at 3 MPa and 980°C. The temperature difference of 20°C between 1000°C and 980°C could be negligibly small to take into consideration.

3.2.3.4 PCS-1000 Gun Description

The inner structure of the PCS-1000 gun is shown in Figure 3.45. The powder tube is placed on the central axial line through the gas chamber. The distance between the outlet of the powder tube and the throat is considerably long. Powder should be inserted at more than 100 mm upstream of the throat to reach the gas temperature. Figure 3.46 shows particle temperatures traveling through the nozzle when powder fed point is changed. When the particle is fed at 50 mm upstream from the throat, the temperature reaches about 750°C, which is 250°C lower than 1000°C of the gas chamber temperature. It is seen in Figure 3.47 that to obtain higher particle temperature, particles will need

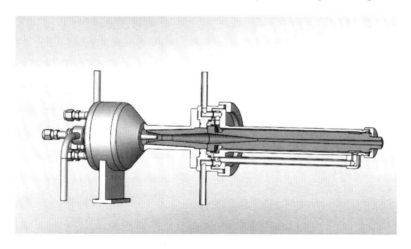

FIGURE 3.45 Schematic illustration of a PCS-1000 gun cross section.

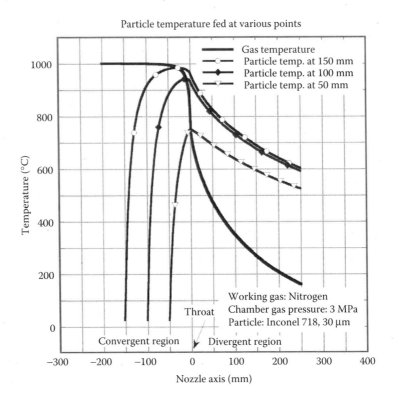

FIGURE 3.46 Particle temperature fed at various points.

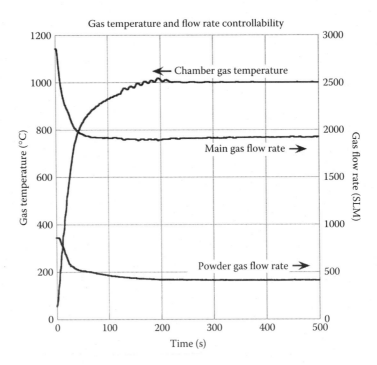

FIGURE 3.47 Controllability of gas temperature and flow rate during operation.

to be fed at upstream points (more than 100 mm from the throat). Higher particle temperature lowers the critical velocity, but makes the deposition efficiency and density higher (Fukanuma et al. 2006). These calculations were conducted under the assumption where the flow is isentropic.

The channels where the water flows for nozzle cooling are also shown in Figure 3.45. Water cooling of the nozzle is effective to reduce the possibility of particle adhering and clogging inside. Keeping the nozzle at below 300°C with water cooling restrains the oxidation of nozzle materials, which allows it to last longer.

The gas temperature and flow rate stability are shown in Figure 3.47. At about 200 seconds after the electric powder is loaded into the heater, the gas temperature becomes constant at 1000°C. Main gas and powder gas flow rates are also significantly stable as are shown in the graph.

3.2.3.5 Powder Feeder

One of the most important units of a CS system is the powder feeder device. If powder feed rate is not steady and uniform during spraying, it is hard to obtain a smooth deposit surface. Several types of feeder systems have been proposed such as vibration feeders, fluidized bed systems, rotary valve, and rotary disc methods (Maev and Leshchynsky 2008; Papyrin et al. 2007).

PCS-1000 system can attach one of the three types of powder feeders: POFC1005, 1025, and 1150. Each of which has 0.5, 2.5, and 15 liter powder container (as shown in Figure 3.48). The feeder uses a volumetric feeding method with a turntable and a trench. The feed rate uniformity depends on flow characteristics of the feedstock. The powder weight in the container also affects feed rates. In general, the heavier the feedstock in the container, the more the feed rate will be. In high-pressure CS systems, the pressure of the powder-feeding hopper becomes 2–5 MPa. Powders tend to be so aggregated due to high pressure that the flow characteristics are aggravated. It is preferable to remove fine particles (less than 5 μm) from the feedstock. An example of feeding characteristics of POFC1025 is shown in Figure 3.49. Water-atomized stainless steel 316 was used as the feedstock. The particle morphology is shown in Figure 3.50. Even with irregular particles, which are considered difficult to feed stably, this figure shows that feed rates can be constant.

POFC1005 is designed for research and development, because it can only handle a small amount of powders of 1–2 kg. POFC1025 is for both research and small production as it can handle 7–8 kg. Lastly, the POFC1150 can be filled to about 50 kg.

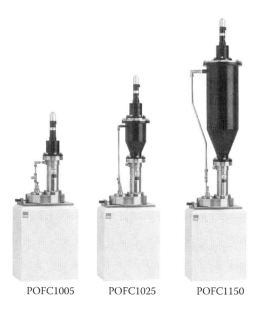

POFC1005   POFC1025   POFC1150

FIGURE 3.48 Types of powder feeder: 0.5, 2.5, and 15 liter capacity.

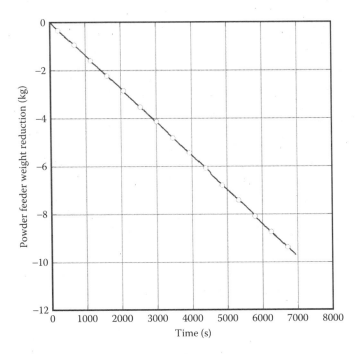

FIGURE 3.49 Flow rates of stainless steel 316L powder.

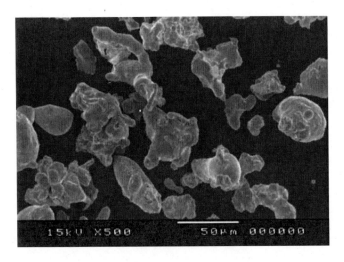

FIGURE 3.50 Particle morphology of stainless steel 316L powder.

Additional powder feeders can be mounted on to the equipment when extra powder gas flow lines are built in the controlling unit. Two or more feeders can be operated at the same time.

3.2.3.6   Examples of Deposits Sprayed with PCS-1000

3.2.3.6.1   Copper Deposits with High Electric Conductivity

Cold-sprayed copper deposits are extensively used in the electric power industry due to its high electric conductivity. CS has advantages in costs, productivity, and quality over other processes such

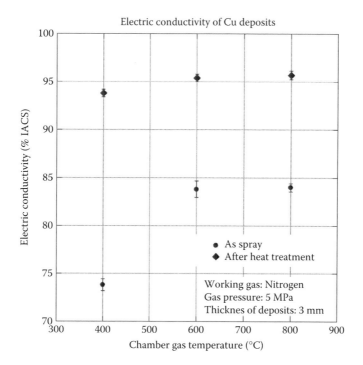

FIGURE 3.51   Electric conductivity of cold-sprayed copper.

as electric plating, thermal spray, and cladding. Some example data on electric conductivity of CS deposits are shown in Figure 3.51. The measurement was carried out by using an AutoSigma 3000 tester provided by GE Inspection Technologies. The specimens are 3-mm-thick Cu deposits sprayed under conditions at 400°C, 600°C, and 800°C of gas temperature and 5 MPa of gas pressure. The graph shows that higher gas temperature provides higher electric conductivity. After heat-treated at 330°C for 2 hours, the conductivities are almost the same.

3.2.3.6.2   Mechanical Properties of Stainless Steel 316L Deposits

Most of the metals and alloys can be easily sprayed due to the recent advancement of CS equipment. Deposits even beyond thickness of 100 mm have been obtained without much difficulty. Mechanical properties of cold-sprayed deposits have not been reported. To open the doors to developing various industrial applications for CS, further studies will be necessary.

Compression tests on stainless steel 316L deposits were conducted and their results are shown in Figure 3.52. The deposits were sprayed under conditions at 1000°C of gas temperature and 5 MPa of gas pressure with nitrogen gas. As shown in the graph, the compression strength of the deposit sprayed reaches 1200 MPa, which is to the limitation of the testing machine. However, ductility was not observed. Further research should be conducted to confirm if deposits have ductility. It is shown in the graph that the deposits heat-treated at 800°C and 1000°C are considerably ductile. The ones heat-treated at 800°C are stronger than the bulk metal with more than 8% strain. The deposit treated at 1000°C is compressed with lower stress.

The specimens were taken with wire cut process from the stainless steel 316L deposit on a tube shown in Figure 3.53. Two types of test pieces vertical and horizontal to the coating layers were cut out of the ring deposit. The data in the graph are from horizontal specimens. The dimensions of the cylindrical specimens were 10 mm in diameter and 15 mm in length (shown in Figure 3.54).

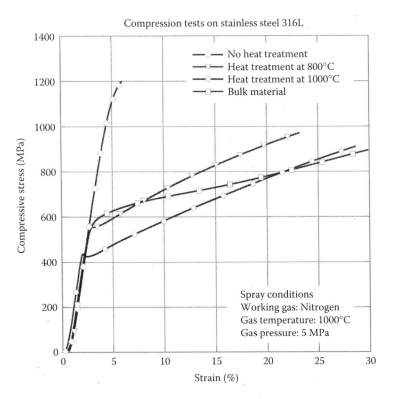

FIGURE 3.52   Stainless steel 316L deposit for compression test specimens.

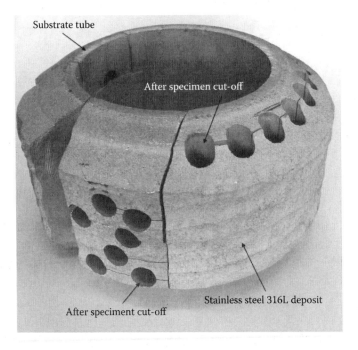

FIGURE 3.53   Photograph of a compression test piece.

FIGURE 3.54 Relations between compression stress and strain of cold-sprayed stainless steel 316L deposit.

CS deposits have properties not only similar to bulk metals, but also interesting and unique ones that would open a variety of possibility for future applications.

3.2.4 Low-Pressure CS System
Roman Gr. Maev

3.2.4.1 Low-Pressure CS Systems Principles

Widespread adoption of CS technologies suggests its potential for successful application in the industrial environment. Therefore, the development of a portable, low-pressure gas dynamic spray (GDS) system is of fundamental importance. It is important, then, to establish the relationship between the operating parameters of a portable GDS machine and the resultant microstructure of the formed spray coatings. To obtain high-quality coatings, the configuration of a portable GDS system and the spray process parameters have to be carefully selected. Due to the sheer number of CS process parameters, however, it is inevitable that a certain degree of trial and error is required to optimize the spray process for each specific coating–substrate combination. Presented below are some specific characteristics related to a portable, low-pressure system. These characteristics are subsequently compared with those of high-pressure CS systems, a more established technology.

The stage related to the particle acceleration in low-pressure CS process is considered to consist of the following three subsequent processes:

1. Gas and powder mixing
2. Acceleration of the particles within the divergent section of the spray nozzle
3. Particle movement within the free jet area (before and after contact with the substrate)

Each of these processes has its own characteristics and may be described by distinct parameters. Both the gas and particle velocities change during each phase of acceleration, as shown in Figure 2.2. One can observe a substantial difference between the velocities of the gas and particle. This is due to the significant disparity in the drag forces that each encounters and must be taken into account when the parameters of the powder-laden jet are calculated.

There are two fundamental aspects that serve to distinguish the low-pressure spray process from that of high-pressure, or stationary, CS:

1. The utilization of low-pressure gas (0.5–1 MPa instead of 2.5–3 MPa)
2. Radial injection of powder instead of axial injection (in most cases)

Through close examination of these differences it can be noted that, due to a decrease in the gas mass flow rate, the particle content in the gas-particle jet is considerably higher in low-pressure systems. In general, the dependence of particle volume concentration on the powder mass flow rate (as shown in Figure 2.2) reveals that the particle concentration is proportional to both the air flow and powder mass flow rates. In low-pressure spray processes both these variables can be compared with those of traditional axial injected high-pressure CS. In Figure 2.3, the powder concentration and deposition efficiency of low-pressure GDS and of different variations of spray systems are presented (as developed by Alkhimov et al., 1994, and Van Steenkiste, 2003). The decrease in the gas flow rate due to the diminishing gas pressure can be observed to cause a sharp increase in the particle concentration of the gas-particle jet. It is known that the solid-phase content in the gas-particle jet in traditional GDS is in the range of 10^{-5}–10^{-6} (by volume), while the content for low-pressure GDS systems is in the range of 10^{-4}–10^{-5}. For this reason, the characteristics of gas-powder flow vary considerably. Hence, the particle impact parameters depend strongly on the type of GDS system that is employed.

As illustrated by Papyrin (2001), Van Steenkiste (2003), and Kay and Karthikeyan (2003), there are various features that characterize typical GDS systems. The main trait of the GDS system presented by Alkhimov et al. (1994) is that the gas and particles form a supersonic jet (with particle velocities ranging from 300 to 1200 m/s) whose temperature remains insufficient to bring about thermal softening. Although the particles are introduced into the gas stream at subsonic velocities, the particle-gas jet is brought to a supersonic speed via the de Laval-type nozzle.

The initial design of the GDS system put forth by Alkhimov et al. (1994) was rather complex. However, Kay and Karthikeyan (2003, 2004) made significant improvements to this model by redesigning the nozzle. This design change allowed the heater mechanism to be removed from the main body of the gun. In this way, the spray gun applications proved to be considerably more flexible; this enabled its use within deposition geometries that would otherwise have been physically precluded in the previous version of the GDS system (Alkhimov et al., 1994). As a consequence, a bulkier and heavier electrical heater is required to heat the large volume of processing gas. Further, moving the gun and heater assembly requires a heavy-duty mechanical manipulator and, as a result, restricts the movement of the spray beam. Thus, the flexibility of the entire spray operation is highly restricted in this arrangement.

Finally, during the spraying of some materials such as aluminum, the powder particles may get deposited inside the nozzle on the walls, blocking the gas flow path. When this happens, the gas flow may be greatly reduced or even stopped, causing an abnormal increase in the temperature and pressure of both the heating element and the gun. Such a sudden increase in temperature and pressure can damage the gun and the heater, and also affect the safety of the operator.

The high-pressure GDS system developed by Van Steenkiste (2003)—or so-called kinetic spray system—is based on the Alkhimov et al. (1994) model. This system is able to produce spray coatings using particle sizes of greater than 50 µm and up to about 106 µm. A further modified version of this device is capable of creating coatings using particles of larger size (up to 120 µm) and is designed to heat a high-pressure air flow up to about 650°C. After introduction of the particles, the heated air-particle jet is directed through a de Laval-type nozzle to achieve particle exit velocities between 300 and 1000 m/s. The primary drawback of this system is that the spray material is injected into the heated gas stream prior to its passage through the de Laval nozzle. This brings about an inherent tendency for clogging, resulting in backpressure and spray gun malfunction. The use of large particles in the Van Steenkiste (2003) design may enhance this problem, requiring additional cleaning of the nozzle.

The second difficulty in CS devices is attributed to the low durability of the convergent throat portions of nozzle. Because the heated main gas stream is under high pressure, the injection of the powder itself requires a high-pressure powder delivery system. Unfortunately, these are quite expensive and are not conducive to the design of portable cold spraying devices. In the Van Steenkiste (2004) design, in an effort to overcome this shortcoming, a specially developed radial method is used for powder injection. This system involves injecting the particles by means of a positive pressure that exceeds that of the main gas pressure. Radial injection takes place within the diverging region of the nozzle, just after the throat. However, the solution of Van Steenkiste (2004) does not eliminate the use of a high-pressure powder delivery system. Furthermore, it introduces an additional difficulty—the necessity to permanently maintain a specific pressure difference between the positive injecting pressure and main gas pressure. Moreover, the value of this difference is strongly dependent on the location of injection point. Thus, in general, an additional adjustment of pressure is needed for different injection points, adding certain complications to the spray system.

The first portable, low-pressure CS system was developed by Kashirin et al. (2002) and referred to as a GDS system. A CS device is composed of a compressed air source that is connected by a gas passage to a heating unit, and in turn fed into a supersonic nozzle. The divergent section of the nozzle is then connected by a pipe to a powder feeder just before the throat. When in operation, the compressed air is delivered to the heating unit where it is heated to approximately 600°C, and subsequently accelerated by the supersonic portion of the nozzle. The powder material is then transferred via the feeder to the supersonic nozzle where it is mixes with the air flow and is accelerated toward the injection point on the nozzle. Within the nozzle the static pressure is maintained below atmosphere, ensuring that powder is effectively drawn in from the powder feeder. This pressure can be maintained only if the cross-sectional area of the supersonic nozzle in this portion exceeds that of the throat. The main difficulty associated with the system of Kashirin et al. (2002) is that the spatial uniformity of powder-air flow is quite low due to particle movement in the long powder passage. Both the particle-wall interparticle friction result in particle accumulation in various places within the powder feeder conduit. The second difficulty is that the precise control of the powder feeding rate is nearly impossible because particle movement within the powder feeder is strongly dependent on the length of the passage pipe, fluidity, and specific weight of powder or powder mixture. In addition, the application of a heating unit where the outlet is connected to the convergent part of the nozzle does not facilitate intensive gas heating in a small coil heating unit.

Therefore, to improve the mechanism of gas heating, as well as the precision with which powder injection may be controlled, the low-pressure portable GDS system needs to be modified. Only after the mentioned shortcomings have been addressed may a truly practical portable gas dynamic spray system become a reality.

3.2.4.2 Powder Feeding for Low-Pressure CS

One of the primary goals of low-pressure portable GDS systems is the development of a powder feeding and metering device for use in delivering powder mixtures to the divergent part of the spray nozzle. A variety of vibratory powder feed systems have been known for many years. Such feeding systems, dating back the 1950s and 1960s, have included both rotationally vibrated bowls and linearly vibrated channels or troughs. In the field of GDS, a distinct need exists for similar portable feed systems.

Not only must this system be portable, it must be reasonably accommodating in that it should facilitate the use of numerous metal and ceramic powders in the spray process. Naturally, each of powder mixtures has its own particle size distribution, specific gravity, coefficient of friction, or other properties that affect its flow characteristics. Therefore, the feed system must ensure consistent particle flow even for powders having significantly different powder flow rates.

Typical designs of powder feed systems may be divided into two groups: (1) fluidized bed hoppers and (2) dispensing hoppers.

A powder feeding system comprising an enclosed hopper and a carrier gas conduit was suggested by Fabel (1976). A schematic of this device is shown in Figure 3.55a. Here the carrier gas supply is connected to a conduit that extends through the lower point of a hopper and continues to the point of powder-carrier gas utilization. To regulate the amount of fluidizing gas that is supplied to the hopper, the pressure at a point in the carrier gas line is monitored, as per the method suggested by

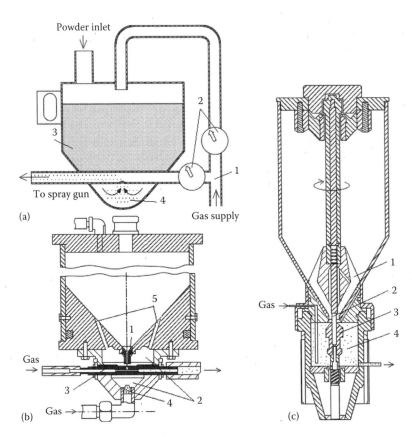

FIGURE 3.55 Fluidizing powder feeding systems design. (a) 3 976 332—Fabel (1976): 1—carrier gas supply system, 2—pressure sensors and valves, 3—hopper, and 4—fluidizing chamber; (b) 4 381 898—Rotolico et al. (1983): 1—powder discharge fitting, 2—fluidizing chamber, 3—powder distribution system, and 4—gas porous insert; and (c) 3 501 097—Dalley (1970): 1—powder feed passages, 2—tie rod, 3—powder chamber valve, and 4—fluidizing chamber.

Dalley (1970). As the pressure within the gas line is dependent on the mass flow rate of solids, the change in pressure may be used to regulate its flow.

Although the feeding system designed by Fabel (1976) has good repeatability, various problems have recently surfaced, particularly with respect to the control and uniformity of the powder feed rate. One such problem is pulsation. This shortcoming, apparently due to a pressure oscillation, results in uneven coating layers. A modified system by Rotolico et al. (1983) (Figure 3.55b) has eliminated this difficulty by inserting an additional fluidizing powder near the conveying gas line.

The use of the fluidized bed hoppers in the high-pressure GDS systems has been proven effective in both academic laboratories and industrial environments alike. The results that have been obtained are summarized in a review by Karthikeyan (2005). However, despite this degree of success, dispensing hoppers with powder feeding rate controls have not yet been applied on a large scale.

The powder feeder designs of Schinella (1971) and Tapphorn and Gabel (2004) use a vibrating structure to move powder from the receiving surface to the discharge channel (Figure 3.56). Schinella (1971) describes the use of a hopper with an outlet channel and hemispherical cup for metering powder and powder flow control. The primary limitation of this powder feeder is that the consistency of powder metering is affected by the agglomeration of powder in the hemispherical cup. In some cases, it may even become plugged.

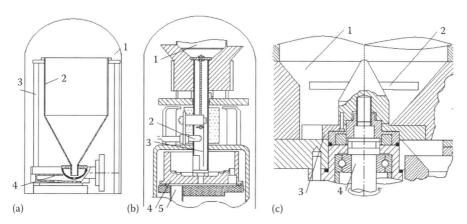

FIGURE 3.56 Powder metering devices. (a) 3 618 828—Schinella (1971): 1—pressurized chamber, 2—feed hopper, 3—hopper frame, and 4—vibratory powder feeder device with outwardly spiral path; (b) 6 715 640—Tapphorn and Gabel (2004): 1—powder hopper, 2—powder feeding rate sensor, 3—flexible metal vane, 4—vibrating spiral bowl, and 5—powder discharge outlet; and (c) 4 808 042—Muehlberger and de la Vega (1989): 1—powder hopper, 2—powder mixer, 3—powder feeder-metering disc, and 4—powder feeder shaft.

An improvement in the design of the powder metering device has been made by Tapphorn and Gabel (2004). In this system, sieve plates are mounted within the hopper for precise metering of the powder as it moves into the vibrating bowl. The powder is metered through the sieve plates by a hopper vibrator that is in turn controlled by a level sensor mounted in the vibrating bowl. Unfortunately, the limitation of this design lies in its repeatability and accuracy. In fact, powder feeding rate measurement by a level sensor is strongly dependent on the agglomeration and fluidity of both powder and, in particular, powder mixtures. Moreover, in the case that the funnel is plugged with a flexible metal vane, the amount of powder flowing from the funnel tube to the receiving surface (and, in turn to the discharge channel) may not be adequately controlled.

3.2.4.3 Portable Low-Pressure CS System Design

A portable gas dynamic CS gun has been developed by the authors (Figure 3.57). This system eliminates many of the inherent difficulties described above by minimizing the scatter of operating parameters and, thus, improving the spray gun efficiency. One advantage of this new system is that the powder flow rate is continuously measured. In this way, the powder flow rate and/or the flow rate of the pressurized gas can be adjusted in real time, leading to a more controlled and efficient deposition by the spray gun. In this design, a gas passageway extends through the spray gun. A gas port supplies pressurized air (or an alternate gas) to the passageway inlet. A nozzle in the passageway forms the pressurized air into a supersonic jet stream. The powder is fed at a controlled rate to the passageway, where it is entrained in the gas and exits the spray gun in the supersonic jet stream. The spray gun further includes a powder flow rate sensor that measures the powder flow rate of the powder. In the example of the spray gun shown in Figure 3.57, the powder flow rate sensor includes a light emitter transmitting light across the duct through which the powder travels. A receiver mounted opposite the light emitter determines the flow rate of the powder based on the measured light intensity. This signal is in turn processed by a specially developed controller (Figure 3.57) that adjusts the gas flow rate and/or the powder flow rate to ensure it is equal to that which is desired.

The portable GDS gun is connected to a pressurized gas source that supplies high-pressure air (or gas) to a heat chamber (1). The converging part (2) and the throat (3) of the nozzle are made of ceramic, while the diverging part (4) of the nozzle is made of steel. The diverging section of the nozzle extends through an outer housing (5) from which it is supplied with powder from a powder container (6). As the pressurized air or other gas passes through the nozzle, it reaches supersonic velocities and draws powder from the feeding hose (7) into the diverging

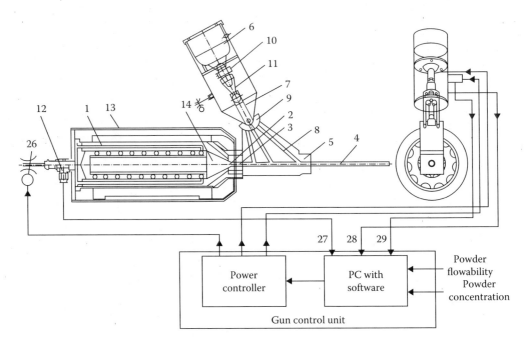

FIGURE 3.57   Low-pressure cold spray gun with a programmed controller.

part of the nozzle. The outer housing has multiple passages (8), each leading to the diverging part
of the nozzle. Powder is selectively supplied to one of these multiple passages in the outer housing using a switch (9). The switch supplies the powder based on the value of negative pressure
at certain points of the air jet. The switch may be set manually or automatically by means of a
controller that is programmed to the expected negative pressure points along the diverging part
of the nozzle. Depending on the pressure of the gas source, the location of negative pressure
point along the diverging part of the nozzle may vary. The switch is set so that the selected passage coincides with the negative pressure point.

The powder from the container is fed to the switch through a vibrating bowl (10) and funnel (11)
into the partial-vacuumed powder passages of the outer housing. The powder then mixes with the jet
of conveyance air and flows through the nozzle where the diverging imparts supersonic velocities.

The jet of conveyance air from the pressurized air source is heated in the heat chamber. The compressed air line contains a variable throttle (12) by which the flow impedance (e.g., the flow cross
section) is regulated from a controller according to a predefined value of either the volumetric flow
of conveyance air and/or the volume concentration of the particles in powder-laden jet. Oftentimes
the controller is a computer having a processor, memory, and additional storage capabilities.

The heat chamber includes a serpentine or helical coil heating element mounted on a ceramic
support and an insulation chamber that is located in an internal chamber housing (13). The air flows
along the helical path defined by the heating element coil. The heated air exits the heater via tapered
chamber (14) which, together with ceramic insert (2), forms the convergent portion of the nozzle.

The powder supply system is shown in more detail in Figure 3.58. The powder supply system
includes a powder container (where the powder to be sprayed is stored in loose particulate form), a
bowl vibration unit (15) for control of the powder flow rate, and a funnel (11) connected to a flexible
hose (7). In addition, two powder container vibration units (bowls) (16, 10) are incorporated into the
powder supply system. Simultaneous control of the two vibration units provides precise and constant control of the powder feeding rate. Powder is fed into the lower vibration unit (10) to ensure
that a certain level of powder is maintained by means of a sensor (17) that controls the operation of
a main powder hopper (not shown). The rate of dispensing powder (powder flow rate) is additionally

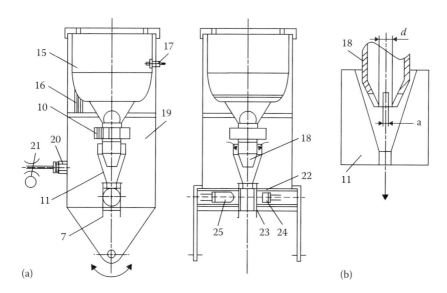

FIGURE 3.58 Low-pressure cold spray gun. The powder supply system: (a) pick up device, and (b) removable nose and funnel.

controlled by the removable lower bowl nose (18) with a variable diameter d of hole and slot size a. The rate of dispensing powder is defined by fluidity of the powder.

The partial vacuum that exists in the partial-vacuum zone in the lower portion (19) of pick-up chamber aspirates air from the atmosphere while being controlled by a flow throttle when passing into the partial-vacuum zone of the chamber. The chamber is fitted with a flow sensor (20) that generates a measurement signal corresponding to the rate of air flow from the atmosphere through the throttle (21) into the partial-vacuum zone of chamber. Hence, the rate of powder passing through the powder passages (8) is controlled.

The pick-up device includes a powder metering unit (24) that detects the flow of powder particles in a measurement duct—this duct can be a simple glass powder transportation tube (23) connecting the funnel (11) to the powder aspirating hose (7) that is in turn attached to the powder switch of the outer housing. The powder metering unit includes an infrared sensor (24) and an infrared emitter or light source (25) disposed within the channel. The infrared sensor can determine the mass flow of powder through the glass tube based on the amount of light from light source that is able to pass through the glass tube to the infrared sensor. Although an infrared light source and infrared sensor are preferred, other wavelengths of light or other waves could also be used. Optionally, an additional powder metering unit can be mounted in the pick-up housing on opposite sides of the funnel (11). This additional powder metering unit, similar in nature to the previously described power meter, measures the powder dispensing rate ω_d from the lower vibrating bowl (10). The powder dispensing rate ω_d can then be compared with the conveyed powder rate ω_p. The amplitudes of the vibration units can be adjusted relative to one another to ensure that the powder dispensing rate ω_d is equal (over some short period of time) to conveyed powder rate ω_p. This prevents clogging in the funnel.

The particle volume concentration significantly affects the deposition efficiency. The particle volume concentration in a powder-laden jet greatly influences the effectiveness of CS process. This is particularly true in the case of radial injection of powder by means of conveyance air. Control of the concentration of particles is achieved through the regulation of both the rate of conveyed powder and the rate of conveyance air. The rate of the conveyed powder ω_p is substantially dependent on the powder dispensing rate ω_d and the rate of conveyance air. Also, the powder rate is approximately proportional to the rate of conveyance air in the partial-vacuum zone of the pick-up housing chamber. Therefore, the conveyance air must be adjusted to obtain the desired particle volume concentration in the powder-laden jet. The controller can automatically set the rate of conveyance air by means of the adjustment motor and throttle (26) (Figure 3.57). On the other hand, the controller can set the powder dispensing rate ω_d by adjusting the amplitudes of the vibration units. In this way, the rate of conveyed powder ω_p may be maintained in permanent balance with that of the dispensing rate, that is, $\omega_d = \omega_p$. The rate of conveyance air may be additionally regulated by changing the location of the injection point via the outer housing switch. This may be done either manually or automatically.

Thus, the controller (Figure 3.57) regulates the powder feeding flow rate, carrier air flow rate, and the powder conveyance air in the partial-vacuum zone of the pick-up chamber. This regulation is based on both measurement signals (27, 28, and 29) and preset specifications and is ultimately managed by means of the vibration units and two throttles. In actuality, the controller receives inputs regarding the desired powder dispensing flow rate ω_d (for instance in g/s), and the necessary volume concentration of powder value C_v. On the other hand, this carrier air flow rate for the air passing through the powder/air duct (1) can be determined as

$$C_V = \frac{\omega_p}{\rho_p \cdot \omega_{\text{air}}} \tag{3.13}$$

where ω_p is the particle feeding flow rate from the funnel (Figure 3.58b), ρ_p is the material density, and ω_{air} is the carrier air flow rate controlled by air pressure and throttle.

4 Fundamentals of Cold Spray Coating Formation

Thomas Van Steenkiste, Keith Kowalsky, Chris Berghorn, Roman Gr. Maev, Wolfram Scharff, Volf Leshchynsky, and Dmitry Dzhurinskiy

CONTENTS

4.1 PARTICLE VELOCITY AND PARTICLE TEMPERATURE PARAMETER EFFECTS ON COATING FORMATION

Thomas Van Steenkiste

Keith Kowalsky

Chris Berghorn

4.1.1 BACKGROUND

The cold spray (CS) process (Alkhimov et al., 1990; Vlcek et al., 2001) is a method for producing a coating by utilizing high-velocity particles. Upon impact with a substrate or previously deposited particles, the particle's kinetic energy is converted into plastic deformation of the particles and ultimately heat. The particles do not melt and can consequently produce coatings with a minimal particle chemistry changes, oxidation, or residual stress. Previous works in the literature have developed the mechanisms by which the interaction of the high-velocity gas streams couple to the particles to produce the high-velocity particles via drag effects (Alkhimov et al., 1990, 1998) The concept of a critical particle velocity for coating formation (Alkhimov et al., 1990, 1994), dependent on the material properties of the coating materials, was developed by Alkhimov, Kosarev, and Papryrin, and has been expanded on by various authors (Alkhimov et al., 1990; Vicek et al., 2001).

Several other new particle coating developments, the KPlaz process developed by Flame Spray Industries Inc. and the Kinetic Spray process, accelerate both larger and smaller diameter particles (Van Steenkiste et al., 2000; Elmoursi et al., 2005) and the radial injection gas dynamic spraying, RIGDS (Maev and Leshchynsky, 2006a, b, and c) processes, for example, also produced similar coatings. RIGDS uses a combination of larger hard particles (materials that do not plastically deform such as ceramics or high-strength materials such as tungsten) with smaller ductile particles to produce coatings. All of these processes rely on the conversion of the particle's kinetic energy to produce either a particle–particle or particle–substrate bonding. The particle's kinetic energy and momentum in the coating formation must be converted into other forms of energy via mechanisms such as plastic deformation (substrate and particle interactions for the initial particles in the coating, as well as the particle–particle interactions as the coating formation builds), which leads to consolidation of the voids between deformed particles and ultimately yields heat. Failure to convert all the incident particle's kinetic energy into substrate or coating formation results in an inelastic collision, that is, the particle will bounce off. This chapter will discuss the roles that the particle temperature and particle velocity as well as the coating material properties play in this coating formation process.

The literature has many references for the mechanisms that influence how particles accelerate in the high-velocity gas stream (Alkhimov et al., 1997, 1998; Dykhuizen and Smith, 1998; Gilmore et al., 1999; Kreye and Stoltenhoff, 2000); however, a brief review is necessary to understand the role that the coating material's physical properties, particle velocity, and particle temperature parameters interact in the CS process to form coatings. A picture of a typical CS system is shown in Figure 4.1a.

The CS process involves preheating the main gas flow and combining it with feedstock particles suspended in a gas mixture from the powder feeder into a mixing chamber. The combination of gas and particles are generally injected axially through a de Laval-type nozzle and the particles accelerate due to drag coupling effects from the high-velocity gas. The particles can also be injected downstream of the throat region in the diverging section of the nozzle (Maev and Leshchynsky, 2006). The conversion of the particle's kinetic energy to mechanical deformation and thermal energy occurs when the particles impact the substrate surface at high velocity. The particle's velocity is influenced by several factors: main gas temperature, molecular weight of the main gas, main gas pressure, main gas flow, powder carrier gas temperature, and carrier gas flow, particle flow rate,

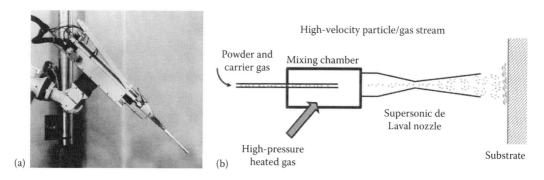

(a)   (b)

FIGURE 4.1   (a) Picture of the CS equipment and (b) Schematic diagram of the CS gun assembly.

particle size, residence time in the high-velocity gas flow, particle shape or morphology, particle density, and if the particles are less than 10 μm the interaction with the bow shock wave at the substrate surface to name a few. The particle velocity inside the nozzle is determined by the coupling of the gas velocity to the particle drag in the main gas stream. The CS process uses a de Laval converging–diverging nozzle to entrain metal and/or composite powder mixtures in a supersonic gas flow. A schematic diagram of the CS machine's key components is shown in Figure 4.1b. The nozzle has an entrance cone (converging region), followed by a throat section, and finally a diverging cross section. Again there are many references (Dykhuizen and Smith, 1998; Alkhimov et al., 2001; Champagne et al., 2004) in the literature for optimizing the nozzle design, exit shape, and other parameter developments. Typically, a CS system will have a pressure sensor port and a main gas thermocouple located in the nozzle assembly. These sensors provide feedback for the control of the gas conditions in the mixing chamber–converging section (before the throat), namely the main gas pressure and the main gas temperature. Similarly, the powder feeder gas pressure, feed rate of the particles and, in some systems, the powder gas temperature are monitored and computer controlled for the powder carrier gas before injection into the mixing chamber. Typically, the main gas temperature is controlled by an electrical resistance heater, although newer designs are heating the main gas using a plasma-based system (the Kplaz system). The primary purpose of heating the main gas is not to heat the particles but rather to increase the gas velocity, which in turn produces high particle velocities via drag coupling. Increasing the particle temperature, however, can provide significant benefits, as will be discussed later.

As stated previously, increased gas velocities generally result in higher particle velocities. The speed of sound for a gas is dependent on both the gas temperature and the molecular weight of the gas. The equation for the speed of sound v is

$$v = \left(\frac{\gamma RT}{M_w} \right)^{1/2} \tag{4.1}$$

where:
γ is the ratio of specific heats (1.4 for air and 1.66 for He)
R is the gas constant (8.314 J/kmole-K)
T is gas temperature
M_w is the molecular weight of the gas

Therefore, to increase a particle's velocity, one needs to increase the gas velocity by increasing the main gas temperature, or use a smaller molecular weight gas such as helium. Once the pressure downstream is 52.8% of the pressure upstream (for air or nitrogen), the flow through the nozzle throat becomes sonic. When the air velocity becomes sonic, further increases in the upstream pressure do

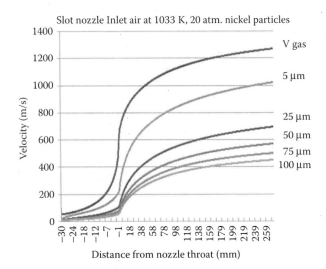

FIGURE 4.2 Computed gas and particle velocities as function of distance along the nozzle. Particle diameters denoted in microns for the particle velocity curves.

not cause a further increase in the gas velocity through the nozzle throat (flow is choked). Increasing the upstream pressure, however, increases the density of the gas, because the mass flow rate is a function of density; the mass flow rate increases linearly with pressure while the gas velocity is constant. This density increase will influence the coupling between the gas velocity and the particle velocity.

Figure 4.2 is a one-dimensional computation of the velocities for the main gas and the particles assuming air as the main gas, inlet gas temperature 1033 K (1400°F), inlet pressure to be 2.0 MPa (300 psi), and nickel as the coating powder. The nozzle has 276 mm diverging section measured from the throat; with a rectangular exit of 5 mm by 12.5 mm. Analytic equations were used to compute the gas velocity and temperature in the nozzle from the gas inlet conditions and the nozzle area versus length (Andersen, 1982). Particle velocities in the nozzle were calculated from the drag forces using correlations in the literature (Henderson, 1976). Particle temperatures were calculated using heat transfer correlations (Carlson and Hoglund, 1964) (using the same conditions as in Figure 4.2) and are shown in Figure 4.3. These simple models, while ignoring boundary layer effects and heat transfer to the nozzle do provide insight into the controlling factors of the particle velocity and particle temperature.

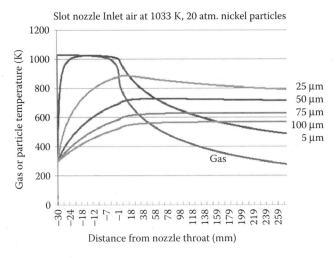

FIGURE 4.3 Computed gas and particle temperatures as a function of distance along the nozzle. Particle diameters denoted in microns for the particle temperature curves.

Figure 4.2 shows that the gas velocity and particle velocity scale roughly as the square root of the absolute main gas temperature. Figure 4.3 shows the heating of the particles as a function of size and distance from the throat. One observes that for small particles, less than 5 μm, the particle velocity and particle temperature curves follow that of the main gas velocity and temperature curves. A rapid increase in particle temperature in the heating section upstream of the nozzle throat is followed by a large increase in particle velocity and simultaneous particle cooling during the trip through the diverging section of the nozzle, while for larger particles (greater than 5 μm), one notes a time-dependent rise and fall of the temperature and velocity related to the size of the particles and its heat capacity. This critical interaction between the particle size and the high-velocity gas allows one to tailor the different spray processes for coating formation. For example, the kinetic spray process (Van Steenkiste et al., 2000; Elmoursi et al., 2005) uses large particles (greater than 50 μm) to produce coatings. These are particles having reduced velocities, but have higher particle temperatures, compared to the typical CS process, which uses higher particle velocities but lower particle temperatures. It is this relationship between particle temperature and particle velocity that is important for coating formation and will be discussed further.

Other factors that can affect the particle's velocity as they travel toward the substrate are the bow shock waves generated by the high-velocity gas stream impinging on the substrate (Alkhimov et al., 1997; Dykhuizen and Smith, 1998; Champagne et al., 2004) particles experience varying degree of deceleration as they pass through the bow shock wave (Han et al., 2004).

Because one is relying on the coupling of the high-velocity gas stream to accelerate the particles, the particle shape (morphology) will determine how effective this interaction is. Long aspect shapes (such as needles or flakes) do not present a large cross-sectional area to promote drag effects. Particle geometry of this type of will tend to orientate into a direction that will present a reduced cross section perpendicular to the gas flow reducing the drag coupling and consequently the final particle velocity. The particle density is another factor to consider (Vicek et al., 2001). Particles with a high density, such as tungsten, will accelerate more slowly than particles with a lower density such as aluminum. High-density materials may also require increased velocity powder carrier gas flows to suspend the particles in the gas flow for the journey to the nozzle. This higher powder carrier gas flow can lead to a reduced mix gas temperature in the nozzle as a consequence of the increased dilution of the cold powder carrier gas flow, if not heated to the main gas temperatures.

Detailed numerical computations of particle deformations have been presented in literature (Dykhuizen et al., 1998); however, the conversion of the particle's kinetic energy is aided by the following simple approximate computations for particle collision stresses and times. We know that plastic deformation is in fact found experimentally, with roughly spherical incident particles flattening into a pancake-like structure with an aspect ratio ranging from 3:1 to 5:1 depending on the particles velocity, temperature, and material composition (Alkhimov et al., 1990; Garetner et al., 2003; Champagne et al., 2004). The collision time, Δt, can be approximated as d/v, where d is the particle diameter and v is the incident velocity of the particle. For a 50 μm diameter particle incident at 500 m/s, the time Δt for the collision process to take place is $\Delta t \cong d/v = 10^{-7}$ s. The impact stress S averaged over the time Δt may be estimated as $\Delta P/A\Delta t$, where A is the cross-sectional area of the (flattened) particle and ΔP is the change in particle momentum as a result of the collision. Here $\Delta P = mv$, the incident particle momentum, and $A \cong \pi d^2$, where an average aspect ratio of 4 has been taken. The time-averaged impact stress S is given by $S \cong \Delta P/A\Delta t = \rho v^2/6$, where ρ is the density of the material of the incident particle. aluminum particles, $\rho = 2.4 \times 10^3$ kg/m³, so $S \cong 100$ MPa. The typical yield point of pure aluminum is 34.5 MPa, and the peak impact stress for copper particles impinging on stainless steel (SS) was found to be 5 GPa (Dykhuizen et al., 1998). Conversion of the particle's kinetic energy at these velocities, temperatures, and sizes provides sufficient energy to plastically deform these materials. Materials that have higher yield stress values will require increased kinetic energy, that is, higher particle velocity.

It is well known that the yield strength of materials, hardness, shear modulus, ultimate tensile strength, and so on, are a function of temperature (Buch, 1999). Reference literatures for various materials and alloys have measured values for these properties as a function of temperature. Generally increasing particle temperatures result in a corresponding decreasing of the mechanical properties. Because cold-sprayed particles utilize plastic deformation in the production of coatings, to convert the particle kinetic energy, increasing the particle's temperature should allow one to achieve the similar coating formation at lower particle velocity. Figure 4.3 shows that as the particle size increases, there is a time-dependent heating and cooling. The larger particles take longer to heat-up once injected into the heated gas stream, but also cool down at a reduced rate in the diverging section of the nozzle and travel time to the substrate before impact. Larger particles arrive at an elevated particle temperature and slower particle velocity compared to smaller size particles.

4.1.2 Coating Formation

For the coating formation to occur, all of the particle's kinetic energy must be transformed into plastic deformation, heat, work hardening, or some other process to the coating and/or substrate. As stated previously, plastic deformation is in fact found experimentally, with roughly spherical incident particles flattening into a pancake-like structure with an aspect ratio ranging from 3:1 to 5:1 depending on the particles velocity, temperature, and material composition. A schematic diagram of the process is shown in Figure 4.4.

The coating buildup is not a simple one particle impact, but rather a series of multiple impacts that transfer the incoming particle's kinetic energy to the substrate initially then to the particles that form the coating. A multistep process has been suggested (Van Steenkiste et al., 2002) consisting of substrate cratering and first layer buildup, followed by particle deformation and realignment, metallurgical bonding, and coating void reduction, and so on, as a possible coating formation mechanism for coating buildup. In Figure 4.5 (etched photo of an aluminum particle after impact), one can clearly see that the once roughly spherical particle at the point of impact a high degree of localized plastic deformation has occurred. One can also observe that the particle grain structure for the majority of that particle that is not in the impact zone is relatively unchanged. Evidence for plastic deformation of previously deposited particles and multiple impacts is also shown in Figure 4.5 as subsequent particle collisions with the existing particles on the substrate surface force material into the voids between the previous deposited particles (see the arrows). The initial first particles impact with the substrate and the bonding produced is presumed to be the result of adiabatic shear (Dykhuizen et al., 1998; Garetner et al., 2003). The original grain structure is preserved, although it is highly deformed, suggesting that bulk melting of the particle does not occur. The measurements of the oxygen levels in the powders and the coatings are determined to be approximately the

Incoming high-velocity particles

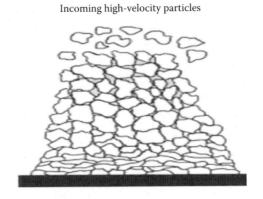

FIGURE 4.4 Schematic diagram showing the formation of coatings using the CS process. Starting with substrate cratering/cleaning & first layer buildup of particles, particle deformation of the previously deposited particles, void reduction, and metallurgical bond formation between particles.

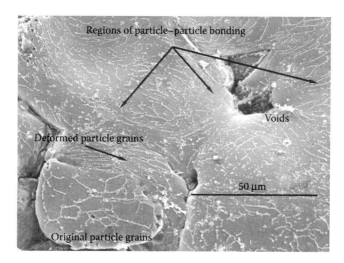

FIGURE 4.5 Electropolished sample of a coating produced at 204°C (500°F) using large particles (greater than 50 μm). Areas of bonding are shown by arrows. Note the degree of plastic deformation that has occurred between particles and the resulting plastic flow of the internal grain structure. The particle in the lower-left side of the picture shows internal plastic flow of the GBs at the impact site; however, the original grain structure was preserved in locations removed from the impact area.

same (Gilmore et al., 1999; Van Steenkiste et al., 1999, 2002) supporting little or no melting of the particles (particle melting would allow for particle oxidation). Evidence has been presented in the literature to support this impact process and numerous micrographs in the literature (Alkimov et al., 1990; Garetner et al., 2003; Champagne et al., 2004; Van Steenkiste and Smith, 2004) demonstrate the plastic deformation of the particles similar to that observed in Figure 4.5.

As stated before, Alkhimov, Kosarev, and Papryrin first demonstrated that there exists a critical velocity region, below which the particles do not form coatings and above which coatings rapidly build (Alkhimov et al., 1990; Garetner et al., 2003; Van Steenkiste and Smith, 2004; Maev and Leshchynsky, 2006). This critical velocity is a function of the physical and mechanical properties of the powder material and their drag interaction with the high-velocity gas stream in the nozzle. Experiments have shown that coatings using materials with higher modulus are more difficult to produce (Alkimov et al., 1990; Garetner et al., 2003; Van Steenkiste and Smith, 2004; Maev and Leshchynsky, 2006) requiring higher main gas temperatures (i.e., increased particle velocity and increased kinetic energy), or incorporation of the materials into a ductile matrix (e.g., ceramic–metal composites) (Elmoursi et al., 2005).

Because the starting powder feed stock is usually a distribution of particles sizes, the critical velocity will be a range of velocities (Gilmore et al., 1999; Alkhimov et al., 2000; Kreye and Stoltenhoff, 2000; Garetner et al., 2003). Individual particles with different diameters or shapes will achieve different velocities due to the drag interactions between the high-velocity gas flows. If the incident particle velocity is too low, the yield strength of the particle material and/or substrate material would not be exceeded (no plastic deformation would occur) and the particle would not adhere. Oxide layer or film on the powder's surface and/or on the substrates surface would also effect the critical velocity needed for coating formation, as oxide films tend to be more brittle than metals requiring its removal before particle plastic deformation could occur.

Substrate properties also contribute to coating formation, and are discussed in the literature (Garetner et al., 2003; Sakaaki et al., 2004; Van Steenkiste and Smith, 2004), but will not be discussed in this chapter. Generally, a roughened surface substrate (i.e., media blasted) will promote higher coating deposition efficiency, because the substrate surface does not have to be cleaned and roughened using the first incoming particles. Contact surface area of the substrate is also increased promoting increased substrate–particle interaction. If the main gas stream is nitrogen or helium, oxidation of the coating and/or substrate can be reduced as the main gas also functions as a shield gas.

Because plastic deformation of the particles must occur for coating formation the yield stress of the particle and/or the substrate must be exceeded during the collision process, and is a part of the necessary energy transformation. For this reason, powders and substrates typically employed in the CS process are metals with relatively high ductility. Ceramic and other high-strength materials are very difficult to deposit and to date the only coatings that have been produced are essentially a few particle layers thick with particles embedding into the substrate and little if any particle–particle bonding. Explosive compaction, however, has produced ceramic coatings, so in theory, if the particles velocity or particle temperature was high enough ceramic coatings could also be produced by the CS process (Murr et al., 1986). Similar results have also been observed in the literature for other brittle materials (Lee et al., 2005).

It has been observed, however, that coatings consisting of mixtures of metals and ceramics or a mixture of a ductile matrix material with a nonductile material can be sprayed successfully (Elmoursi et al., 2005; Lee et al., 2005). Many composite coatings, using ceramic magnetic materials such as Terfenol-D ([$Tb_{0.3}Dy_{0.7}$]$Fe_{1.9}$), AlNiCo, $SmFe_2$, and other rare earth compounds combined with copper and aluminum matrix have been successfully produced (Pinkerton et al., 2002).

Figure 4.6 is an aluminum/silicon carbide coating. The silicon carbide particles do not plastically deform and become incorporated into the ductile aluminum matrix during the coating formation. The aluminum matrix incorporates the material that does not plastically deform. Deposition efficiencies are generally lower because the amount of incorporation of the nonductile components into the coatings is usually less than the original starting percentage. The RIGDS method (Maev and Leshchynsky, 2006) uses mixtures of metals and ceramics to produce coatings; here the energy of the nondeforming material help in the coating formation of the ductile materials. An analogy would be a hammer and anvil method used to form steels, where some of the incoming particles are sandwiched between the substrate and the nondeforming particles.

Reactive coatings (combination of two or more materials that are cold sprayed without reacting together) are another new development possible with CS. A CS composite coating of tin or tin alloys with copper particles, for example, could be used as a reactive coating. The copper particles would react with the tin particles, if this coating were exposed to a high enough temperature, but would not react below this temperature. These types of coatings could be used in soldering, brazing, or other applications.

200 µm

FIGURE 4.6   Cross-sectional SEM of the Al/silicon carbide composite coating on a brass substrate (bottom). No evidence of plastic deformation of the silicon carbide particles is observed.

However, the critical velocity for coating formation is only one of the critical aspects used to determine the coating formation. Equally important is the particle temperature. These two parameters ultimately determine whether a particle will plastically deform and produce a coating or not. The particle velocity will determine the amount of kinetic energy available for plastic deformation. The particle temperature will determine the mechanical properties of that particle at the moment of impact.

4.1.3 Particle Velocity

From Equation 4.1, one observes that the particle velocity is dependent on main gas temperature, molecule weight of the main gas, and the drag coupling between the particles and main gas stream. The simplest solution to increase the particle velocity is to increase the main gas temperature. However, there are physical limitations on the heater elements if one wants to maintain a heater having small weight, size, and electrical consumption. Generally, for electrical resistance heaters a practical upper bound would be around 800°C. A new CS process heater developed by Flame Spray Industries Inc. (Zhao et al., 2005) utilizes a plasma to heat the high-pressure main gas to temperatures exceeding those achievable by conventional electric heaters (Figure 4.7).

Substituting a lower molecular weight gas such as helium or a helium gas mixture will also increase the main gas velocity and consequently increase the particle velocity. For most processes, however, this option is economically expensive unless one can reclaim the helium. Typically, most of the CS systems in industry use nitrogen or air as the main gas.

The drag coupling between the particles and main gas stream is a function of particle size, particle shape, particle density, and total kinetic energy of the gas flow, and the drag coupling between the particles and main gas stream (Han et al., 2004) have shown that the particle acceleration potential from the main gas flow is proportional to the density and velocity, respectively, of the gas in the nozzle and the local flow area of the nozzle. Because particle acceleration potential is the product of the mass flow rate and the gas velocity, it is directly influenced by the total mass flow rate. The CGT Kinetiks 4000 system, for example, operates at higher pressures, compared to other CS systems, with an additional nozzle heater to increase the particle velocity from Figure 4.2, one observes that smaller particles can quickly accelerate to near the gas velocities while traveling through the nozzle.

Decreasing the initial powder feed stock size will generate high particle velocities and result in increased kinetic energy as shown in Figure 4.2. Lower density materials will accelerate faster. Ultimately, the particle velocity optimization is a function of several variables and will depend on the type of heating system, main gas, powder material, particle size, particle shape, type of coating (one material or composite), cost factors, and so on of the final coating product.

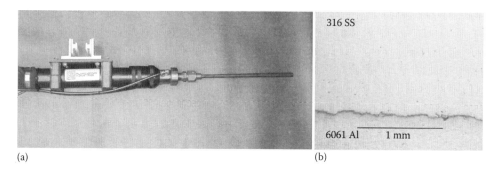

(a) (b)

FIGURE 4.7 (a) Picture of a new Kplaz gun developed by Flame Spray Industries. Main gas is heated by plasma. (b) Micrograph of coating produced using the Kplaz spray gun.

4.1.4 Particle Temperature

Particle temperature is primarily a function of the powder material properties, powder size, main gas temperature, and the residence time in contact with the main gas stream and/or a heated powder carrier gas stream. Optimizing the particle temperature can be accomplished using a combination of the variables above; however, limitation of the spray systems, coating material, coating properties, and so on, may determine what final particle temperature is best.

The simplest method to increase the particle temperature is to increase the main gas temperature. This method results in increased particle velocity and particle temperature. As shown in Figure 4.1b, the gas mixture injected into the nozzle is composed of a main gas flow and powder carrier gas flow. Changing the ratio of the powder feed gas that is injected into the mixing chamber with the main gas (at elevated temperature) can cause an increase in the particle temperature because the powder feed gas is generally at room temperature and carries the powder into the heated main gas flow. This dilution of the elevated temperature main gas reduces the temperature of the combine gas flows into the de Laval nozzle. Particle temperature increases on the order of 50 K have been reported by changing this ratio (Han et al., 2004). However, for high-temperature, high-yield strength materials, this may not be sufficient for coating formation based on the limitations of the cold systems.

One observes in the literature that comparable coatings can be produced using the kinetic spray process (Van Steenkiste et al., 2000; Elmoursi et al., 2005). The process uses larger particles, traveling slower (300–500 m/s), but are at higher particle temperatures (several hundred K), compared to the CS process (Alkhimov et al., 1990; Vicek et al., 2001). These larger particles also exhibit a lower critical velocity range, compared to the critical velocity range of the smaller CS particles, but also produce comparable coatings. Larger particle feedstock powders provide opportunity for higher particle temperatures because the residence time in the main gas flow is longer, (slower velocity due to mass effects and negligible cooling in the diverging section of the de Laval nozzle, see Figures 4.2 and 4.3) and have been used successfully by the kinetic spray process (Van Steenkiste et al., 2000; Van Steenkiste and Smith, 2004; Elmoursi et al., 2005) to form coatings.

The effect of temperature on the yield strength of various materials is shown in Figure 4.8. Generally, one observes a decrease in the mechanical properties with increasing temperature. This could explain the lower critical particle velocity for the kinetic spray processes using relatively large, above 50 μm, particles.

Increasing the residence time by lengthening the mixing chamber before the nozzle has also been successfully used to increase the particle temperature (Han et al., 2004; Zhao et al., 2005) and is an optional part of the new Kinetiks 4000 CS system. These devices are designed to minimize the disturbance to main gas flow prior to the particles entry into the convergent section of the nozzle,

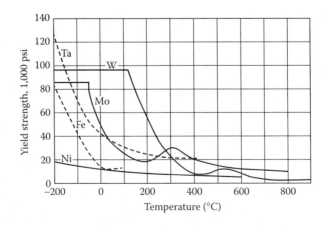

FIGURE 4.8 Effect of temperature on the yield strength of body-centered cubic Ta, W, Mo, Fe, and face-centered cubic Ni. (From Bechtold, J. H., *Acta Metall.*, 3, 252, 1955. With permission.)

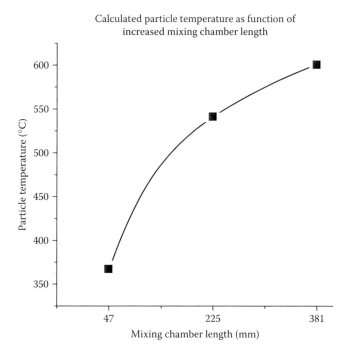

FIGURE 4.9 Calculated particle temperature as a function of mixing chamber length for a 90 μm diameter nickel particle.

while increasing the residence time and particle temperature. Figure 4.9 shows calculated particle temperature as a function of increased mixing chamber length for a 90 μm nickel particle. As the mixing chamber length increases the particle temperature increases. A calculated particle temperature increase of approximately 250°C is possible for 234 mm chamber extension. Increasing the length of the convergent section of the nozzle has also been used to increase the particle temperature (Sakaki and Shimizu, 2001).

The use of additional heaters on the powder feeder lines is another option that heats the powder carrier gas and subsequently heats the particles suspended in the elevated temperature gas flow. Figure 4.10 is schematic diagram and picture of an inline powder heater assembly. Depending on the material properties of the powder particles, the particle delivery at elevated temperatures can become more complicated and harder to maintain a precise mass flow of particles, because hot particles generally become *sticky* clumping together limiting their ability to flow well. Nozzle clogging and sticking to interior walls of the nozzle assembly and/or powder feed lines is possible without properly designed components and attention to details with heated particles (Han et al., 2004). Oxidation of the particle's surface is another factor to be concerned at higher particle temperatures when the carrier gas is air. Experiments (Van Steenkiste and Smith, 2004). have measured a doubling of the oxide levels with the use of increased powder gas heating, when using air carrier gas. Nitrogen, helium, or other inert processing gases would be required with elevated particle temperatures where oxidation would occur.

Experiments, using a nickel composite consisting of 10% AlNiCo$_8$–80% Ni–10% Fe, by volume, as the coating material, were completed to observe what effect particle temperature would have on coating formation. Powder line heating, mixing chamber length, and larger diameter particles were used to increase the particle temperatures. Figure 4.10 shows the experimental setup using resistance powder heater assembly for these experiments. The temperature of the powder feeder line was measured using a K-type thermocouple attached to the outside of the metal feed line approximately 50 mm upstream of the point of injection into the nozzle. The kinetic spray nozzle 300 mm long

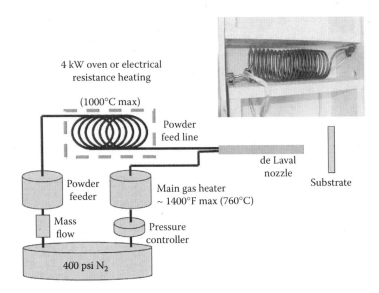

FIGURE 4.10   A schematic diagram of a typical CS system with an external heater on the powder feed line. The external powder feed line heater increases the residence time of the particles in contact with elevated temperature powder carrier gas. A picture shows electrical resistance heater developed by Flame Spray Industries.

with an exit area of 12.5 × 5 mm. The throat diameter was 2.8 mm. The $AlNiCo_8$ powder particles (received from Dexter Magnetics) had a particle diameter of 51–230 μm. The nickel particle size range was 63–90 μm in diameter (F. J. Brodmann Inc.). The iron powder was 45–250 μm size range (Quebec Metal Powders Limited). The main gas temperature was 732°C (1350°F).

   The mixing chamber extension lengths were 381, 272, 197, 150, 122, and 47 mm. The mixing chamber extension length devices were used with and without the additional powder line heaters. The powder feed rate was 2.5 g/s. The traverse speed was 0.8 mm/s for the experiments.

   Figure 4.11 is a graph of powder line temperature (all other spray parameters fixed) as a function of single pass coating thickness. By adjusting only the temperature of the powder line, which by default adjusts the particle temperature, the coating thickness increased from 1 mm to over 2.5 mm for an approximate 200°C powder line temperature increase. As the powder traverses through the heated length of the powder line its temperature increases due to contact with the heated walls and gas in the line.

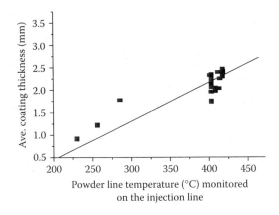

FIGURE 4.11   Single pass coating thickness of a nickel composite coating as a function of increasing power feeder line temperature. As the powder line temperature increases the particle temperatures increase as a consequence of increased residence time with the hotter powder feeder carrier gas.

The particle's increased residence time, in contact with elevated powder feeder carrier gas temperatures, results in increased particle temperatures (with some increased particle velocity addition from the increased mixed gas temperature). The material properties of the nickel particles at elevated temperature were modified in such a way that the ensemble was able to bond more efficiently, resulting in thicker coatings and increased deposition.

Figure 4.12 is graph of main gas temperature as a function of mixing chamber length. The temperature of the gas is measured in the converging section of the nozzle using a K-type thermocouple. If the main gas temperature and flow rate into the mixing chamber is held constant, the increased heat loss, from increasing the mixing chamber length (due to increased thermal losses from the additional mass, increased surface area, etc.), will result in an overall temperature decrease in the exiting main gas before it enters into the nozzle. From Figure 4.12, one observes a decreasing main gas temperature with increasing mixing chamber length. However, the additional mixing chamber length allows for an increase residence time for the large-size particles resulting in increased particle temperatures (see calculated values in Figure 4.9). Optimizing the mixing chamber length for the particle feedstock size provides the particles sufficient time to heat up while not deceasing the main gas temperature and particle velocity, sufficiently to prevent coating formation.

Figure 4.13 is a graph of deposition efficiency as a function of increasing mixing chamber length. One can clearly observe that increasing the mixing chamber length (increasing particle temperature) results in an increase in deposition efficiency. The largest deposition efficiency recorded in Figure 4.13 is when an optimizing mixing chamber length was used in conjunction with the preheated powder feeder lines. The yield strength and hardness of nickel decreases rapidly with temperatures of 200°C and higher as shown in Figure 4.8 and in the literature (Buch, 1999). Presumably, the increased mixing chamber lengths and powder feeder line heaters are increasing the large-size nickel particle's temperature to around this range when the powder feeder line

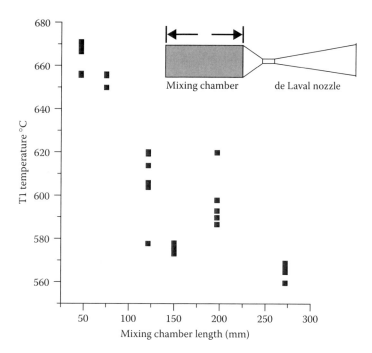

FIGURE 4.12 Main gas temperature as a function of the mixing chamber length. Temperature of the gas is measured in the converging section of the nozzle. Because the main gas temperature into the mixing chamber is constant, the increase heat loss, from increasing the mixing chamber length, results in a temperature decrease in the exiting main gas for increased mixing chamber length; however, the increase residence time for the large-size particles results in increased particle temperatures.

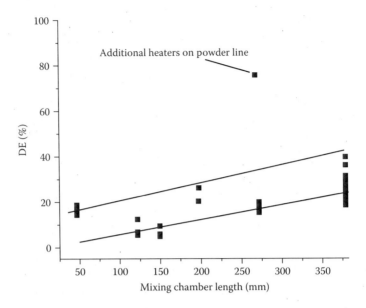

FIGURE 4.13 Deposition efficiency as a function of mixing chamber length. Largest deposition efficiency, approximately 76%, was observed when increased mixing chamber length was used in conjunction with an external powder feeder line heating.

temperature was above 250°C. Figure 4.13 shows that the optimum combination of mixing chamber extension of 272 mm with powder line heating provided the best deposition efficiency for the 10% $AlNiCo_8$–80% Ni–10% Fe powder mixture. Figure 4.14 is a cross-sectional micrograph of the coating produced using the optimum spraying parameters above. The micrograph shows a structure consisting of a matrix of plastically deformed nickel and iron particles surrounding the brittle $AlNiCo_8$ particles on an aluminum substrate.

The particle temperature optimization is again a function of several process and material variables as demonstrated in the experiments above and will depend on the powder material and type of equipment used. The required properties of the coating may also ultimately determine what combination of particle temperature/particle velocity is best suited for cold spraying equipment and starting powder available.

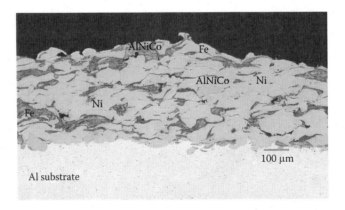

FIGURE 4.14 Cross-sectional micrograph of composite coating using a starting 10% $AlNiCo_8$–80% Ni–10% Fe powder mixture. The coating was produced using the optimum mixing chamber length combined with powder line heater.

4.1.5 Concluding Remarks

Optimizing both the particle velocity and the particle temperature are the most critical factors that contribute to the formation of CS coatings. These two parameters closely interact with each other to produce coatings. The particle velocity determines the amount of kinetic energy available for plastic deformation and bonding and the particle temperature determines the physical properties of the particle at the moment of impact. Increasing the particle temperature, generally can results in reducing the critical velocity, while decreasing the particle temperature will generally result in a need for higher particle velocity. CS coating formation of high-strength, low-ductility, higher service temperature materials will push one to operate with higher main gas heater temperatures (to achieve increased critical velocities) in conjunction with higher particle temperature heating methods. The balance between the kinetic energy and momentum of each individual particle that must be 100% transformed in conjunction with the material properties will allow one to build coatings using the CS process. New developments in plasma heating of the main gas stream (see Figure 4.7) could in theory produce increase particle velocities and increase particle temperatures that could open an operating window of spray parameters for the production of ceramic coatings and high-temperature materials. The benefits of the CS process that one realizes from not melting particles and not changing feedstock chemistry have demonstrated that the CS process is a viable alternative to some of the more traditional coating processes, such as thermal spray.

4.2 COATING MICROSTRUCTURE AND PROPERTIES

Roman Gr. Maev

Wolfram Scharff

4.2.1 Classification: Structure–Property Relationship

In this chapter, a so-called PSPP (processing–structure–properties–performance) approach, modified after Olsson (1997), is used to describe the structure–properties relationship for Cold Spray (CS) coatings (Figure 4.15). The following three links need to be analyzed to achieve an integral model for the wear performance of a surface:

1. Influence of the surface properties on the wear processes
2. The link between the material properties and the microstructure
3. Influence of the surface manufacturing process on the microstructure

Advanced surface engineering offers many possibilities to modify the properties of surfaces to achieve the better exploitation performance using various surface treatments, deposition of thin layers or processing of thick surface coatings as shown by Holmberg and Matthews (2009). Thin coatings, like physical vapor deposition (PVD) and chemical vapor deposition (CVD) coatings, are

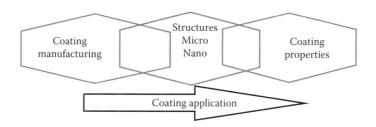

FIGURE 4.15 The PSPP approach to establish links between coating manufacturing, structure, and properties.

excellent in many tribological applications, and they are being used largely today. However, they may be helpless, in particular, in harsh, high-load and high-temperature conditions due to both material and structural limitations originating from their tiny thickness, which is typically in the range of 1–3 μm and even less. The use of thick composite coatings is another solution to tailor surface properties of the component. As shown in Chapter 1, the typical processing methods for thick coatings are thermal spraying and laser powder cladding. These methods offer a flexible route to produce the composite-structured, rather thick coatings, in a typical thickness range of 150 μm to 3 mm, on a substrate material selected according to other criteria, like price, extent of alloying elements and other additives, mechanical strength and low weight (Davis, 2004). CS technology opens the wide possibilities to produce the thick metal and composite coatings in the wider thickness range (100 μm to 10 mm) with high-dimensional accuracy due to coating formation at the relatively low temperatures.

In general, a composition, structure, and surface characteristics determine the properties of materials. The surface properties of various components have a strong influence on their behavior. In the most of cases, the component surface is exposed by a combination of various factors such as mechanical, tribological, thermal, corrosion, and so on. These factors can cause degradation of material properties or even its damage. Coating can reduce the environmental effects in many cases (Figure 4.16).

Over the last decades, myriad hybrid coating materials suitable for thick coatings—mainly mixtures and composites of ceramics, metals, and polymers—has been developed. Entirely new possibilities for developing high-performance materials have opened up in the 2000s, with the development of new manufacturing methods, widespread adoption of nanotechnology, and more in-depth understanding through new process diagnostics and higher modeling capacity (Holmberg et al., 2014). Particle-reinforced composite materials consisting of a metal matrix and hard dispersed particles offer a potential solution for increased wear resistance demands. An improvement in wear properties is often counterbalanced by the deterioration of other properties, such as impact resistance or corrosion resistance. It is important to tailor and optimize the material for the particular application, taking all the requirements into consideration.

Thick composite coatings, such as the CS coatings, need to be characterized and classified in terms of their complex composite microstructure, including interparticle boundaries, grains, pores, phases, and so on. Figure 4.3 shows a typical microstructure of SS–titanium CS coating (Figure 4.17a) and general coating characterization based on its microstructure (Figure 4.17b).

Dispersion phases in the coatings can influence the coating properties and as a rule improves them. The combination of the coating matrix (base material) and dispersion phases results in significant change of the properties, and so careful selection of this combination is very important (Steinhauser and Wielage, 1997). Mainly, various compounds in the form of particles or, in small amounts, short fibers, are used as the reinforcement phase. These second phases are incorporated during deposition process or created later within the coating using post-processing treatment. The classification of the main structure formation processes is presented on the diagram (Figure 4.18).

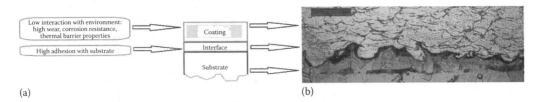

(a) (b)

FIGURE 4.16 Functions of the coatings. (a) The main possible functions of the coating and interface and (b) an example of the CS Copper coating structure on the steel substrate for long-life corrosion protection.

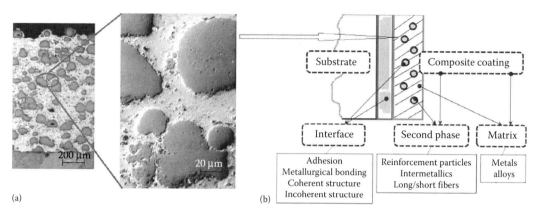

FIGURE 4.17 Schematic depiction of CS MMC coating. (a) Microstructure of SS–titanium CS coating (unetched) and (b) MMC coating microstructure schematics.

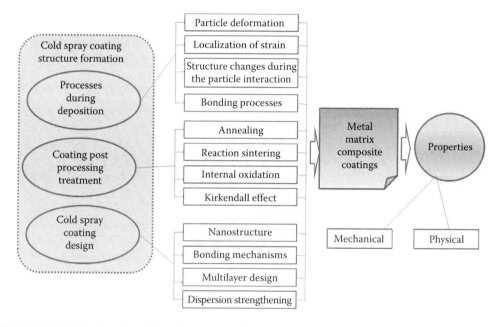

FIGURE 4.18 Classification of the main structure formation processes.

The main structure formation processes during CS deposition are the following:

1. Severe deformation of the particles
2. Adiabatic shear band formation at the particle–substrate and particle–particle interface resulting in metallurgical bonding
3. Shock consolidation of the particulate aggregates, and structural changes due to hardening and softening processes at the particle impact

The specific features of these processes were described in Chapter 2. The typical structures of the coatings and their link with CS technology parameters are analyzed in detail in Section 4.2.

In spite of widespread dissemination of the CS technology during the last decade, main materials that are being applied for CS are soft metals that have ability for severe plastic deformation (SPD). The advanced material science needs the new composite materials that consist of a soft matrix and hardening phases such as carbides, intermetallics, and so on. The deposition of these composite materials by CS and structure formation processes during CS and following processing steps have not been deeply described yet. The main goal of this chapter is to analyze the specific features of cold spraying process of composite coatings obtained by deposition of powder mixtures or separate layers of soft and hard materials, and to define their influence on composite structure and properties to obtain and to develop the new composite coating applications.

The properties of composite materials depend on their phase composition and structure of various phases. For this reason, it seems to be reasonable to manage the cold-sprayed material structure by following heat treatment. Such a combined technology opens the new opportunities to control the phase composition, dislocation, and grain structure of materials. In this case, the new possibilities to obtain various phase composition and properties of materials are opened. That is why from application point of view, development of the combined CS–heat treatment technology is very important for composite materials (coatings) obtained by CS.

The main processes controlling CS coating structure formation during annealing (Figure 4.18) are the following:

1. Recrystallization, phase transformation during coating annealing
2. Reaction sintering processes with some compounds and intermetallic phases formation
3. Internal oxidation processes

Some specific effects such as the Kirkendall effect, pores, and cracks initiation and growth have to be taken into account.

In Chapter 1, the general description of CS in the family of thermal spray coating technologies has been made. It is shown that CS has its own advantages and drawbacks compared with other technologies. The main advantages of CS are known to be the formation of coatings due to severe deformation of the particles without melting at low temperature, and low level of internal stresses in the coatings due to absence of crystallization of melted phase. It allows us to achieve mechanical properties of the coatings much higher than for those obtained by plasma spray, high velocity oxygen fuel, and other flame spraying techniques. Therefore, the utilization of the main advantages of the process to obtain the new materials and composites is of great importance.

It is well known that one of the main problems of the development of CS process is a profound understanding of the mechanisms governing the particle impact and the subsequent particle–substrate and particle–particle bonding. It is now widely known that the bonding in the CS is formed due to SPD occurring at the particle/substrate interface and the localization of plastic deformation in a thin material layer. It is believed that the jetting at the local intensively deformed zones is useful for breaking and transformation of the oxide films which promotes the bonding process. The localization of deformation is induced by the adiabatic shear band formation in the impacting particles. Although the impact times are very small, the formation of adiabatic shear bands allows us to achieve the intensive diffusion and bonding. That is why, application of the combined CS–heat treatment technology allows us to carefully control and to design the micro- and nanostructure of the coatings to obtain their desired properties.

Thermal treatment of Al alloy, SS and other coatings causes precipitation hardening and therefore an improvement in coating properties. Dispersed phases can also be incorporated into cavities within the coatings (e.g., in pores or microcracks). In this way, the coating properties can be improved. Multilayer and nanostructured CS coatings also fall into the category of composite coatings. Such systems are useful as deposits on automotive parts, tools and other components. Gradient coatings can be produced by CS, similar to those produced by other thermal spraying technologies. A great variety of coatings and treatment processes now exists. Most published articles and books (Champagne, Papyrin) have discussed cold spraying technology of composite coatings but recently much work has been done on thermal treatment of the CS composite coatings. Cold-sprayed composite coatings can be applied at harsh tribological conditions and highly corrosive environment, for example, for energy generation, steel manufacture, and metal forming (Champagne, 2007).

Search and establishment of structure–property relations for the CS composite coatings is important task to develop engineering procedure of coating design for certain applications. It is well known that the composite coatings consist of matrix and dispersion phases (Figure 4.17). The interactions of different constituent phases define the composite properties. The CS composite coatings can be applied for various purposes, including:

- Increase of strength and providing the effect reinforcement (by occurrence of high-strength phases in a soft and tough matrix)
- Formation of zones able to absorb an energy, relax stress, and prevent the crack initiation and growth
- Improvement of tribological properties (diminishing friction coefficient and increase of wear resistance)
- Improvement of corrosion resistance
- Improvement of thermal barrier properties (thermal insulation)

Some of the possible matrix and second phase materials for metal-matrix composite (MMC) coatings are shown in Table 4.1.

It can be stated that composite coatings offer favorable options for the improvement of the various exploitation properties of materials, and the main features of structure formation listed on diagram (Figure 4.18) will be analyzed in the following paragraphs.

4.2.2 METHODOLOGY

4.2.2.1 Modeling Approach

To carefully analyze the PSPP (Olsson) relationships a lot of manufacturing, structure, and properties parameters have to be taken into account which are difficult to be defined at present time. For example, Holmberg et al. (2003) demonstrated how the properties can be linked by finite element method (FEM) modeling to wear performance for tribological contacts with components coated by thin homogenous PVD and CVD coatings which are relatively homogeneous. However, it is very complex to build a computational model covering the whole PSPP chain for CS coatings that are not homogenous, and so there is a need to characterize the microstructure in the relevant material model. The real exploitation conditions need to be used to define the material properties, and the risk for deformation, cracking, fracture, wear, or corrosion damage can be estimated. At this case the simulations would show the interaction between the different manufacturing, structure, properties, and exploitation parameters and help to find the main ones to be optimized. At present, time the computational modeling and simulation approach of CS coating structure formation is

TABLE 4.1
The Possible Matrix and Second Phase Materials for MMC Coatings

Material Type	Matrix Phase	Second Phase
Metals	Ni, Cu, Fe, Cr, Co, Ag, Au, Zn, Sn, Cd, Pb, Pd, Al	W, Ti, Ni, Fe
Alloys	Ni–Cr, Ni–Ag, Cu–Ni, Cu–Ag, NiCrBSi, steels	Al-Si, steels, Ni-Cr
Carbides	–	SiC, TiC, Cr_3C_2, WC, B_4C, ZrC, HfC
Oxides	–	Al_2O_3, SiO_2, TiO_2, Cr_2O_3, ZrO_2, BeO
Borides/nitrides	–	Cr_3B_2, TiB_2, ZrB_2, HfB_2, BN, B_4N, TiN, AlN
Intermetallics	–	Fe–Al, Ni–Al, Ti–Al
Other compounds	–	Graphite, CaF_2, MoS_2, WS_2, NbS_2

comparatively laborious and requires a good fundamental understanding of the powder materials and CS phenomena to be modeled. The computational modeling and simulation approach includes several steps, which combine exploitation conditions analysis, software development, material characterization, empirical testing and numerical analysis. The methodology to develop an accurate and generic equation for optimization of the wear resistance can be characterized by the following main steps for the tribological coating applications (Holmberg et al., 2014):

1. Identify the relevant dimensional scale level for the tribo-contact to be optimized
2. Analyze the contact conditions and identify possible influencing variables and parameters
3. Construct the contact and material model by, for example, FEM or molecular dynamics techniques
4. Determine the relevant material properties of the material pair, for example, by macro-, micro-, or nano-level material characterization
5. Carry out computer simulations for the chosen material pairs and contact conditions, to show the prevailing stresses, strains, and deformations

The steps of coating microstructure and properties characterization are principal at the start of modeling. From this viewpoint, an identification of the relevant dimensional scale level of the application case is crucial (Figure 4.19).

Determination of the relevant scale level for the real application process requires deep understanding of surface interaction mechanisms (wear, corrosion, heat transfer, etc.). For example, identification of the scale level for tribology applications (Figure 4.19) allows us to define the real tribology parameters (friction coefficient, lubrication regime, wear rate, etc.) influencing on the wear process. It allows us to define the real material structure, and mechanical and physical properties of the composite coating needed to the certain application case. To get accurate input values for the models, there is a need to determine the mechanical properties of the surface material on a macroscale level but also to determine very detailed microscale properties of the composite structures.

Examples of cold-sprayed composite structures are shown in Figure 4.20 (Bielousova, 2013). The results reveal that a higher magnification needs to be applied to define the microstructural features of CS coating formation such as particle deformation, dynamic recrystallization (DRX), twinning, phase transformations, and so on. As shown in the basic monographs of Papyrin (2001)

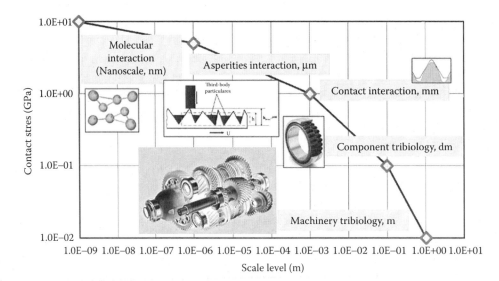

FIGURE 4.19 Identification of the relevant scale level for the tribology case.

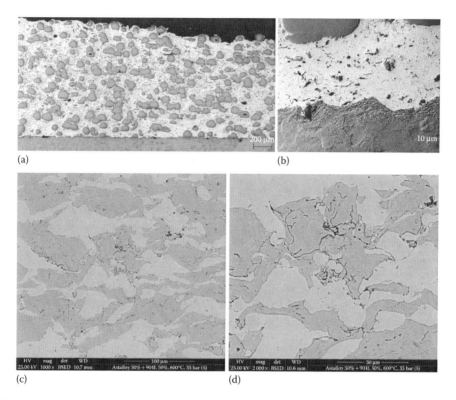

(a)

(b)

(c)

(d)

FIGURE 4.20 Microstructures of CS composite coatings: (a, b) Optical micrographs of AISI 316L (40v%) + aluminum coating on mild steel substrate, (a) magnification 100×, (b) magnification 100×; and (c, d) SEM micrographs of Fe (50%) + AISI 904L (50%) composite coating, (c) magnification 1000×, (d) magnification 2000×.

and (Champagne 2007), one of the main coating structure formation mechanisms is the creation of shear bands under adiabatic conditions associated with high loading rates encountered during the particle impact. The mushroom shape of the particles in the CS coatings is known to be the evidence of this particle deformation mechanism (Figure 4.20c and d). As shown in fundamental review of Antolovich and Armstrong (2014), the primary requirement of adiabatic shear band formation is that no heat should be exchanged between the shear band and the surroundings. It is possible because metals and their alloys have a relatively low heat capacity, which during deformation promotes an increase of the high local temperature.

The presence of adiabatic shear band areas at the particle interface of CS coatings is now well established on the basis of examination of transformed microstructures and modeling of the particle deformation during impact. A dislocation-based approach is crucial to explain the high temperature increase in the area of strain localization. This approach allows us to analyze the localization effects on the basis of the interaction between the physical properties of a metal, its microstructure (e.g., grain size) and a dislocation avalanche that results in highly localized strain (Antolovich and Armstrong, 2014).

The one-particle impact simulations presented in Chapter 2 can be extended to predict microstructures of a coating, as shown by Schmidt et al. (2009) for different materials. Depending on the process parameters and material properties, different particle strains and resulting porosities can be predicted. As a result, the prediction of mechanical properties of the coatings may be performed based on the respective strains and shear instabilities characterization. To highlight differences in attainable microstructures and mechanical properties, the microstructures of Cu and AISI 904 SS coatings are presented in Figure 4.21a and b. The mushroom shape of the Cu particles are clearly

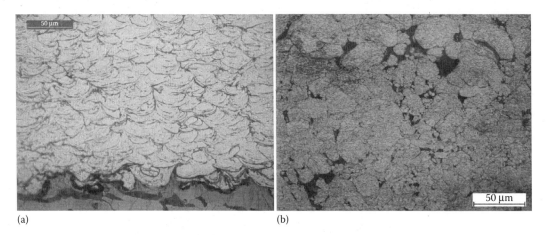

(a)   (b)

FIGURE 4.21 Microstructures of Cu and AISI 904 SS coatings. (a) Cu particle velocity is about 700 m/s and (b) AISI 904 particle velocity is about 600 m/s.

seen in Figure 4.21b because of high particle velocity (about 700 m/s), while the shape of AISI 904 particles is distorted in less extent because the particle velocity ($\approx$600 m/s) is lower than critical. Moreover, porosity of this coating is high as well. The main reason of these results is a strong coating structure dependence on particle velocity. If the Cu particle velocity is sufficient ($>V_{cr}$, Figure 4.21a), the Cu coating microstructure exhibits adiabatic shear band formation features as compared with that of SS coating (Figure 4.21b).

So, problems of the cold spraying should be faced on and resolved by microstructure analysis and particle impact modeling. The particle and substrate deformation characteristics and bonding behaviors are shown to be complex, and the processes such as impact-induced interface melting, adiabatic shear instability, jetting of materials, and DRX in the interfacial regions have to be analyzed. The bonding mechanism between the particles and the substrate or among the particles is not well understood. So, current developments on the CS process are aimed for designing/improving the equipment, optimizing the cold spraying parameters, exploring possible bonding mechanisms, or demonstrating the applicability of various spray pure and composite materials.

4.2.2.2 Structure and Phase Transformations Characterization Methods

Since the early 1860s, Prof. Henry C. Sorby of Sheffield developed a microscopy examination technique (Higgis, 1971). The great progress results in wide application of optical microscopy, scanning electron microscopy (SEM), and transmission electron microscopy (TEM), differential thermal analysis (DTA), dilatometry, X-ray diffraction (XRD). They are the major characterization techniques used in characterization of the coating structure.

The low-magnification microstructural analysis of the samples is done using various optical microscopes (Olympus-PME3, Center Valley, PA, Leica CTR4000, etc.) with magnification ranging from 50× to 2000×.

The SEM (e.g., Quanta, FEI, Hillsboro, OR, Jeol, etc.) with an energy dispersive spectroscopy (EDS) attachment are used to analyze the microstructures, determine local chemical distributions and the porosity of the coatings. In our own studies, the FEI Quanta 200 FEG scanning electron microscope was used for microstructural analysis of the various coatings and substrates, including characterization of both the substrate and the coating, as well as adhesion and cohesion interactions. The used SEM allows for high-resolution imaging with resolution up to 5 nm due to special field emission gun (filament), excellent backscatter and secondary electron images, elemental analysis with EDS and elemental mapping, and cathodoluminescence of trace elements. The main microstructural characteristics observed through SEM are morphology of the coatings, thickness of the layers, porosity, oxide particles, and cracks. Because a high-quality contact between the coating and the substrate is a fundamental factor for good adhesion, the coating/substrate interface microstructure needs to be carefully evaluated to optimize the CS parameters.

The phase transformation temperature (Tg) and the crystallization temperature (Tc) of the samples are being determined by DTA (e.g., DTA 7, Perkin Elmer, Norwalk, CT). First, the samples are grounded and screened less than 25 μm. About 25 mg of the powder sample is placed into an alumina crucible. The DTA measurement is being performed by flowing atmospheric air at the rate of 25 cm³/min, and the sample is heated from 25°C to 1350°C at the rate of 10°C per minute. The powdered alumina sample is used as a reference material. Tg and Tc values are determined from the tangent intersection of the DTA curves. The melting temperatures (Tm) are also being obtained from the peak temperature of the endothermic peak.

The thermal expansion coefficient and the glass transition temperature measurements of the coatings and the bulk materials are performed using a dilatometer (e.g., Orton Dilatometer, Model 1000D Orton Ceramic Foundation, Westerville, OH). The samples are being prepared for the length of one inch and then placed in a dilatometer. Each sample is being then heated to 1000°C with the rate of 3°C per minute. The sample length measurements are being recorded automatically for every 10°C increment during the heating cycle by a linear variable displacement transducer.

The phase analysis of the samples is being done by XRD, using, for example, a Scintag XDS 2000 (XDS 2000, Scintag, Cupertino, CA). The XRD system has 35 kV of voltage and 20 mA of current. The Cu-Kα radiation ($\lambda = 1.5405$ Å) is usually used to obtain XRD patterns. Diffraction scans are run from 20° to 80° with a rate of 1° per minute at room temperature. Because XRD provides the content of overall phases, the local chemical distributions are being obtained using EDS.

4.2.2.3   X-Ray Diffraction-Based Residual Stress Measurement

The residual stress measurements in the real CS coatings are very important for coating characterization. The residual stress measurements may be implemented on Proto LXRD stress analyzer by the XRD technique. It uses the distance between crystallographic planes, that is, d-spacing, as a strain gage. When the material is in tension, the d-spacing increases and when the material is under compression the d-spacing decreases. This method can only be applied to crystalline, polycrystalline, and semicrystalline materials. The diffraction angle 2θ is measured experimentally and the d-spacing is then calculated using Bragg's law. The presence of residual stresses in the material produces a shift in the XRD peak angular position which is directly measured by the detector. The determination of the shift in the Bragg angle is thus interpreted in terms of shifts in the lattice spacing. The commonly known Bragg's law forms the fundamental basis of the XRD theory:

$$n\lambda = 2d\sin\theta \qquad\qquad (4.2)$$

where:
  λ is the wavelength of the X-Ray radiation
  n is an integer (a whole number of wavelengths)

Once the d-spacing is measured for unstressed (d_0) and stressed (d) conditions, the strain can be calculated using the following relationship:

$$\varepsilon = \frac{(d - d_0)}{d_0} \qquad\qquad (4.3)$$

Strain along any tilt direction can be measured using the $\sin^2\psi$ method. For the $\sin^2\psi$ method where a number of d-spacing are measured, stresses are calculated from an equation derived from Hooke's law for isotropic, homogeneous materials:

$$\varepsilon_{\varphi\psi} = \frac{1+\nu}{E_{\text{eff}}} \sigma_\varphi \sin^2 \psi - \frac{\nu}{E_{\text{eff}}} (\sigma_{11} + \sigma_{22}) \tag{4.4}$$

where:

$(1 + \nu)/E_{\text{eff}}$ is the X-Ray elastic parameter of the material

σ_φ is the stress in the direction of the measurements

σ_{11} and σ_{22} are the stresses along x and y directions

ψ is the angle subtended by the bisector of the incident and diffracted beam and the surface normal

$\varepsilon_{\varphi\psi}$ is the strain at a given ψ tilt

Residual stresses can be measured using two main techniques. The first is the single exposure technique (SET), whereby a stress measurement is performed using only one tilt angle. This technique gives the user a very quick and efficient method to perform a stress measurement and is particularly suited for the need to take many measurements very quickly. The second is the multiple exposure technique, whereby multiple tilts are used in the analysis. This method is more revealing in materials for which the d versus $\sin^2\psi$ relationship is not linear, as assumed in Equation 4.3, but takes much longer than the SET.

4.2.2.4 Coating Adhesion Strength

Adhesive strength of coatings prepared by CS determines its applications in the industrial field. Therefore, many researchers have been focused on the bonding mechanism in the last few years. Huang and Fukanuma (2012) reviewed the studies of an influence of powder, substrate, and heat treatment on the adhesive strength, and numerical simulation results that helped to explore the bonding mechanism. These studies on the bonding mechanism of CS suggested that the adhesive strength is mainly affected by the mechanical interlock and diffusion bonding based on the shear instability or metallurgical bonding caused by molten impact. So far, the underlying mechanism of bonding of CS has not been well clarified yet. So, the adhesion strength measurement is considered to be mandatory at coating characterization.

The ASTM C633 testing standard defines the procedure for measuring coating adhesion to the substrate in a tensile testing machine. The procedure is as follows:

1. A 25 mm diameter A516 low carbon steel cylinder end is to be coated with the required coating.
2. Another cylinder with the same diameter is glued on top of this, using FM 1000 adhesive (curing temperature 175°C) with published tensile shear strength of about 60 MPa (Figure 4.8a).
3. After positioning this assembly into the tensile testing machine they are pulled apart.
4. Maximum pull stress is recorded.

In some cases, the adhesion strength of the new advanced coatings is higher than the tensile strength of the used adhesive (the adhesive with the highest tensile strength was selected from several available adhesives); therefore, the adhesion value could not be determined from the ASTM C633-01(2008) standard test and an alternative approach may be applied (see Figure 4.22b). Tensile testing according to ASTM E8/E8M-09 is a fundamental material science test in which a sample is subjected to uniaxial tension until failure. A tensile specimen with coating is cut out to have two shoulders and a gauge (section) in between. The shoulders are relatively large so they can be readily gripped, whereas the gauge section has a smaller cross section so that the deformation and failure

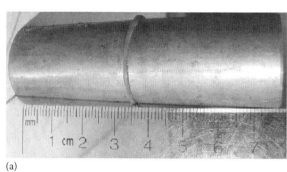
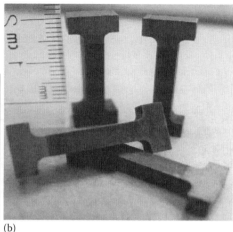

(a)   (b)

FIGURE 4.22   (a) Coated sample (ASTM C633) glued on top using FM 1000 adhesive (curing temperature 175°C) with tensile shear strength of about 60 MPa and (b) *dog bone*-shaped samples with brazed Cu insert for convenient gripping.

can occur in this area. To apply the tensile test to determining the adhesion strength of the deposited Cu coatings the following steps are being been carried out:

- *Dog bone*-shaped samples (Figure 4.22b) are cut out across the coating substrate interface, so that one of the sample shoulders is cut out of the deposited Cu coating, while the second shoulder and the gage would be cut out of substrate
- To make the Cu shoulder of the sample larger for a more convenient gripping a small copper insert is brazed to the coating surface
- A special sample holder was designed and fabricated to fit the dimensions of the *dog bone*-shaped sample with the working length of 20 mm

4.2.3   STRUCTURE FORMATION PROCESSES DURING DEPOSITION

4.2.3.1   CS Copper Coating Structure Formation

4.2.3.1.1   Introduction

In this section, the analysis of structure formation coating is made on the basis of authors' own experiments with cold spraying of pure copper. Cold gas dynamic spray is shown in previous chapters as a surface coating technique in which a coating is formed due to bombardment of a metallic substrate by metallic particles accelerated to high velocities and consolidation of these particles due to impact. This process can generate nanocrystalline (nc) layers in the impact zone of both the particles impinging upon the substrate and the substrate itself. Understanding the physical mechanisms underlying the severe deformation of the impact zones and grain refinement of material being deposited by CS is crucial for the optimization and improvement of the CS technology.

Localized shear deformation and the form of SPD in a narrow band generated during dynamic deformation at high strain rates has been a topic of great interest for decades. Localized shear is shown (Section 2.2.2) to be an important mode of deformation which leads to superhigh strains and development of interparticle bonding in various powder composites. The CS technology is known to be the process of metallic particles impacting the substrate with a velocity of about 600–1000 m/s that results in particle deformation at strain rates in the range of 10^3–10^9 s^{-1}. That is why it is well recognized that localization of deformation induced by an adiabatic shear band formation in the impacting particles is one of the dominant mechanisms for successful bonding in CS

(Surinach et al., 1996; Gedevanishvili and Deevi, 2002). However, there is a lack of experimental data demonstrating the contribution of SPD to CS coating bonding as well as numerical simulation of the process. This is due to unknown characteristics of each impinging particle and the substrate under very high strain conditions during multiparticle impact, resulting in the severe deformation of both the particle and the substrate, complexity of the dislocation structure formation processes at high temperatures such as DRX, and so on.

In this section, an experimental study of properties and structure formation of the CS deposited Cu coatings is presented together with the finite element (FE) simulations of multiparticle impact to characterize the severe deformation in the areas impacted by the impinging particles; the influence of SPD on the nc structure of the formed coatings–substrate interface is shown too.

The CS process was used for depositing Cu coatings on A516 low carbon steel substrates. Commercially available Cu powder, with particles ranging in size from 5 to 100 μm, has been used. The particles were accelerated to a high velocity by injecting them into a stream of high-pressure carrier gas which was subsequently expanded to supersonic velocity through a converging–diverging de Laval nozzle. After exiting the nozzle, the particles were impacted onto a substrate, where the solid particles deformed and created a bond with the substrate. As the process went on, particles continued to impact and form bonds with the previously consolidated material resulting in a uniform deposit with very little porosity and high bond strength.

The A516 Gr.70 low carbon steel substrates were grit blasted (blasting air pressure 0.75 MPa, stand-off distance 150 mm, traverse speed 20–40 mm/s) with alumina (mesh 36) to remove the scale layer formed due to the hot rolling process used to produce the substrate material. Commercially available low oxygen Cu powder (Tafa, Inc., Concord, NH) with an average diameter of about 22 μm was used. To form a Cu coating, the following CS parameters were utilized with Plasma Giken CS machine: propellant gas—nitrogen, gas pressure—5 Mpa, gas temperature—800°C; powder feed rate—100 g/min, and nozzle traverse speed (NTS)—500 mm/s.

To determine the adhesion strength of the deposited Cu coatings, tensile testing was carried out according to ASTM E8/E8M-09 (Figure 4.8). The surface roughness of the grid-based A516 Gr.70 low carbon steel substrates was measured by Seimitsu 2800 E roughness measurement machine. The profile diagram is shown in Figure 4.23.

4.2.3.1.2   FEM Simulation

To study the Cu particle deformation during deposition with the CS process, a dynamic simulation of multiple impact of the Cu particles impinging upon the carbon steel substrate was run, using commercial Lagrangian Finite Element software Abaqus/Explicit Version 6.9 (Section 2.3.1). A 3D FEM analysis was performed for 25 μm and 50 μm particle size configurations. Figure 2.28 shows the scheme of the simulated model. The bottom surface of the simulated substrate was fixed and additionally the horizontal movement of the substrate in two perpendicular directions was limited by fixing the substrate sides. For both the 25 μm and 50 μm Cu particles impinging upon the substrate, their temperature was $T_p = 80°C$, the substrate temperature was $T_S = 20°C$ and the impact velocity was $V_p = 800$ m/s.

The numerical simulation results are shown in Figures 4.24 and 4.25. Figure 4.24a and c demonstrates the Mises stress distribution in the particles and substrate during impact of two parallel particle columns (shown in Figure 4.24b and d). It can be seen from the images in Figure 4.24a and c that the plastic deformation of the substrate due to impact is considerable in spite of the fact that the yield strength of carbon steel substrate is higher than that of Copper particles.

The calculated particle penetration depth into the substrate is about 32.1 μm for the case shown in Figures 4.24a, and 14.8 μm (Figure 4.24b). A comparison of Figures 4.24a and c shows the influence of the distance between columns and eccentricity of the particle location on the plastic deformation of both the particles and the substrate. In the case of the gap between particle columns in the range of 5 μm, (Figure 4.25a) the substrate material is severely deformed in the area between

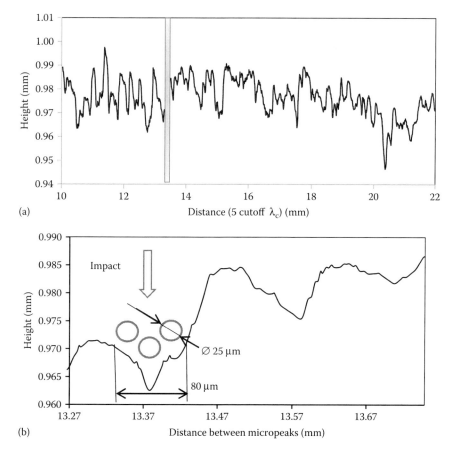

FIGURE 4.23 Roughness of the substrate after grit blasting with Al_2O_3 0.7 mm grit. (a) Roughness profile of the cut off distance 10 mm and (b) the profile of a shaded area (length–400 μm). The impinging upon the substrate particles are shown as grey circles.

particles, which results in the formation of substrate material jets. Another specific feature of multiparticle impact is nonuniformity of the particle deformation in the case of nonsymmetric impact (Figure 4.24c).

To characterize the particle deformation process during multiparticle impact, the equivalent strains in various nodes (Figure 4.25a) on the particle surface were calculated. The results are shown in Figure 4.25b. One can note that the total strain of the nodes #6–10 varies in the range of $\varepsilon = 15$–25, which is related to the severe strains due to localization of deformation at the interface. The strains of the nodes in the particle core are in the range of $\varepsilon = 2$–5 in the case of multiparticle impact. Thus, the main conclusion of numerical modeling is that CS technology is characterized by severe deformation of both the particles and the substrate, which depends on the technological parameters of CS. The localization of deformation, severe deformation and formation of material jets seem to be the main processes that lead to obtaining nanostructured coatings with high mechanical properties.

4.2.3.1.3 Structure Formation

The morphology and structure of the CS deposited coatings in general depend on the substrate roughness is usually reflected in the topography of the formed coating. The morphology also depends on the mechanical properties of the substrate and the impact conditions of the powder material sprayed at high velocity. It also depends on the coating/substrate interface formation mechanism. In the case of Cu coating/A516 Gr.70 low carbon steel substrate, the interface is formed by the deposition of Cu

Material type	Matrix phase	Second phase
Metals	Ni, Cu, Fe, Cr, Co, Ag, Au, Zn, Sn, Cd, Pb, Pd, Al	W, Ti, Ni, Fe
Alloys	Ni-Cr, Ni-Ag, Cu-Ni, Cu-Ag, NiCrBSi, steels	Al-Si, steels, Ni-Cr
Carbides	–	SiC, TiC, Cr_3C_2, WC, B_4C, ZrC, HfC
Oxides	–	Al_2O_3, SiO_2, TiO_2, Cr_2O_3, ZrO_2, BeO
Borides/nitrides	–	Cr_3B_2, TiB_2, ZrB_2, HfB_2, BN, B_4N, TiN, AlN
Intermetallics	–	Fe-Al, Ni-Al, Ti-Al
Other compounds	–	Graphite, CaF_2, MoS_2, WS_2, NbS_2

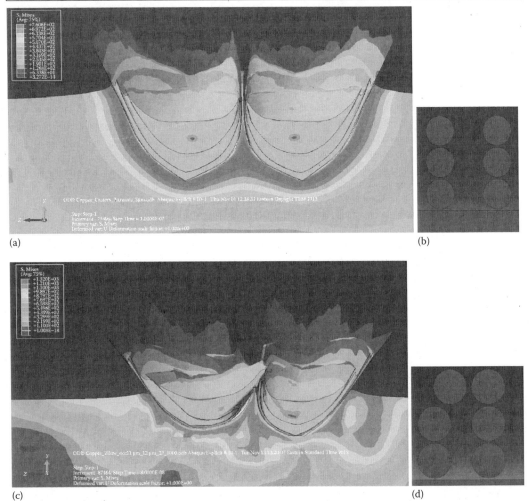

(a)   (b)

(c)   (d)

FIGURE 4.24 Numerical simulation results of the multiparticle impact with the carbon steel substrate. (a, c) Mises stress distribution for symmetric (b) and nonsymmetric (d) particle impact. Particles—copper, diameter of 20 μm; impact velocity—1000 m/s; eccentricity—11 μm (d); gap between particle columns 5 μm (b) and 12 μm (d); impact time—80 ns; depth of penetration—32.1 μm (a) and 14.8 um (b); crater diameter—34.5 μm (a) and 36.4 μm (c).

by high-pressure nitrogen carrier gas which accelerated Cu particles at a velocity around 800 m/s. It was found that the initial substrate surface roughness at the interface produced by alumina grit-blasting (Figure 4.13) was considerably altered by the accelerated Cu particles impacting the grit-blasted substratesurface. As can be seen in Figure 4.26, the formed interface structure is comprised of micropeaks up to 50 μm high and craters with a mean diameter of 20–40 μm and a depth of 10–15 μm.

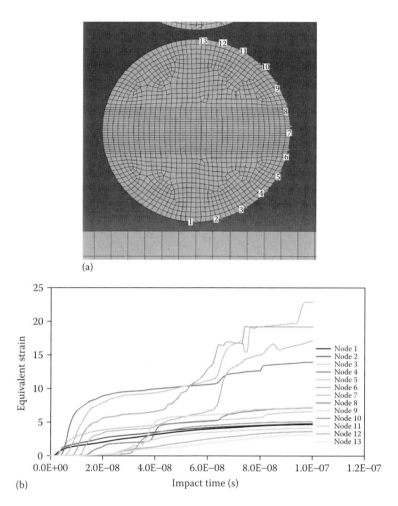

FIGURE 4.25 Equivalent strains in the various nodes of the bottom particle. (a) Nodes location and (b) equivalent strains at the various times of impact.

The alumina grit size increase, which results in higher surface roughness, leads to obtaining the mechanical interlocking of Cu particles at the interface. The presence of the interlocked Cu localized in the substrate craters can be seen (Figure 4.27). The substrate surface roughness considerably influences the vortex structure of the Cu layer. The role of the substrate surface roughness micropeaks in mechanical interlocking of the impinging copper particles and forming the vortex structure of the Cu layer is well observed (Figure 4.28) in prepared 45°-angled cross sections.

The severe deformation of the deposited material and strain localization result in formation of the vortexes that are presented in Figures 4.26 through 4.28. It is interesting to note that the vortexes have a 3D structure, which is shown in the microstructure of coatings cut parallel to the substrate surface (Figure 4.29).

The vortexes seen consist of small grains being generated during recrystallization of Cu because of high temperature and strain created at the interface due to multiple particle impact. The influence of the NTS on the grain structure of copper coating is shown in Figure 4.30. The increase of the NTS from 300 to 500 mm/s results in diminishing of a single pass thickness, which leads to an additional strain accumulation and increase of total strain of the Cu grains. This leads to submicro and nanograins similar to those shown on TEM micrograph (Figure 4.31).

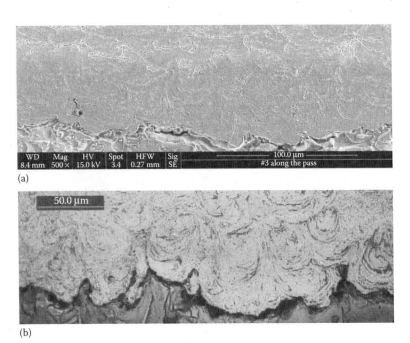

(a)

(b)

FIGURE 4.26 Vortex morphology of at the interface area (substrate preprocessing grit size: (a) 0.4 mm and (b) 0.7 mm).

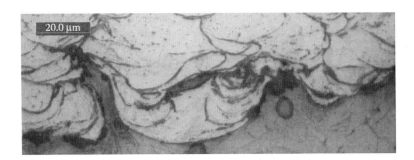

FIGURE 4.27 Structure of coating/substrate interface.

The separate recrystallized nanograins free from dislocations and similar to those described by Borchers et al. (2003), can be seen. The presence of low-angle boundaries confirms the assumption about the influence of SPD taking place during the CS process on the nc structure formation at the coating–substrate interface.

To single out the separate mechanisms contributing to the Cu layer/low carbon steel substrate bonding, the adhesion tension tests were performed in accordance with the procedure described above. The results are shown in Figure 4.32.

The microstructures of the Cu–carbon steel interface shown in Figures 4.16 through 4.28 exhibit the effect of interlocking. Moreover, fracture topography (image is not shown) examination shows that the fractures take place in the localized *necks* within the interlocked particles. In other words, a partly ductile fracturing of Cu grains is taking place in some areas of the formed interface.

The interlocking effect is associated with the effect of nanostructuring, which allows us to achieve the proper bonding between Cu and Fe. Both effects caused by SPD result in considerable increase of adhesion strength. As can be seen in Figure 4.32, the adhesion strength reaches approximately 150–160 Mpa which is 2.5–3 times higher than that of thermal spray coatings.

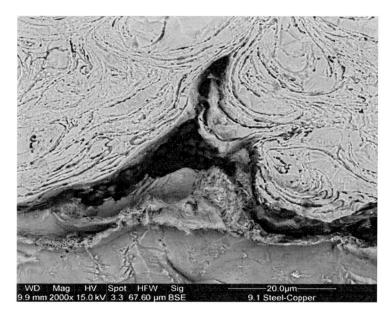

FIGURE 4.28 The *turbulent* structure of the sublayer in angled cross sections. Deep etching with Nital.

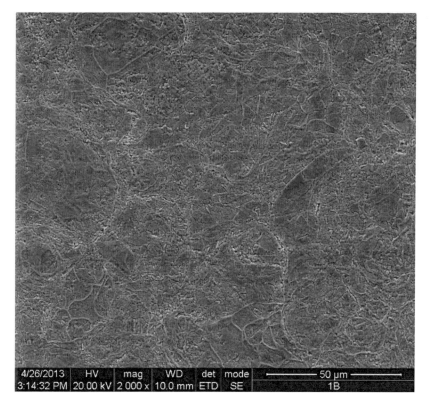

FIGURE 4.29 Microstructure of Cu coating in the plane parallel to a substrate surface.

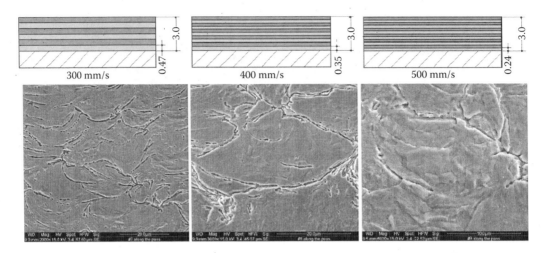

FIGURE 4.30 Effect of secondary deformation during the following pass deposition on the coating structure (core of coating).

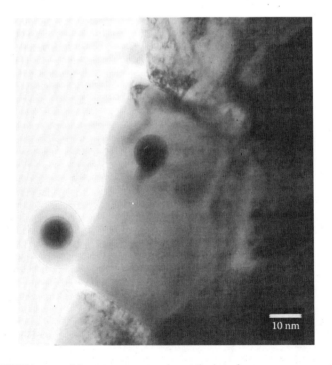

FIGURE 4.31 HRTEM image of the copper nanograins at the interface.

The experimental and simulation study of nanostructured Cu coating formation by CS allows us to make the following conclusions:

- Severe deformation of both particles and substrate is shown by numerical modeling of multipaticle impact. The total strains of the nodes at the particle interface are found to be in the range of $\varepsilon = 15$–25.
- The localization of substrate deformation peaks is found and linked to the possibility of interlocking effects between the impacting particles and substrate.

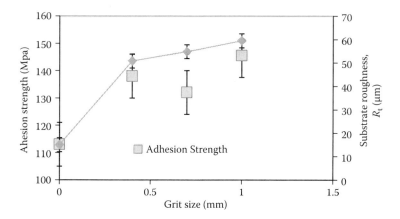

FIGURE 4.32 Dependence of roughness on adhesion strength of copper coatings deposited on carbon steel substrates.

- Considerable vortex formation process is experimentally demonstrated by the proper selection of the CS process parameters.
- Severe deformation of copper particles results in effective nanostructuring processes. TEM microscopy confirms the presence of nanograined in the copper—carbon steel interface.
- Coupling of nanostructuring and interlocking effects due to strain localization and severe deformation at CS results in obtaining a coating with high adhesion strength.

4.2.3.2 CS Metal-Matrix Composite Coating Structure Formation

4.2.3.2.1 Definition of Structure Parameters

The main factors that need to be established to fully characterize the structure of particulate-reinforced MMCs are the size and the volume fraction of the particles (Maev and Leshchynsky, 2007). The size parameters of a material's microstructure can exert a strong influence on its mechanical properties even when all other parameters remain unchanged. An example of this influence can be seen as metallic matrixes are strengthened by particles where the generation and movement of dislocations are controlled. Kouzeli and Mortensen (2002) showed that the analysis of such interaction phenomena has led to the development of a number of dislocation models which can adequately describe the correlation between mechanical properties and microstructural parameters in dispersion or precipitation hardened alloys.

The case of strengthening in particle reinforced MMCs has been extensively studied in the past. Many dislocation models have been designed to account for strengthening in this class of materials. However, most of these models tend to underestimate the strengthening increment in the metallic matrix due to the presence of In reinforcement particles. Kouzeli and Mortensen (2002) state that this underestimation is due to the neglect of another, very different, contribution to composite strengthening, namely load sharing. Unlike precipitation or dispersion hardened alloys, the volume fraction of the reinforcing phase in GDS composites is relatively high. Also, interparticle bonding areas fail to create a continuous network, as compared with regular MMC produced by infiltrating. In this case, the matrix sheds load to the reinforcement during straining. This concept of *load sharing* is central to composite continuum mechanics, which predicts strengthening in particle-reinforced composites based on knowledge of the bulk metal-matrix properties, the volume fraction, aspect ratio, and spatial arrangement of the particles. Analyses of GDS MMC properties from this viewpoint have not yet been performed; neither have the volume fraction of the reinforcement particles of the spatial arrangement of the particles been adequately examined.

The volume fraction, ε_j, of particles can be calculated from the weights of the ingredients making up the composites by means of the following equation:

$$\varepsilon_j = \frac{m_s \rho_m}{\left(m_s \rho_m + m_m \rho_s \right)} \qquad (4.5)$$

where:
    m_s is the mass of reinforcement particles
    m_m the mass of matrix particles
    ρ_s is the density of the particles
    ρ_m is the density of the matrix

As well as knowing the average volume fraction of particles, it is essential to also determine if there are any local variations in the composition due to factors such as the settling of particles during spraying. This can be monitored by polishing random sections taken from different areas of the coating.

The spatial arrangement of the particles in MMC is characterized by a mean free interparticle distance, λ. According to Underwood (1985), measurements of the mean free interparticle distance, λ, may be conducted by superposing random lines on micrographs of the as-deposited composites. λ may be determined from the number of particle intercepts per unit length of the test line, N_L, and the volume fraction of particles, ε_j, from

$$\lambda = \frac{1 - \varepsilon_j}{N_L} \qquad (4.6)$$

As-deposited composite microstructures feature a near homogeneous distribution of particles both in aluminum and nickel matrixes with relatively low porosity (Figure 4.33).

The basic microstructural features of the Al- and Ni-based composite coatings made by low-pressure CS are summarized in the Table 4.2 (Maev and Leshchynsky, 2007). In all these composites, the reinforcement interparticle distance varies in the range of 10–200 μm and is comparable with the soft matrix particle size (about 45 μm). Thus, the interparticle boundaries seem to exert a similar influence on dislocation creation and motion in the matrix as compared with the reinforcement particles.

4.2.3.2.2 Topography and Microstructure of Fe-Based MMC Coatings

The SPD of particles during CS processes results in both the consolidation and strengthening of the resultant coating. As shown previously, the extensive plastic flow of a particulate material is the main process which governs the structural formation of the coating. The soft metals are well-known

100 μm

FIGURE 4.33 SEM structure of CS Ni matrix composite coating.

TABLE 4.2

Al- and Ni-Based Microstructure Characteristics

Composite Designation	Particle Designation	Average Particle Size (μm)	Porosity (%)	Volume Fraction of Reinforcement (%)	Interparticle Distance (mm)
Al-Al$_2$O$_3$-1	Al$_2$O$_3$	10	3.1	10	187.5
Al-Al$_2$O$_3$-2	Al$_2$O$_3$	10	3.5	15	100.
Al-Al$_2$O$_3$-3	Al$_2$O$_3$	10	4.3	30	44.6
Al-Al$_2$O$_3$-4	Al$_2$O$_3$	45	4.9	50	10
Ni-TiC-1	TiC	20	7.5	10	90
Ni-TiC-2	TiC	20	8.1	20	57.1
Ni-TiC-3	TiC	20	9.5	30	33.3

materials which in many cases exhibit excellent ductility and low flow stress. Metal particles with high kinetic energy build up a dense coating due to deformation upon impact. Thus, the macroscopic mechanical properties of composites are strongly conditioned by the bulk properties of the constituents; they are also influenced by the mechanical behavior of the matrix/reinforcement interfaces.

The deformation behavior of the material is also dependent on its crystal structure, because the dislocation behavior depends on the same. Strain hardening of body-centered cubic (BCC) materials is usually much more sensitive to strain rate than those of face-centered cubic (FCC) which remain fairly ductile over a wide range of strain rate, while BCC metals tend to lose their ductility and generally undergo a brittle transition.

An overview of structure formation processes in CS, as well as some microstructures, has been presented above. A detailed analysis of the bonding process has also been undertaken by Borchers et al. (2004). As shown in this study of the interfaces in high-pressure CS process, interparticle bonding areas do not form a complete bonding network. In fact, some of the bonding areas exhibit microstructural features such as jets. Using TEM, several remarkable features are noted at the interface. In some locations within Al there appears a tangled structure, where the interface has bifurcations; here, a thin strip of material with a thickness of some nanometers seems to be interlaced. The Al particles interface is in fact somewhat wavy on the order of some 100 nm, with the grains having low dislocation density. The grain size near the interface is about 500 nm. CS Cu coating also exhibits a nonequilibrium grain boundary (GB), which is characterized by ultrahigh dislocation densities adjacent to the GBs (Figure 4.17). The microstructural features of CS Ni coatings are similar to those of Cu. The particle–particle interfaces consist of nano-sized grains and large grains with coffee-bean-like contrast, which is typical for shock consolidated structures (Meyers et al., 1999).

The described microstructure features clearly show that effective interparticle bonding takes place only in local areas. The similar conclusions were made in Chapter 2 based on the numerical simulation results.

At present time, there are two primary issues of CS that have not yet been completely defined: (1) the types of interparticle bonds, and (2) the surface areas occupied by the definite type of interparticle bonds that are created by the appropriate powder consolidation technique. The primary bonding structures (described above) may be characterized by definite macroscopic mechanical properties. Thus, based on the analysis of the mechanical properties, the structure of interparticle bonds and the contact surface area data may be evaluated. It may in fact be possible to estimate the contribution of each bonding mechanism to the mechanical strength of the composite as a whole. Therefore, having characterized the surface topography and microstructure formation mechanisms, it is possible to determine the primary features of processes governing the buildup of Fe-based MMC coatings. ARMCO iron and SS-based CS coatings were examined. The SEM images of the

surface topography of ARMCO iron and SS-based CS coatings (Figures 4.34 through 4.42) show
the following specific features (Bielousova, 2013):

- Mechanism of coating buildup depends on the hardness and ductility of powder and type
  of crystal lattice
- Powder particles (BCC) with small hardness and high ductility (ARMCO iron) exhibit
  the intensive bulk deformation during the first particle impact and following impacts by
  another particles (Figure 4.34)
- FCC SS particles with higher hardness and smaller ductility (AISI 904) impinge with the
  coating top surface with jetting (shown with arrows in Figure 4.35) that reveals about
  severe surface deformation due to adiabatic shear band formation at the interface

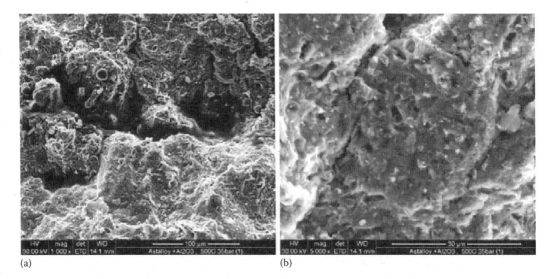

(a)   (b)

FIGURE 4.34   (a, b) SEM of the surface topography: Fe + Al$_2$O$_3$ 20% coating, spraying temperature 500°C.

(a)   (b)

FIGURE 4.35   (a, b) SEM of the surface topography: AISI 904L coating, spraying temperature 450°C.

- The formation of the Fe-AISI 904 composite coating is achieved by both mechanisms: hard AISI904 particles penetrate into soft ARMCO iron particles, and mutual deformation of both types of the particles (Figure 4.36)
- Metal-matrix Fe-SiC coating formation proceeds by severe deformation of ARMCO iron particles and embedding SiC particles into Fe surface (Figure 4.34b)

The SEM observation of the ARMCO iron and SS-based CS coating microstructures in the areas of deposit and interface (Figures 4.37 through 4.42) demonstrated that the examined deposits form large particle aggregates. An increase of the particle velocity by raise of propellant gas temperature results in change of the coating microstructure due to change of particle consolidation mechanism. FCC AISI 904 austenitic coating sprayed with gas temperature of 450°C ($V_p \approx 500$ m/s) consists of rather equal axial particles with grains of 3–5 μm. The distribution of strains within the particle is mostly uniform (Figure 4.37). The flattening ratio (Equation 4.2) of the grains at a periphery area of the particles is in the range of $R_1 = 2, ..., 2.5$. The increase of gas temperature up to 550°C

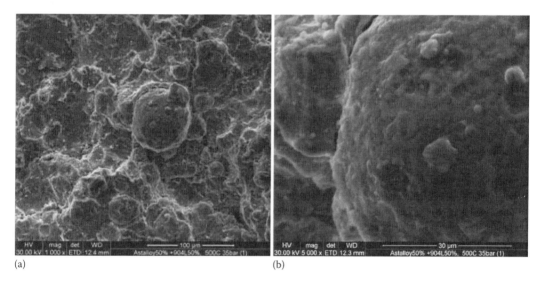

(a)   (b)

FIGURE 4.36   (a, b) SEM of the surface topography: Fe 50% + AISI 904L 50% 904L coating, spraying temperature 500°C.

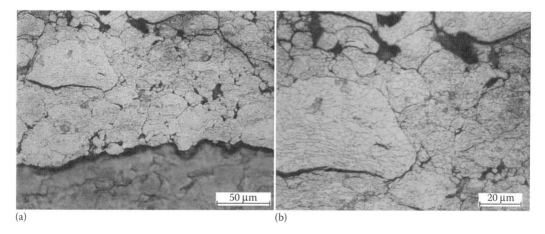

(a)   (b)

FIGURE 4.37   (a, b) Optical microstructure of AISI 904L coating, spraying temperature 450°C.

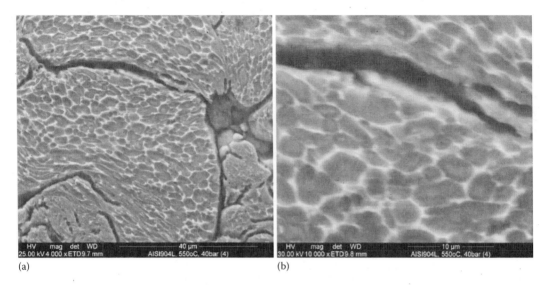

FIGURE 4.38   (a, b) SEM microstructure of AISI 904L coating, spraying temperature 550°C.

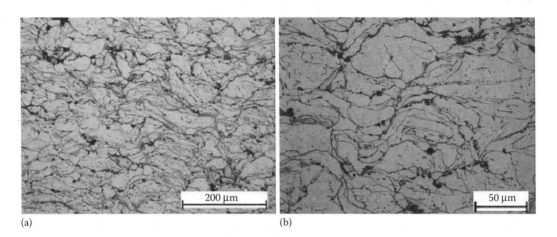

FIGURE 4.39   (a, b) Optical microstructure of Fe + Al_2O_3 20% coating, spraying temperature 550°C.

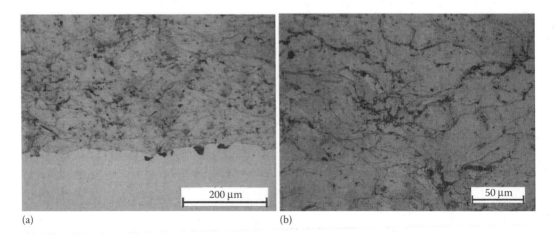

FIGURE 4.40   (a, b) Optical microstructure of Fe + Al_2O_3 20% coating, spraying temperature 600°C.

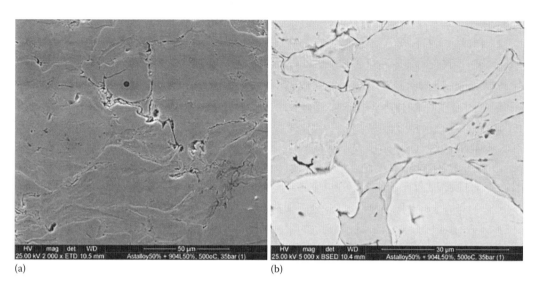

FIGURE 4.41   (a, b) SEM microstructure of Fe 50% + 904L 50% coating, spraying temperature 500°C.

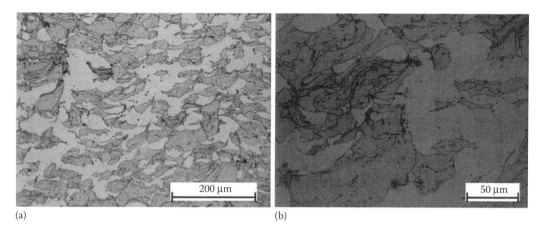

(a)   (b)

FIGURE 4.42   (a, b) Optical microstructure of Fe 50% + 904L 50% coating, spraying temperature 600°C.

($V_p \approx 600$ m/s) leads to the strain localization at the particle/particle interface which is clearly seen on the SEM micrographs (Figure 4.38a). Flattening ratio of grains near to interface is about $R_1 = 4, \ldots, 7.5$ (shown by arrows in Figure 4.38a), while that of another area of the particle (Figure 4.38b) is about $R_1 = 2.5, \ldots, 3.0$.

Influence of the particle velocity on structure formation of BCC Fe + Al_2O_3 20% composite coating is seen on micrographs (Figures 4.39 and 4.40). The interparticle boundaries of the coating sprayed with $V_p \approx 600$ m/s are clearly seen after etching by 5% Nital coating sprayed with $V_p \approx 650$ m/s. The slight increase of the particle velocity up to $V_p \approx 650$ m/s results in considerable change of microstructure because a thickness of the interparticle boundaries became small (Figure 4.40). The possible reason of this effect is intensification of diffusion processes at the interface after the particle impact that leads to annihilation of defects at the interface area and thinning of the boundaries. In some cases (when the strain localization is likely to be developed) the structure of interface is difficult to etch.

The microstructural changes observed at the same micrographs lead to the hypothesis that small Al_2O_3 particles are deposited at the surface of ARMCO iron particles. While the Fe + Al_2O_3 20% powder mixture contains the Al_2O_3 particles of 10 μm size, a lot of the smaller Al_2O_3 particles may be observed that reveals about fracture of Al_2O_3 particles during impact with the substrate.

The two phase CS composite coatings consisting of BCC and FCC metals attract attention because they may be applied for manufacturing of hybrid parts, joining dissimilar metals, and so on. Microstructures of the Fe 50% + 904L 50% coatings sprayed with two particle velocities (550 and 650 m/s) are shown in Figures 4.41 and 4.42. Increase of particle velocity results in a raise of ARMCO iron particle strains (flattening ratio $R_1 = 4.5, ..., 5.0$) and appearance of mushroom shape of the AISI 904L particles.

The main features of the composite coating microstructure analysis are the following:

- Some areas of severe deformation of the AISI 904L grains are found at the local areas of interface.
- Small grains of 2–4 μm size obtained during powder manufacturing (by atomizing) are preserved in CS coating structure which proves the CS capability to retain the nanostructure due to low particle temperatures and high strain rates.
- Al_2O_3 particles are distributed at the interparticle boundaries and possess the hardening of MMC.

4.2.3.2.3   Structure Formation of Al-based MMC Composite Coatings

The CS Al-based composite coatings are widely applied in industry at present time (Champagne, 2006). Some specific features of the Al-based composite coatings structure formation are shown below (Leshchynsky et al., 2011).

The CS Al-based composite coating microstructure is characterized by change of the soft Al particles deformation mechanism due to reinforcing of the interface with SiC particles. Some activation of surface layers at the SiC–Al interface due to embedding of SiC particles is believed to enhance the interparticle bonding. Other factors are also affecting on the particle consolidation process. Among them are: (i) interparticle friction and (ii) fracture of the ceramic particles during impingement, pore formation at the Al–SiC interface, and so on.

As shown in Chapter 2, the adiabacity of the process is a very important factor particle–substrate interaction during the particle impact. As it is observed that with its increase, the heat evolved during the plastic deformation is not transferred to the neighboring regions, which leads to deformation localization. The deformed region heats up and loses strength, which results in further strain localization and better interparticle bonding. This, in turn, results in higher density and strength of CS coating.

To compare the effect of matrix type on the structure formation of composite coatings the deposition of Cu + SiC and Cu + Al_2O_3 powder mixtures were made. The deposition of the Cu + SiC and Cu + Al_2O_3 mixtures is characterized by the similar mechanisms as that of Al-based mixtures. The microstructure of the deposited coating made of Al + Al_2O_3 and Al + SiC powder mixtures is shown in Figure 4.43. The data reveal about a specific mechanism of the fracture of ceramic particles during the deposition of these powder mixtures. It is known that the fracture toughness of SiC is about of 4.6 MPa $m^{1/2}$ while that of Al_2O_3 is 3.5 MPa $m^{1/2}$. Thus, the SiC particles fracture toughness is higher than that of Al_2O_3, and there are a lot of small SiC-particles sustaining an impact load at CS and appearing at the microstructure without failure. Moreover, the big SiC particle has large unfractured areas (Figure 4.43b). Contrary, upon impact, a significant amount of small cracks appears in the Al_2O_3 particles with a size about 60 μm, these cracks tend to unite with one another and also have tendency to the further break-off of some volume of the material (image is not shown). Diminishing the particle size decreases the probability of the ceramic particles destruction. Additionally, porosity is seen at the soft matrix/reinforcement particle interface (Figure 4.43b).

We should note that the failure of the ceramic particles during CS of the Cu powder mixtures happened more than that in Al-based mixtures. The possible reason of such a behavior is the higher values of Young's modulus and yield strength of a copper matrix, which results in a higher stresses at the interface during the ceramic particle impact that leads to fracture of the ceramic particles. This process is mostly influenced by the gas and particle temperature and velocity. Figure 4.44b

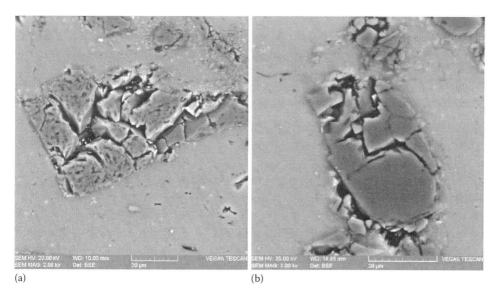

(a)   (b)

FIGURE 4.43 Cross-sectional view of fractured Al_2O_3 and SiC particles. Content of Al_2O_3 and SiC in the coatings is about 43%–45%. (a) Al 30–45 µm and Al_2O_3 + 63 Al_2O_3 µm (powder mixture 50% × 50% by volume) and (b) Al 30–45 µm and SiC +63 µm (powder mixture 50% × 50% by volume).

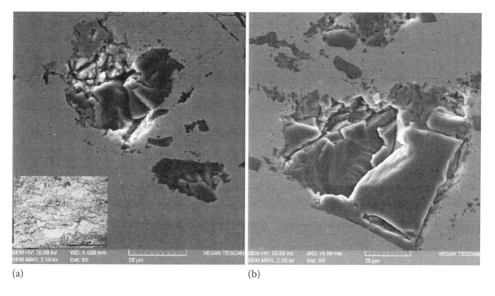

(a)   (b)

FIGURE 4.44 View of fractured Al_2O_3 particles. (a) Cu –53 µm and Al_2O_3 63–100 µm (powder mixture 50% × 50% by vol.), gas temperature $T = 300°C$ and (b) Cu –53 µm and Al_2O_3 63–100 µm (powder mixture 50% × 50% by vol.), gas temperature $T = 500°C$ (content of Al_2O_3 in the coating is about 12%–15%).

shows that the Al_2O_3 ceramic particle is only destroyed partially when sprayed at the 500°C temperature, despite of its higher kinetic energy than that of the smaller particle shown in Figure 4.44a. The higher gas temperature (Figure 4.44b, gas temperature $T = 500°C$) leads to an increase of the Cu matrix temperature and, consequently, to the softening of the Cu matrix that diminishes a probability of the ceramic particles failure during impact.

The experimental data demonstrate the CS deposition process is characterized by a significant rebound of ceramic particles comparing to that of the soft particles of the basic powder (Al and

Cu deposition efficiency is in the range of 90%–95%). Measurement results reveal that after deposition of the 50% × 50% Al–Al$_2$O$_3$ and Al–SiC powder mixtures, the concentration of ceramic particles in Al–Al$_2$O$_3$ and Al–SiC coatings lies in a range of 43%–45% while Cu-based coatings contain only 12%–15% of the ceramic phase. The main reason of this fact is believed to be the different conditions of the ceramic particles interaction with deposited coating during their impingement with the substrate. While, upon the impact, the soft particle deforms intensively with a formation of adiabatic shear bands and adheres to the coating, the conditions of the adhesion of a ceramic particle will be defined by the matrix strain. The soft matrix should be deformed to achieve a ceramic particle adhesion.

The mechanisms of adhesive bonding between the ceramic particles and the soft matrix at CS are not clearly understood yet, and study of the microstructural mechanisms of soft matrix severe deformation by hard particles during impact seems to be very important. From this viewpoint let's analyze the structure formation of aluminum–SS and aluminum–titanium CS coatings made by CS of 50% × 50% powder mixtures (Leshchynsky et al., 2011).

The CS of 50% Al–50% SS powder mixture is characterized by forming of SS particle clusters (Figures 4.45 and 4.46a) which consist of 2–5 particles. The analysis of microstructure shows the presence of a thin Al layers between SS particles. This effect seems to be a result of particle cluster formation during the gas–powder jet acceleration in the de Laval nozzle. One have to note that the content of the SS particles in the coating is about 33%–35% (defined by image analysis), while the composition of sprayed powder mixture is 50% Al–50% SS by volume. Thus, the rebound of SS particles is about 15%. The microanalysis results reveal that the average strains of the SS particles are within 15%–25%, and there is no evidence of the strain localization (mushroom shape of the particles, jetting occurrence) of these particles. It is clearly seen the formation of the dense coating in the result of Al particles severe deformation during their impact with substrate and impact of SS particles with Al layers previously formed at CS.

It is interesting to note that severe deformation and high local temperature at the areas of strain localization lead to reactions and formation of intermetallics at the Al–SS particles interface. The areas of the reactions are seen in Figure 4.46b. Chemical composition of various fields at the interface defined by EDS is shown in the Table 4.3.

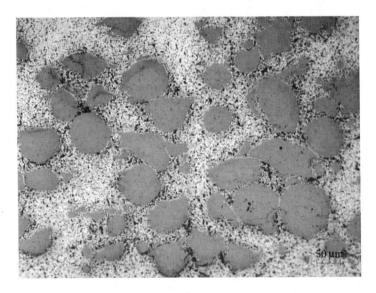

FIGURE 4.45 Optical micrograph of coating Al 30–45 μm + AISI 316L 80–100 μm (50% × 50% by volume). Content of SS in Al matrix is about 35%.

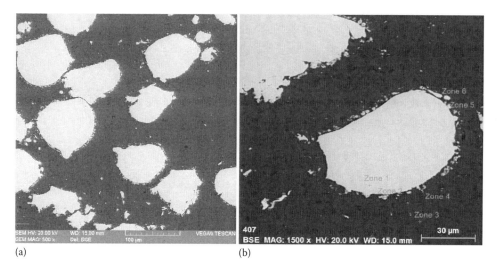

(a)   (b)

FIGURE 4.46   (a, b) SEM images of Al 30–45 μm and SS 316L 80–100 μm (50% × 50% by volume).

TABLE 4.3

Chemical Composition of Al-SS Coating Areas (Zone Numbers are Shown in Figure 4.46b)

	Chemical Composition (%)				
Area	Al	Cr	Fe	Mo	Si
1	–	7.02	90.49	1.06	1.43
2	56.34	–	43.66	–	–
3	100	–	–	–	–
4	100	–	–	–	–
5	71.15	1.77	26.86	0.23	–
6	3.69	5.09	90.53	0.69	–

The data reveal about formation of Fe–Al intermetallic compounds with some content of main alloying elements of the AISI 316L SS (Cr and Mo). The SEM images (Figure 4.46) exhibit the great effect of SS dissolution in Al matrix. Microstructure analysis of the cold-sprayed coatings reveals that a significant part of titanium particles rebounds from the surface during the deposition of Al + 50% Ti powder mixture, that is why the content of titanium particles in the obtained coating determined by image analysis, is about 19% (Figure 4.47).

It can be seen that the shape of titanium spherical particles is altered to elliptic. There is no specific change of particle morphology caused by an intense shock wave generated by the impact of a high-velocity particle with the substrate passing through the powder. It means the deformation of titanium particles occurs above all in a static way. The short duration, large amplitude stress waves propagating through the material during the particle impact cause compaction and bonding of the Al particles. The shock compaction features with a formation of adiabatic shear bands is only occurring in a soft aluminum matrix. This fact indicates that the velocities achieved at the CS deposition at these conditions are lower than critical velocities needed for the formation of adiabatic shear bands in the surface layers of the titanium particles. The results of critical velocity measurements reveal about particle velocity in the range of 600 m/s, while the critical velocity of Ti powder is about 860 m/s (Papyrin et al., 2007).

These SEM and EDS examination results reveal about presence of the reactions at the titanium/aluminum interface (Figure 4.48). Reaction products are seen at the area adjacent to Ti–Al particles

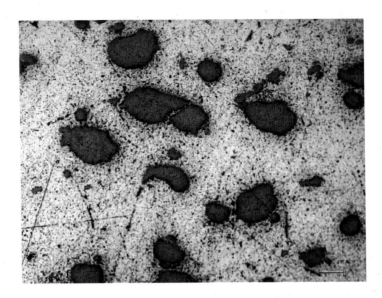

FIGURE 4.47 Optical micrograph of the coating cross section from Al (30–45 μm)-Ti (80–100 μm) (powder mixture 50% × 50% by volume).

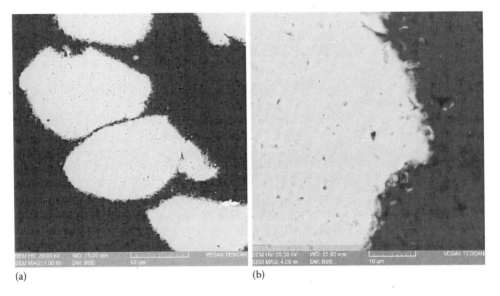

(a)   (b)

FIGURE 4.48 SEM images of Al–Ti coating microstructure. (a) Magnification 1000× and (b) Magnification 4000×. (From Leshchynsky V. et al., Impact particle behaviour in cold spray of composite coatings, in *Proceeding of OTSC*, Hamburg, Germany, 2011, pp. 1076–1081. With permission.)

interface as shown on the Figure 4.48b. It can be observed that the dissolution of titanium in aluminum matrix is taking place, and results in substantial increase of the titanium particles roughness. The roughness of Ti particles before spraying is about 0.01 μm while that of Ti particles in the coating is about 2–5 μm (Figure 4.48b). The EDS data (not shown) demonstrate that the concentration of titanium in aluminum matrix about the particle interface is changing in a wide range of 35%–55%. As a result of using the air as a carrier gas, the oxidation reactions of titanium and aluminum are taking place during the deposition. However, the EDS analysis results show that the concentration of the oxygen in the aluminum matrix does not exceed the level of 5%–7%, which means a relatively low oxidation degree during the deposition. From this view point, CS technology has some certain advantages.

Based on above described results the following conclusions may be made:

- Impact breaking of the ceramic particles during CS MMC coating formation depends on particle velocity particle fracture toughness and type of metal matrix.
- The particle rebound effect depends on modulus of elasticity and yield strength of metal matrix.
- It is defined a reinforcement of Al and Cu matrix with micro- and nanoparticles of ceramic (Al_2O_3 and SiC) phase due to attrition of ceramic particles during impact and their following embedding to the coating.
- Occurrence of the interface reactions due to impact of SS and Ti particles with Al particles at the CS process results in creation of Fe–Al intermetallic compounds.

4.2.4 COATING POSTPROCESSING TREATMENT

4.2.4.1 Annealing/Sintering Metal Coatings

4.2.4.1.1 Introduction

One of the characteristics and useful features of metal and alloys is their ability to undergo. The heat treatment of steel and other metals and alloys is the best-known example of the modification of their internal structure without appreciable change of the parts' shape and dimensions postprocessing technology, and solid-state reactions can be caused to occur in any metal-based coating. The CS metal and MMC coatings have enormous potential for further heat treatment and sintering technologies application because the CS coatings are obtained by shock consolidation from pure powders at low temperature without big particle oxidation. Before discussing the sintering and structure transformation reactions, it seems to be reasonable to consider the simpler cases in which heating alone is sufficient to cause marked changes in metallic structures and properties. The essential purpose of an annealing treatment is to permit the alloy to approach its normal or equilibrium condition. For example, nonuniformities in chemical composition remaining from the casting process can be eliminated by an homogenization annealing. Another type of nonequilibrium state of the CS coating is the cold-worked structure of the coating described in Chapter 2. The phenomena of recovery, subgrain formation, and recrystallization, which are discussed below, occur at tempering and annealing of the cold-worked CS coating structure.

The essential mechanism by which annealing and sintering processes occur is the diffusion of the atoms of the alloy. Several references have already been made to diffusion phenomena: the changes necessary to maintain equilibrium during the formation of a solid and the addition of various metals and compounds for the purpose of coating properties getting. In this section, the annealing/sintering effects in the CS coatings will be referred. In the following section the quantitative aspects of this subject will be developed for application also to the reaction sintering of MMC composite coatings.

As shown in Chapter 2, the CS coating formation is being achieved due to particle shock consolidation when the particles are accelerated to velocities above a material-dependent critical velocity. In this case, the metal particles will bond to the substrate and form a dense, well-adhered deposit (Papyrin et al., 2007). The CS technology is capable of making deposits with low-oxide content at low temperatures. Depending on particle diameter and density, process gas properties, and inlet conditions as well as the nozzle geometry, particle velocities range from 500 to 1500 m/s. The process gas inlet temperature is about 300–900°C which is significantly lower than the melting temperature of the spray material and the gas cools down rapidly during its expansion. So, the particles remain solid until impact, and negative effects like oxidation and phase changes can be avoided, which makes cold spraying particularly suitable for oxidation sensitive materials (Stoltenhoff et al., 2006). For this reason cold spraying of aluminum, copper, titanium, nickel, and zinc is very attractive for many applications. Some of the Al applications, such as spray forming, require the cold-sprayed material to exhibit a modest amount of ductility (Hall et al., 2006). In their as-sprayed state, cold-sprayed metals tend to be brittle because the particles are heavily cold worked during the deposition process and interparticle bonding tends to be incomplete. Thus, using of annealing affecting the ductility of cold-sprayed aluminum deposits is very important. This will allow new applications for CS processing such as spray forming.

With respect to demands from industry, cold-sprayed copper coatings can cover a broad range of applications. The low oxygen content and thus the high thermal and electrical conductivity in combination with the focused spray spot allow spraying narrow conducting lines or contacting segments for automotive and electrical applications. The focused spot size, the high spray rates, and the high deposition efficiencies in cold spraying result in comparably thick layers attainable in one single pass. For example, the CS coating technique was attempted to fabricate a 10 mm copper shell of canisters for spent nuclear fuel disposal. Additionally, the method could be also used for the repair of cylinders or chills in the printing and casting industry (Stoltenhoff et al., 2006). Based on this, a little analysis of the Al and Cu coatings structure formation and their properties is presented below.

4.2.4.1.2 Al Coating

Al coating was deposited by high-pressure cold spraying with nitrogen (pressure 40 bar, gas temperature 350°C). SEM micrograph of the as-sprayed Al deposits (coating top layer) presented in Figure 4.49 shows that CS Al coating have some porosity and grain distortion in the areas of the interparticle boundaries. Annealing at 400°C during 1 hour results in obtaining uniform grain structure of the coating in the areas of particle/particle interfaces. There is no evidence of abnormal grain growth in the annealed structures up to 400°C (Figure 4.49). The effect of annealing temperature on the grain and dislocation structure in the areas of particle/particle interfaces is seen on TEM images presented in Figures 4.49 and 4.50. The processes of dislocation annihilation and recombination, and static recrystallization at the annealing temperatures >0.4 Tm are clearly seen. However, the concentration of stalking faults in the come subgrains remains relatively high (Figure 4.50), and the size of grains is in the range of 200–500 μm due to severe deformation of particle surface layers.

Presence of oxide films and uncompleted bonding at the particle/particle interface greatly influence on the coating ductility, and influence of annealing temperature on the ductility parameter is of great importance. The fracture surfaces of the annealed deposits revealed features that are consistent with the observed increase in ductility analyzed in Section 4.3.1. While the as-sprayed samples exhibit brittle failure with a relatively smooth and blocky fracture surface, the annealed fracture surfaces exhibit a typical ductile failure with coalescence of microvoids, similar to data of Hall et al. (2006) (Figure 4.51 and 4.52).

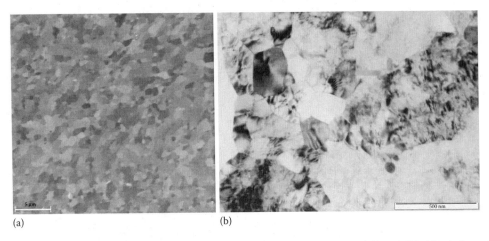

(a)   (b)

FIGURE 4.49 TEM images of the grain (a) and dislocation structures and (b) of CS Al coating after annealing at 400°C, 1 hour.

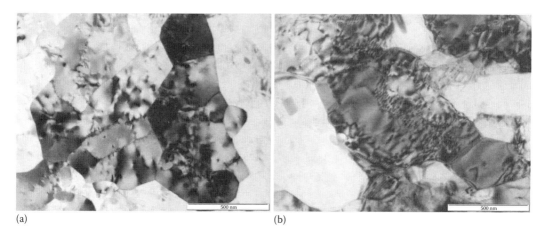

(a)   (b)

FIGURE 4.50 TEM images of the dislocation structure of CS Al coating after annealing at 200°C, 1 hour (a) and at 300°C, 1 hour (b).

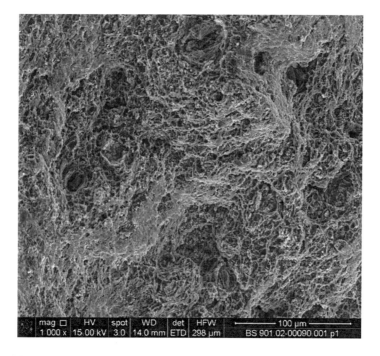

FIGURE 4.51 SEM topography of fracture surface of the CS Al coating annealed at 400°C.

Micrographs of the annealed samples showed considerable elongation and flow of the individual grains. The fracture appears to be of a mixed mode with intergranular failure and a large amount of voids formed during deformation of the tension sample (Figure 4.52). Thus, optimization of CS Al coating annealing/sintering parameters allows us to considerably improve the interparticle bonding and, consequently, coating mechanical properties.

4.2.4.1.3   Cu Coating

As shown in Chapter 2, the CS coatings are formed by particle shock consolidation and its structure consists of particles of mushroom-like shape. The particle shock consolidation process results in preferably solid-state bonding between the particles that leads to an incompleteness of interface

FIGURE 4.52 SEM topography of fracture surface of the CS Al coating annealed at 400°C. Closed view at magnification 20,000×.

bonding and relatively low mechanical strength. Because of the low thermal stability of the severe deformed grains in the area of interfaces in cold-sprayed coating, the annealing considerably modifies the CS Cu coating structure and properties (Stoltenhoff et al., 2006; Wen-Yua Li et al., 2006; Seo et al., 2012). These papers have demonstrated the effect of annealing on the structure and properties of cold-sprayed coatings. However, detailed studies of annealing temperature effect on the Cu coating structure and properties have not been made yet. It would be reasonable to expect healing the incompleteness of interface bonding and adequate improvement of coating mechanical properties.

The SEM images of Cu coating after annealing (Figure 4.53) demonstrate the effect of annealing temperature on the healing the interface bonding. While the total length of the interparticle boundaries in the sample annealed at 200°C during 3 hours is similar to that of as-sprayed samples (Figure 4.53a), annealing at the 400°C at the same time leads to 30% diminishing of interparticle length (defined by image analysis) (Figure 4.53b).

Increasing the annealing temperature up to 600°C results in complete recrystallization of Cu coating structure (Figure 4.54). The equiaxed grains were grown in the coating instead of elongated grains in the as-sprayed coating. The grain size in the coating annealed at 600°C for 12 hours (Figure 4.54b) was larger than that of the Cu powder (Figure 4.54c). However, this difference is about 5–10 μm, contrary to results of Wen-Yua Li et al. (2006). Some twins occur in these annealed structures coating. Moreover, particle boundaries were no longer observed from the cross section and the GBs were observed. It is reported by Wen-Yua Li et al. (2006) that under solid-state diffusion bonding, the oxide films between particles will also be coalesced as the grain grows during annealing.

As shown by detailed study of Stoltenhoff et al. (2006), after annealing at 600°C, the microstructures of cold-sprayed coatings processed with nitrogen or helium are mainly affected by

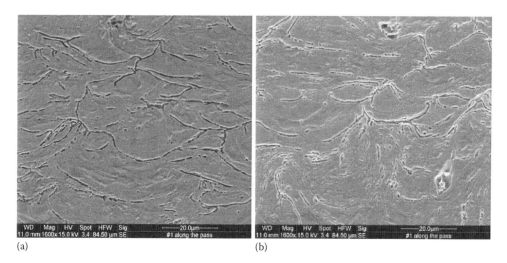

(a)   (b)

FIGURE 4.53 SEM images of annealed Cu coating microstructure after etching. (a) Annealing at 200°C during 3 hours and (b) at 300°C during 3 hours.

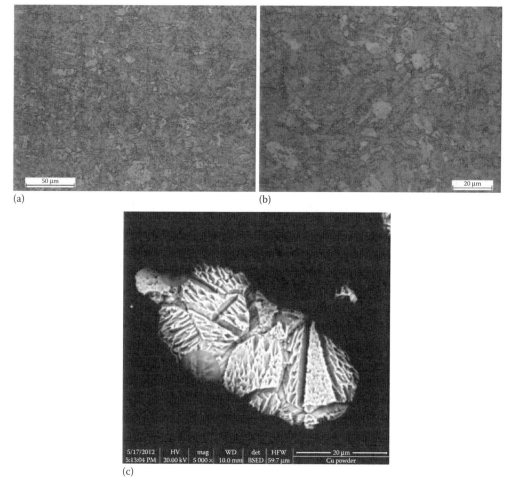

(a)   (b)

(c)

FIGURE 4.54 Optical microstructure of Cu coating after annealing at 600°C: (a) magnification 250×; (b) magnification 500×; and (c) SEM image of microstructure of Cu powder particle.

recrystallization and the spheroidization of oxides. Diffusion not only allows microstructural rearrangements, but also significantly improves the adhesion to the substrate. The facts that Cu coatings show spheroidized oxides decorating former particle–particle boundaries and that Cu coatings sprayed with helium show a more arbitrary oxide distribution and oxide clusters indicate different diffusion mechanisms within particle–particle interfaces. Authors of this paper suggested the explanation based on behavior of highly strained areas of the particles. In coatings sprayed with nitrogen as process gas, the comparatively high amount of just cold forged interfaces and, respectively, not well-arranged large-angle GBs should result in a higher mobility of oxygen and copper than in normal GBs that promote formation of oxides in these cold forged interfaces. Pieces of evidence of this hypothesis are obtained by Seo et al. (2012) TEM study. We obtained the similar results during the TEM examination of Cu coating annealed at 600°C (Figure 4.55). The arrows shows the Cu oxides and the dotted line—the possible former interparticle boundary.

4.2.4.2 Reaction Sintering Technology

4.2.4.2.1 Introduction

Reaction sintering of the cold-sprayed deposits will be analyzed on the basis of iron aluminides which are known to be the products of iron and aluminum reaction during sintering. Iron aluminides and other compounds have been a subject of renewed attention in the recent past. The desired oxidation and corrosion resistance, relatively low thermal diffusivity, high melting point, low material cost, and low density characterize these materials as potential candidates for several high-temperature structural applications in heat treating, automotive, and power generation industries. So, their synthesis development attracts attention in the past decades.

Sintered iron aluminides are very promising novel structural materials mainly because of their attractive physical and mechanical properties, but also because of the low cost of raw materials used for their manufacture (Gedevanishvili and Deevi, 2002). Powder metallurgy eliminates many of the limitations during the fabrication process (Jordan and Deevi, 2003), for example, it enables forming complex shapes and choosing random compositions, deals with problems of grain overgrowth, gives better homogeneity of the chemical composition and structure, and is an energy-saving process (Joslin et al., 1995). However, the use of this method is limited by the necessity of controlling the self-propagating high-temperature synthesis (SHS) (Joslin et al., 1995) reaction utilized as one of the sources of heat during the sintering process. The SHS is very rapid and strongly exothermic which causes unwanted porosity of the sinters (Gedevanishvili and Deevi, 2002). The studies (Joslin et al., 1995; Gedevanishvili and Deevi, 2002) show the great complexity of the phenomena taking place during the presintering process correlated with diffusion and phase transformations. The researchers concluded that the mechanism of exothermic phase transformations is strongly

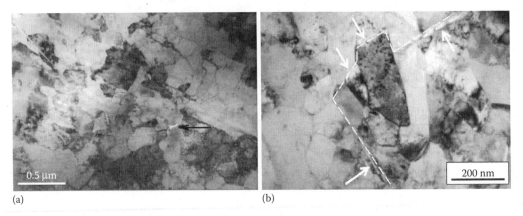

(a)   (b)

FIGURE 4.55 TEM images Cu coating after annealing at 600°C. (a) magnification 50,000× and (b) 150,000×.

correlated with many factors such as elemental powder particle size, atmosphere, heating rate, and the chemical composition of the sample.

Fe–Al intermetallics have been well studied as structural materials (Suryanarayana, 2001). To this extent, Fe–Al intermetallics were expected to replace stainless steels in many engineering applications. However, this effort has been less successful, mainly because the brittle fracture and low ductility of Fe–Al at ambient temperatures limit their application. The combined CS–sintering processing technique may overcome this problem because of high ductility of Al particles to be sprayed, high density of CS Al–Fe-based coatings, and careful temperature control of the following Al–Fe reactive sintering process. The sequence of solid-state transformations in the synthesis of Fe–Al intermetallics according to the reactions $Fe + Al = FeAl$, $FeAl + 2Fe = Fe_3Al$, and $2FeAl + 3Al = Fe_2Al_5$ in powder mixtures was studied in Gialanella (1995) and Jozwiak et al. (2010). The results indicate that the process involves the formation of atomic configurations that may become nuclei of stable or metastable intermediate phases. Prolonged milling of powder mixtures leads to homogenization of the synthesis product and the formation of a solid solution or an intermetallic phase with a low degree of long range order. For this reason, cold spraying of the Al–Fe powder mixtures seems to be very important.

One has to note the additional important specific feature of Fe–Al intermetallics—ductility at the high temperature. Authors (Bednarczyk et al., 2008) show that the microstructure of Fe_3Al–5Cr alloy following the straining at 800°C and 900°C displays elongated primary grains with clear boundary migration. The creation of subgrains with single dislocations and the effects of cell formation are apparent in the substructure (Bednarczyk et al., 2008). It means the Fe–Al-based coating structure has much better resistance for cracks nucleation and grow than ceramic Yttria-Stabilized Zirconia coatings.

So the main goal of the combined CS—sintering technology development is to obtain alternative ductile intermetallics-based coatings with effective thermal barrier properties and high thermal expansion coefficient. The task of this section is to examine the structure changes of the Al–SS composite coating due to formation of Fe–Al intermetallics during reaction sintering (Bielousova, 2013).

4.2.4.2.2 Al–SS Coating Microstructure and Microhardness: Kirkendall Effect

The intermetallic synthesis of Al–SS coatings was performed at 600°C, during 3 hours. The microstructure of Al–SS composite coating after annealing is depicted in Figures 4.56 and 4.57.

One can note that the above-described intermetallics reactions during annealing result in formation of the intermetallic compound layer at the SS steel particle surface. The thickness of the intermetallic layer depends on conditions of Al matrix dissolution in the austenite of SS steel particles. Because of higher diffusion rate of Al atoms in austenite than that of Fe atoms in Al-based solid solution, the vacancies are being accumulated at the SS steel particle–Al matrix interface—the Kirkendall effect. The Kirkendall effect results in pore formation and loosing contact between Al and SS steel particles. That is why, further consuming of Al atoms is interrupted and reaction diffusion is stopped. The large pores formed due to the Kirkendall effect are clearly seen in Figure 4.57. It is interesting to note that the pores are located in the areas with high concentration of SS steel particles. It results in low thickness of Al matrix bridges between SS steel particles. It seems the localization of vacancies accumulation in the areas of Al bridges occurs due to better diffusion

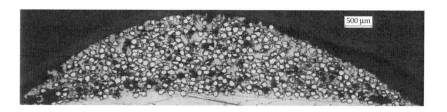

FIGURE 4.56 Optical image of microstructure of sintered Al–SS coating (general view at low magnification).

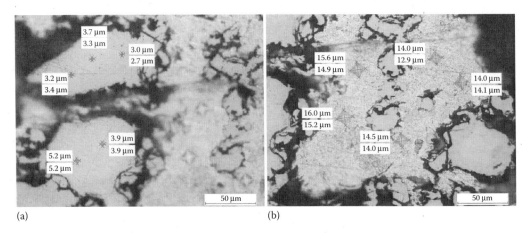

(a)   (b)

FIGURE 4.57   Optical images of microstructure of sintered Al–SS coatings with indents at various loads. (a) SS steel particles with indents, 50 mN load and (b) Al matrix with indents, 200 mN load.

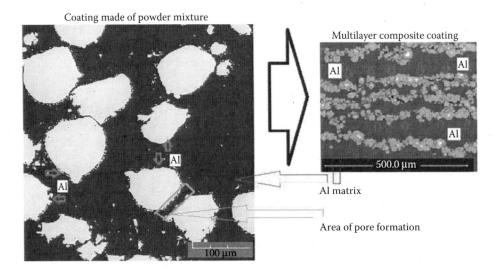

FIGURE 4.58   Schematic of pore formation process due to the Kirkendall effect at annealing of Al–SS composite coating.

coefficient of Al atoms in these severe deformed areas (shown by arrows) as compared to that of Al matrix (Figure 4.58).

To avoid such uncontrolled diffusion flow of both vacancies and Al atoms, and formation of big pores in composite coating, the layered composite material was made by CS (Figures 4.58b and 4.59a). In this case, uniform diffusion flow of Al atoms to the interface is organized, that results in considerable decrease of the Kirkendall porosity. This effect seems to be achieved by the creation of a one–two particle layer structure by CS as shown in Figure 4.59a. It allows us to diminish a number of Al bridges between SS steel particles and decrease the effect of pore zones formation during sintering.

Application of multilayer composite coating approach allows us to obtain good bonding between intermetallics synthesized during annealing and Al matrix. To illustrate this effect, the SEM image of fracture topography of Al–SS composite coating after annealing at 575°C is shown in Figure 4.59b. The separate particles are well bonded with Al matrix after annealing. In spite of the fact that the temperature of annealing was 575°C, approximately 50% of the SS steel particle volume is converted into intermetallic phase.

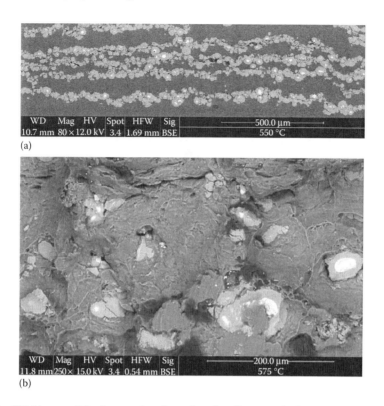

FIGURE 4.59 SEM image of the fractured surfaces from bending test of Al–SS coating annealed at 575°C. (a) Magnification 80× and (b) Magnification 250×.

4.2.4.2.3 Intermetallic Growth Kinetics

On the basis of assumption that intermetallic layer thickness bears a direct relation with growth kinetics, the average thickness of aluminide layer on the embedded particles may be used to calculate the kinetic parameters of reaction diffusion for compositions studied with the procedure developed in Jindal et al. (2006). The measurement results of aluminide layer thickness on the embedded particles at various sintering (annealing) conditions are shown in Figure 4.60.

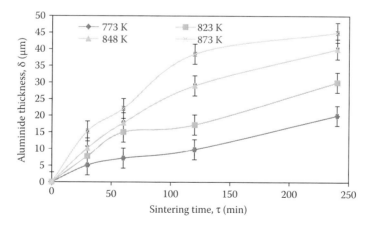

FIGURE 4.60 Growth of the aluminide intermetallic layer on the SS steel particles embedded into Al matrix at various time and temperature.

The measurement results reveal about classical dependence of aluminide thickness on temperature and time in the solid-state diffusion. The SS particles exhibit SHS reactions with Al matrix at the temperature 873 K with formation of typical melted phase. This transition from solid-state diffusion to SHS reaction is seen at the graph Figure 4.60. The image of coating structure after SHS reactions at 873 K is shown in Figure 4.61.

The intermetallic layer dependence δ on the holding time τ for solid-state diffusion seems to be described by differential Equation 4.7:

$$\frac{d\delta}{d\tau} = \frac{k}{\delta} \tag{4.7}$$

where k is the proportionality constant which is known as parabolic rate constant (Bouayad et al., 2003; Jindal et al., 2006). The integration of Equation 4.7,

$$\int \delta \cdot d\delta = \int k \cdot d\tau$$

gives Equation 4.8:

$$\frac{\delta^2}{2} = k\tau \tag{4.8}$$

Thus, in accordance with Equation 4.8, parabolic relationship between intermetallic layer thickness and sintering time may be observed. The experimental results of sintering SHS 717 particles with Al matrix are presented in Figure 4.62 in the coordinates $\delta = f(\tau^{0.5})$. Linear approximations are shown to describe the parabolic function $\delta = f(\tau^{0.5})$ with simulation veracity of $R^2 = 0.96, \ldots, 0.99$. Thus, the growth kinetics conforms well to parabolic rate (Equation 4.7). It allows us to define the parabolic rate constants k for different solid-state sintering temperatures (Figure 4.62). The results are shown in Figure 4.63 in the coordinates $\ln k = f(1/T)$ to define an adequacy of kinetics to the Arrhenius law:

$$k = k_0 \exp\left(\frac{-Q_a}{RT}\right) \tag{4.9}$$

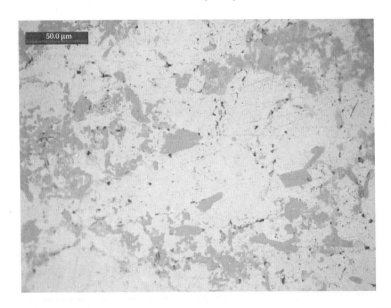

FIGURE 4.61 OM image of the microstructure of intermetallics layer after sintering at 873 K.

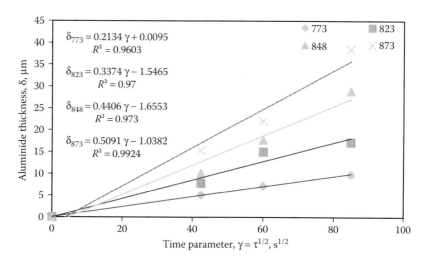

FIGURE 4.62 The experimental results of sintering SHS 717 particles with Al matrix in the coordinates $\delta = f(\tau^{0.5})$.

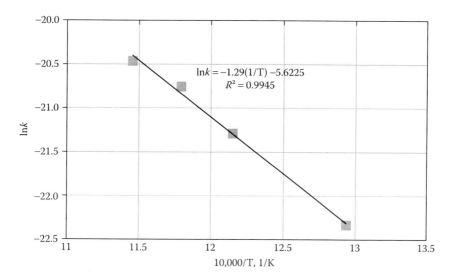

FIGURE 4.63 Parabolic rate constant ($\ln k$) dependence on reciprocal of absolute temperature ($1/T$).

where:

k_0 is pre-exponential factor
Q_a is activation energy of the aluminide layer growth
R is universal gas constant

It can be seen, the data are located well on the corresponding approximation line. Thus, the constants of the Arrhenius equation (4.9) may be determined from relationship

$$\ln k = \ln k_0 - \left(\frac{Q_a}{RT}\right) \tag{4.10}$$

The evaluation provides $k_0 = 3.6 \times 10^{-3}$ cm²/s and $Q_a = 107$ kJ/mol. We have to note that the pre-exponential factor for our case is higher by an order than in the handbook of Massalski (1990),

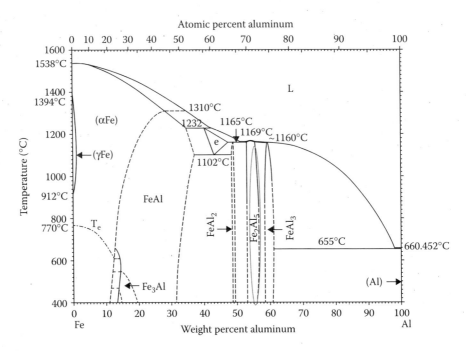

FIGURE 4.64 Binary phase diagram Al–Fe (Massalski, 1990).

which reveals about easier nucleation of the intermetallic phase on the SHS 717 particle surface. It seems to be the effect of SHS 717 particle penetration process into Al matrix that results in severe deformation and activation of Al matrix at the SHS 717 particle–substrate interface.

In accordance with binary phase diagram (Massalski, 1990), there are several intermetallic compounds in Al–Fe system (Figure 4.64). However, Ansara et al. (1998) showed that thermodynamically $FeAl_2$ has a lowest free energy of formation followed by Fe_2Al_5 and Fe_4Al_{13}. So, Jindal et al. (2006) suggested that the first phase to be formed will be Fe_4Al_{13}. At the later stage of the diffusion process Fe_4Al_{13} and Fe phases will react with each other to form a phase with composition between that of the interacting phases and closest to that of the lowest eutectic composition, that is, Fe_2Al_5 (see the red circle area on the phase diagram Figure 4.64). The activation energies for the growth of Fe_2Al_5 layer reported in literature shown in the work Jindal et al. (2006) are presented in Table 4.4.

Comparison of the data of Table 4.4 indicates about reaction of Fe_2Al_5 compound formation during sintering process. The high content of alloying elements in SHS 717 particles results in change of intermetallic composition formed during solid-state sintering.

The obtained parabolic constants (Figure 4.63) for different temperatures are compared with those available in literature for respective temperatures. The results are shown in Figure 4.65. One can note that obtained data are in the range of 10^{-14}–10^{-13} m²/s (points 1 in Figure 4.65), whereas $k = 10^{-10},...,10^{-11}$ m²/s for liquid (Al)–solid (Fe) diffusion couple. It seems to be influence of SHS

TABLE 4.4
Values of Activation Energy for Diffusion Growth of the Intermetallic Layer

Reference	Jindal et al. (2006)	Bouayad et al. (2003)	Denner and Jones (1977)	Eggeler et al. (1986)	Present Work
Q_a (kJ/mol)	85	74.1	155	134	107

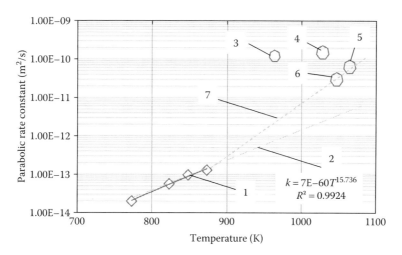

FIGURE 4.65 Parabolic constants comparison. 1—present work; 2—linear approximation of data 1; 3—Heumann et al. data, 4—Denner and Jones data; 5—Bouche et al. data; 6—Bouayad et al. data; 7—possible linear approximation of liquid (Al)-solid (Fe) reaction.

synthesis reaction with sharp increase of the temperature. Our data are similar to those presented by authors (Jindal et al., 2006) at the solid-state diffusion conditions.

The above analysis of reaction kinetics of Al–Fe intermetallic compounds clearly shows the real opportunity of controlled solid-state intermetallic synthesis of separate SHS 717 particles at the temperatures of 550°C–600°C.

4.3 MECHANICAL BEHAVIOR OF COLD-SPRAYED MATERIALS

V. Leshchynsky

D. Dzhurinskiy

4.3.1 GENERAL PRINCIPLES OF THE MICROSTRUCTURAL EVOLUTION UNDER COLD SPRAY

The structural evolution of powder material during the CS particle impact is being achieved due to plastic deformation, recrystallization, joining, and strengthening mechanisms, which results in certain mechanical properties of CS coatings. The high-velocity deformation and consolidation of the particulate body are known to be the main processes of the material structure evolution under CS, and definition of the evolution general principles needs to be performed. As clearly shown by Hansen and Jensen (2011), the new discoveries and advanced experimental techniques have created paradigm shifts in the way deformation and annealing structures are characterized and analyzed. The shifts have also enabled quantification of important structural parameters, which form the basis for analytical modeling and simulation and in turn have led to improved and new structural property models.

CS coatings are characterized with high particles' interface density and SPD of the particles that promotes intense microstructure refinement. From a microstructural perspective, multiphase powder systems can be classified (Raabe et al., 2010) as either particle-like alloys or as lamellar or filament-type micro- or nanostructured materials. From a chemical perspective, these composite coating systems can be classified as immiscible pure-metal–metal-matrix compounds, intermetallic–metal-matrix compounds, or carbide–MMCs. In cases where extreme strains are imposed, such as in CS, severe deformation of the particles can lead to complex curling (Raabe et al., 2010), where the minority phase forms into filaments that are bent about their longitudinal axis. In general, the different deformation processes during CS may lead to differences in nanostructure, mechanical alloying and other reactions. The most essential criteria to identify the coating structure formation are the certain micro- and nanostructure parameters, maximum attainable strain, the mutual solubility of the elements in each phase, mechanical properties of the composite coatings.

The task of this section is to give an overview of the microstructure specific features and the resulting mechanical properties obtained by extreme straining due to CS (true strains of 2–4, in some cases up to 10). The aim is to identify microstructure features that are common to a number of different material combinations and deformation processing conditions, including not only cold spraying but also forging operations, mechanical alloying, and frictional joining, as they reveal similar severe strains (Raabe et al., 2010).

For the purpose of discussion throughout this chapter, based on Kumar et al. (2003), we define nanocrystalline structures of metals and alloys as those with an average grain size and range of grain sizes smaller than 100 nm. Ultrafine crystalline (ufc) metals and alloys are defined to be those with an average grain size in the 100–1000 nm range, and microcrystalline (mc) structures have average grain dimension of a micrometer or larger. It is obvious the mechanisms of deformation and the properties of the crystalline materials not only depend on the average grain size, but are also strongly influenced by the grain size distribution and the GB structure (e.g., low-angle versus high-angle GBs). The variation of flow stress as a function of grain size from the mc to the nc regime is schematically shown in Figure 4.66.

In many mc and ufc metals and alloys with average grain size of 100 nm or larger, strengthening with grain refinement has traditionally been described on the basis of the Hall–Petch law (Kumar et al., 2003):

$$\sigma_s = \sigma_s^o + k_y d^{-1/2} \tag{4.11}$$

In this case, a pile-up of dislocations at GBs is approved as a key dislocation generation process resulting in dependence of flow stress on the grain size. As the microstructure is being transformed to the nc scale, the above dependence breaks down, and the flow stress versus grain size function differs considerably from that at higher grain sizes. With further grain refinement, the yield stress peaks in many cases at an average grain size value on the order of 10 nm or so. Further decrease in grain size can cause weakening of the metal. Although there is a lot of pieces of experimental evidence of the nc materials mechanical behavior, the underlying mechanisms are not well understood.

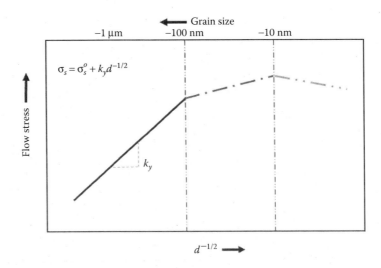

FIGURE 4.66 Hall–Petch flow stress dependence on grain size.

As previously shown, the structure formation at CS process consists of particle deformation and particle–particle or particle–substrate joining. That is why the unified description of the structural evolution during cold spraying can be described by the following two processes at submicroscale and microscale:

1. *Microscale*: A particle–particle or particle–substrate heterophase interface structure formation due to dislocation penetration through the interfaces (even among noncoherent phases), DRX and localization effects across the interfaces by micro- or shear bands (Ohsaki et al., 2007; Raabe et al., 2010)
2. *Submicroscale*: Structural subdivision by dislocations of the grains and forming hierarchical structures on a finer and finer scale (Hansen and Jensen, 2011) at severe strains of the particles due to their high-velocity impact.

The specific feature of CS-formed coating structures is that extremely strained heterophase interparticle areas are involved in all cases. In these regions, severe strains, DRX, and other processes of high-velocity deformation can be observed that may finally lead to interface related plasticity, structural transitions (e.g., amorphization), atomic-scale mechanical alloying, phase formation, and phase decomposition. It results in formation of an impact deformation-driven microstructure hierarchy and the corresponding microstructure–property relations. Numerical simulation of CS coating formation demonstrates producing a particle gradient structure which is characterized with microstructural gradient at a macroscopic scale. Figure 4.67a and b shows the typical microstructure of high-pressure cold spraying Cu coating (etched with reactive containing 50 ml 3% H_2O_2 and 50 ml NH_3, Figure 4.67a) and microscale schematics of Cu coating structure elements such as particle core, bonded, and stick interparticle boundaries. It is possible to approximately evaluate location of the areas of good metallurgical bonding between particles due to differences of etching kinetics of interparticle and GBs. The stick interparticle boundaries are known to contain pores and other surface defects. That is why these areas are etched in higher extent than those of metallurgical bonding.

The similar features may be seen at microstructure of low-pressure cold-sprayed AA 6022 coating (Figure 4.68a) after etching with Keller reactive (0.5 g NaF; 1.0 ml HNO_3; 2.4 ml HCl; 100 ml H_2O). The stick interparticle boundaries are thick and black while the GBs look as thin white lines. The submicroscale features of CS coating structure formation are shown in Figure 4.68b and c. The TEM images of the Al particle core (Figure 4.54b and c) demonstrate the subdivision of gains and reduction of cell size.

Evolution of the particle microstructure at submicroscale due to its impact at CS may be evaluated on the basis of strain gradient plasticity (SGP) theory (Molinari and Ravichandran, 2005). Based on microstructural observation of the particle deformation (Figure 4.68b and c), it may be recognized that, in accordance with Molinari and Ravichandran (2005), one of the key features of deformation is the reduction of cell size with increasing of strain. The model of the submicrostructure evolution during CS particle deformation is assumed to be similar to that of Molinari and Ravichandran (2005) (see Figure 4.69).

It has been shown by many researchers that the grains are divided into cell blocks separated by geometrically necessary boundaries (GNBs) that are hardly penetrable by dislocations and contribute to internal stress, while the cell boundaries are more permeable to dislocations and contribute to the hardening of metals (Hansen and Jensen, 2011).

The second parameter in the phenomenological constitutive model of a response of metals over a wide range of loading conditions is an internal variable δ related to characteristic microstructural length scale (Molinari and Ravichandran, 2005). The cell size δ_c (Figure 4.69) is known to be important characteristic length for microstructure description, and $\delta_c = K/\sqrt{\rho_t}$, where K is material constant of order 10–20 and ρ_t is the total dislocation density. The effective length scale δ_c cannot be quantitatively related to the cell size δ_c. Some GNBs are seen on TEM images of LPCS microstructure (Figure 4.68b and c).

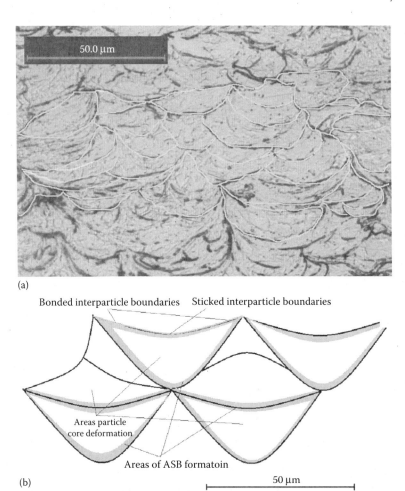

FIGURE 4.67 Microstructure of high-pressure cold-sprayed Cu coating (a) with sharing interparticle boundaries (thick—stick boundaries, which have a big etching capability, and thin —bonded boundaries obtained due to possible DRX in adiabatic shear band formation areas) and (b) Schematics of coating microstructure.

As previously shown in Chapter 2, the formation of the adiabatic shear bands and extreme deformation in these local areas result in large plastic strains at the particle–particle or particle–substrate interfaces. Thus, during the particle impact, the strains are being developed in two phases of the heterophase interface. Details of Cu–Fe interface after Cu particle impingement on steel substrate are illustrated with a high-resolution TEM image (Figure 4.70). An array of dislocations at the Cu–Fe interface is shown by dotted line. This dislocation structure demonstrates that Cu–Fe semicoherent interface may be obtained.

To illustrate microstructure features at microscale, the cross sections AA and BB (insets) are shown in Figure 4.71a and b. It is seen that some interparticle boundaries are not well bonded, which results in severe penetration of etchant in these areas and bigger etching effects. The cross section BB shows the interparticle boundaries preferably in the areas of week bonding.

Analysis of the CS composite coating microstructures (Figure 4.71a) clearly demonstrates particle spreading in the plane perpendicular impact direction. The schematics of a representative unit cell of the particle in the CS coating structure (Figure 4.72) may be assumed taking into account that the horizontal cross section of the spherical particle after deformation is circled, and vertical cross section is elliptical. In this case, an axisymmetric scheme of the particle deformation may be approved. The particle deformation during impact is characterized by strain localization at the

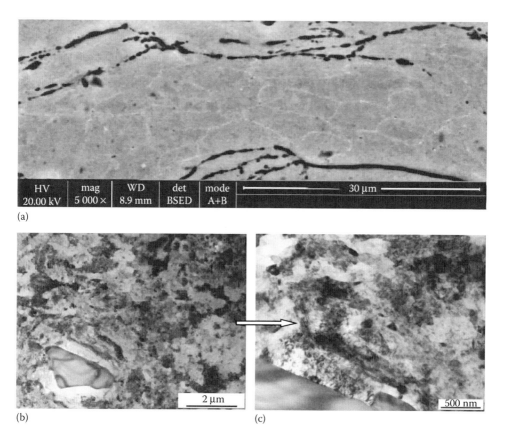

FIGURE 4.68 Microstructure of low-pressure cold-sprayed AA 6022 alloy composite coating at different magnifications. (a) SEM image ×5000, etched and (b, c) TEM images ×80,000, reinforcement particle is seen.

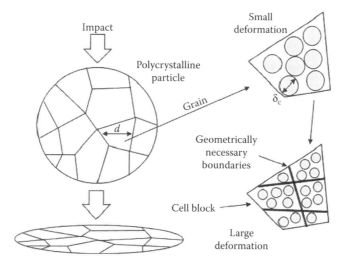

FIGURE 4.69 Schematic of microstructural evolution within a particle's grain during CS deformation (d is the initial grain size, and δ_c is the cell size). GNBs denote the geometrically necessary boundaries that divide the cell blocks at large strains (after Molinari and Ravichandran, 2005).

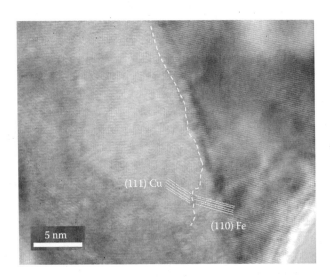

FIGURE 4.70 HRTEM image of Cu–Fe particle–substrate interface after high-pressure cold spraying Cu powder on the carbon steel substrate.

interparticle contacts. Following the above considerations, the idealized representative structure of the particle will contain the particle walls of thickness $w/2$ and the particle core. The particle wall structure differs from that of the particle core because of strain localization effects. Thus, it may be assumed that the walls form a topologically continuous skeleton structure within which isolated islands of the particle core are encapsulated, similar to Estrin et al. (1998).

Figure 4.72 depicts the coating structure which consists of one representative unit cell particle after impact and other deformed particles (the interparticle boundaries are shown). The particle consolidation during impact may be described by continuous development of the interparticle walls with thickness of w. As shown in Chapter 2 by numerical simulation of the particle impact, it is difficult to precisely define the areas and thickness of interparticle contacts occupied by adiabatic shear bands, and other areas of severe strains resulting in certain bonding strength. That is why we would suggest to characterize the coating structure evolution with a volume fraction of the walls which can be calculated using the unit cell shown in Figure 4.72. If the surfaces of the deformed particles are the spheres of radius R_d, diameter of the deformed particle is $2R_f$, and its height $-2h_x$,

$$2R_d = \frac{R_f^2 + h_x^2}{h_x}$$

The volume fraction of interparticle walls may be defined as

$$f = \frac{(S_d \cdot w/2) \cdot 2}{4/3\pi R_p^3} \tag{4.12}$$

Here, S_d is the surface area of spherical segment with radius R_d and height h_x: $S_d = 2\pi R_d h_x$. R_d may be defined on the basis of an incompressibility rule that means volumes of the particle before and after deformation are equal. A volume of the spherical segment of the height h_x (Figure 4.72) is $V_d = (2/3)\pi R_d^2 h_x$. Based on the assumption that the deformed particle consists of two segments, the incompressible condition is the following:

$$\left(\frac{2}{3}\pi R_d^2 h_x\right) \cdot 2 = \frac{4}{3}\pi R_p^3 \quad \text{and} \quad R_d = \sqrt{\frac{R_p^3}{h_x}} \tag{4.13}$$

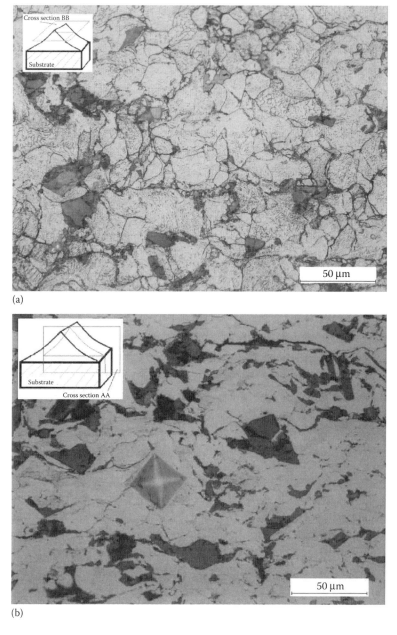

(a)

(b)

FIGURE 4.71 CS SS steel –SiC coating optical microstructure. (a) Cross section BB, interface after etching and (b) cross section AA, unetched interface.

Taking into account that $S_d = 2\pi R_d h_x$, the volume fraction f is

$$f = \frac{3w}{2R_p}\left(\frac{h_x}{R_p}\right)^{1/2} \tag{4.14}$$

The main goal of this analysis is to define the dependence of the volume fraction of the interparticle walls on geometry parameters of the particle deformed upon impact (radius R_f and height h_x). To analyze the process of particle deformation during impingement, the classical energetic approach

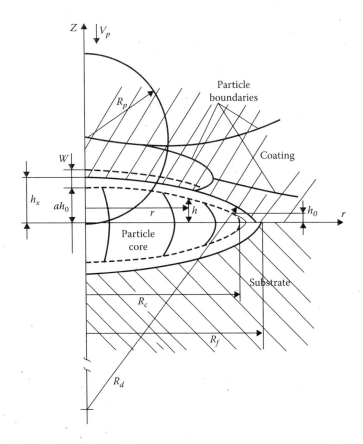

FIGURE 4.72   CS coating structure model.

(Avitzur, 1983) is applied. For the case of the spherical particle impact, the energy A_d consumed for the particle strain is spent on overcoming internal resistance of material

$$dA_d = \iiint_V \int^w \sigma_S \, d\epsilon_i dV \tag{4.15}$$

The task of determination of deformation energy A_d may be solved on the basis of the analysis of deformation of the particle core and thin layer of localized strain which have the less flow stress than that of the core. The shape of the particle core is assumed to be similar than that of the particle. This surface may be described by the following equation (Evstratov, 1995):

$$h = \pm\left[h_0 + ah_0\left(1 - r^2 / R_c^2\right)\right] \tag{4.16}$$

where:
  h and r are hot fixes of the spherical surface
  a, h_0, and R_c are parameters defining the geometry of the particle core (Figure 4.72)

The radial displacements of particle points during the particle deformation under impact may be described as

$$u_r = br\left(1 - z^2/h^2\right)^2 \tag{4.17}$$

where parameter b may be defined on the basis of the incompressibility rule:

$$\pi r^2 \Delta h = \int_0^h 2\pi r u_r dz \tag{4.18}$$

After inserting Equation 4.17 into Equation 4.18 and integration, the radial displacement equation is (Evstratov, 1995):

$$u_r = 15 \Delta hr \left(1 - z^2 / h^2\right)^2 / 16h \tag{4.19}$$

The second component of displacement u_z may be defined on the basis of incompressibility equation for axisymmetrical scheme (Drucker, 1967):

$$\frac{\partial u_z}{\partial r} + \frac{u_r}{r} + \frac{\partial u_z}{\partial z} = 0 \tag{4.20}$$

So, displacement u_z is presented by Equation 4.21:

$$u_z = \frac{30 \Delta hz}{16h} \left(1 - \frac{2z^2}{3h^2} + \frac{z^4}{5h^4}\right) \tag{4.21}$$

Based on Equations 4.9 through 4.11 and 4.19, the strain components ε_r, ε_θ, ε_z, and equivalent strain ε_i are defined, and deformation energy and particle deformation load may be calculated using Equation 4.15. Taking into account two-layer representation of the deformed particle (Figure 4.72), the full deformation load P_d of the particle may be described by the following equation:

$$P_d = \pi R_c^2 \overline{p}_{\text{core}} \sigma_{\text{s core}} + \pi \left(R_f^2 - R_c^2\right)_w^2 \overline{p}_{\text{layer}} \sigma_{\text{s layer}} \tag{4.22}$$

where $\overline{p}$ is dimensionless parameter normal stress $\overline{p} = \left(p/\sigma_s\right)$.

Because $P_d \cdot \Delta u_z = dA_d$ and the kinetic energy of the spherical particle of the certain size is proportional to $\rho_p V_p^2$ Equation 4.22 may be rewritten

$$\pi R_p^3 \rho_p V_p^2 = 2\Delta u_z \left[\pi R_c^2 \overline{p}_{\text{core}} \sigma_{\text{s core}} + \pi \left(R_f^2 - R_c^2\right)_w^2 \overline{p}_{\text{layer}} \sigma_{\text{s layer}}\right] \tag{4.23}$$

Here $\left(R_f^2 - R_c^2\right)_w^2 = \left(w/2\right)\left(2R_c + \left(w/2\right)\right)$. After transformation and replacement with $a = \Delta u_z \overline{p}_{\text{layer}} \sigma_{\text{s layer}}$ and $b = \Delta u_z \overline{p}_{\text{core}} \sigma_{\text{s core}}$ we will obtain from Equation 4.23:

$$R_p^3 \rho_p V_p^2 = a \frac{w^2}{2} + 2aw \left(R_f - \frac{w}{2}\right) + 2b \left(R_f - \frac{w}{2}\right)^2 \tag{4.24}$$

Taking into account function $\beta = \left(h_x / R_p\right)^{1/2}$ and Equation 4.22,

$$R_p^w \rho_p V_p^2 = 2a \frac{f}{\beta} + \frac{4}{3} a \left[\left(\frac{R_f}{R_p}\right) \frac{f}{\beta} - \frac{1}{3} \left(\frac{f}{\beta}\right)^2\right] + 2b \left[\left(\frac{R_f}{R_p}\right)^2 - \frac{1}{9} \left(\frac{f}{\beta}\right)^2\right] \tag{4.25}$$

Taking into account spherical segment approximation of the deformed particle shape (Figure 4.13) and incompressibility condition, the ratio β may be presented as $\beta = \left(h_x / R_p\right)^{1/2} = \left(R_p / R_d\right) = \left(1/\gamma\right)$, where $\gamma = \left(R_d/R_p \right)$ and $R_d = \left(R_f^2 + h_x^2 \right)/2h_x$. As shown in Chapter 2, the flattening ratio is $\left(R_f/R_p \right) = \delta$. Equation 4.23 may be presented after rearrangements as

$$R_p^w \rho_p V_p^2 = 2af\gamma + \frac{4}{3}a\left[\delta\gamma f - \frac{1}{3}\gamma^2 f^2 \right] + 2b\left[\delta^2 - \frac{1}{9}\gamma^2 \right] \tag{4.26}$$

$$R_p^w \rho_p V_p^2 = 2af\gamma + \frac{4}{3}a\left[\delta\gamma f - \frac{1}{3}\gamma^2 f^2 \right] + 2b\left[\delta^2 - \frac{1}{9}\gamma^2 \right]$$

$$\frac{4}{9}a\gamma^2 f^2 + a\left(2\gamma - \frac{4}{3}\delta\gamma \right)f + 2b\left[\delta^2 - \frac{1}{9}\gamma^2 \right] - R_p^w \rho_p V_p^2 = 0$$

$$\frac{4}{9}\frac{a}{b}\gamma^2 f^2 + \frac{a}{b}\left(2\gamma - \frac{4}{3}\delta\gamma \right)f + 2\left[\delta^2 - \frac{1}{9}\gamma^2 \right] - \frac{R_p^w \rho_p V_p^2}{b} = 0 \tag{4.27}$$

Taking into account $a/b = q$ and assuming $\left(R_p^w \rho_p V_p^2/b \right) \approx 2$, Equation 4.27 is transformed to a quadratic equation (4.28) with solutions of $f_{1,2}$ (Equation 4.29).

$$\frac{4}{9}q\gamma^2 f^2 + q\left(2\gamma - \frac{4}{3}\delta\gamma \right)f + 2\left[\delta^2 - \frac{1}{9}\gamma^2 \right] - 2 = 0 \tag{4.28}$$

$$f_{1,2} = \frac{-q\left(2\gamma - (4/3)\delta\gamma \right) \pm \sqrt{\left(2\gamma - (4/3)\delta\gamma \right)^2 - 4(4/9)q\gamma^2 2\left(\delta^2 - (1/9)\gamma^2 - 1 \right)}}{2(4/9)q\gamma^2} \tag{4.29}$$

In Equation 4.28, the coefficients γ, δ are functions of the particle velocity, which are defined based on our experimental results of the cold-sprayed SS and carbon steel coatings microstructure examination and SS and carbon steel and Cu particles impact modeling shown in Chapter 2.

It is difficult to exactly define the function $q = q(V_p)$ because $q = (a/b) = \left(p_{\text{layer}}/p_{\text{core}} \right) = \left(\sigma_{\text{s layer}}/\sigma_{\text{s core}} \right)$. It is believed to define the flow stress ratio dependence on the particle velocity based on the Gao and Zhang (2010) constitutive model describing properly the dynamic response of a material under a deformation with large temperature variation, high strain rate and large strain. Taking into account particle impact modeling results (Chapter 2), the strain rates of the particle core and interparticle layer (Figure 4.72) deformation are evaluated. Based on these data, the $\sigma_{\text{s layer}}$, $\sigma_{\text{s core}}$ and q ratio are determined. The results of γ, δ, and q evaluation are shown in Figure 4.73 as the γ, δ, and q dependences on particle velocity are presented.

Based on experimental data and modeling results (Chapter 2), the functions $\gamma(V_p)$ and $\delta(V_p)$ may be presented with polynomial approximations (Figure 4.73). Based on modeling data (flow stress dependences on strain and strain rate, Chapter 2), the yield stress ratio $q(V_p)$ dependence is well approximated with linear function up to particle velocities of 800 m/s. A slow increase of the q ratio with the increase particle velocity may be explained by an alteration of the particle deformation mechanism. In the areas of adiabatic shear band formation at the particle velocities more than 800 m/s, the yield strength will fall that will result in localization of the plastic deformation at high strain rates.

The function of volume fraction of interparticle walls $f = f(V_p)$ exhibits maximum (Figure 4.74) which can be explained by extensive strain localization in the area of interparticle walls. The interparticle walls thickness $w/2$ (Figure 4.72) seems to decrease due to the localization process that results in diminishing of volume fraction f. One can note that adiabatic shear band formation occurs

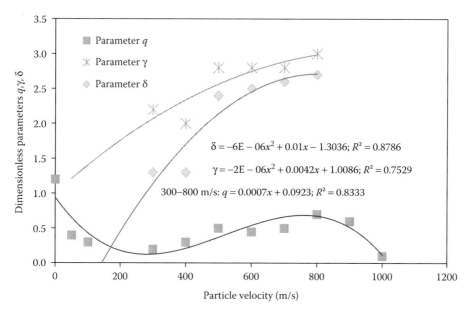

FIGURE 4.73 Evaluation of the parameters of Equation 4.28 for Cu particles.

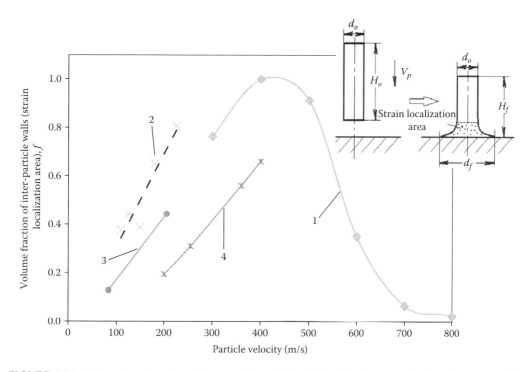

FIGURE 4.74 The volume fraction of interparticle walls/strain localization area fraction f versus particle velocity. 1—f calculation results in accordance with Equation 4.22; 2—Gust (1982) experimental results of Cu cylinders impact deformation; 3—Eakins and Thadhani (2006) experimental results of Cu cylinders impact deformation; 4—Gust (1982) experimental results of Al cylinders impact deformation.

at the narrow area of approximately 1 μm thickness, which leads to considerable decrease of f at the velocities of 700–800 m/s. These particle velocity values correspond to numerous critical velocity data (Assadi et al., 2003).

4.3.2 EFFECT OF NANOSTRUCTURE ON THE PROPERTIES

4.3.2.1 Interface Analysis and Its Influence on the Strength and Ductility

As shown in the previous section, the presence of interfaces of various origin and its interaction with dislocations and other lattice defects control the structure formation of CS coatings. At present time, the deformation and fracture mechanisms of the polycrystalline, sintered, and deposited materials are shown to be controlled by the presence of interfaces which constitute either barriers to dislocation motion, easy paths for deformation, nucleation sites for dislocations, or twins and/or weak sites for the initiation of cracking (Pardoen and Massart, 2012). Arzt et al. (2001) and others demonstrate that interfaces can be much more dominant in setting the strength and ductility of metals than chemical composition and crystallography. The ultrafine grain (UFG) and nanostructured materials exhibit the highest sensitivity to these structure features (Meyers et al., 2006). The main reason of this effect seems to be creation of barriers for dislocation movement and interaction. The constraint exerted by the interfaces at the nanoscale leads to local gradients of deformation which are physically accommodated by geometrically necessary dislocations (GNDs), for example, Fleck and Hutchinson (1997). The interfaces of submicron and micron size create the barriers for dislocation movement and generation and areas of dislocation and other defects drainage. The above effects depend on the dimensions of the microstructure elements, which, first of all, is characterized by the interface spacing and interfaces properties (Pardoen and Massart, 2012). SGP theory (e.g., Fleck and Hutchinson, 1997; Needleman and Sevillano, 2003) develops a theoretical framework to evaluate the effect of interfaces on the plastic flow. Based on classification of Pardoen and Massart (2012), the types of interfaces and its interactions with dislocations are represented taking into account microscale interfaces shown in Figure 4.75b.

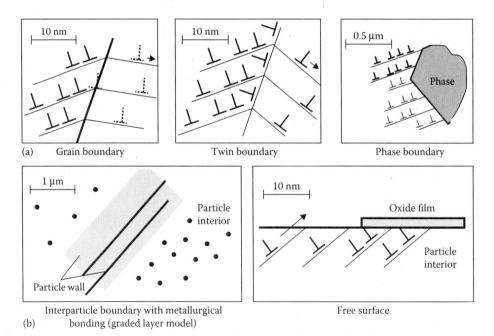

FIGURE 4.75 Different types of interfaces occurring at particles deformation and consolidation at CS in accordance with Pardoen and Massart, 2012. (a) Particle deformation interfaces and (b) particle consolidation interfaces.

Figure 4.75 depicts the classification of interfaces from the viewpoint of its size, structure and behavior taking into account the particle deformation and consolidation during CS:

- GBs produce different levels of constraint on dislocation motion, from fully impenetrable if the local stress is small enough to partly transparent if the stress is large enough to allow dislocations to fully or partly transmit. Reducing the grain size results in an increase of the strength at missing of the ductility. The grain size influences on yield stress (the Hall–Petch law, Equation 4.11).
- The twin boundaries are created during dynamic loading of the particle during CS impact. As shown by Zhu et al. (2011), the interactions between the twin boundary and moving dislocation involve a combination of dislocation blocking with the creation of sessile and/or partial interface dislocations. Additionally, generation of the new dislocations may occur. So, the strength of the interface depends on strain and other structure specific features. Deformation twins constitute dynamic interfaces that multiply during deformation leading to an evolution of the microstructure characteristic lengths that have a much greater effect on strain hardening than static twins (Christian and Mahajan, 1995; Idrissi et al., 2010).
- The two types of the phase boundary interfaces (Figure 4.75a) occur: (1) static interfaces, which form during crystallization and thermal phase transformations; and (2) dynamic ones, which form due to stress and strain assisted phase transformation. The static phase boundary interfaces are impenetrable even at high stress levels due to its incoherency and the lattice mismatch between the primary and second phases. An optimization of the volume fraction, size and morphology of the second phases is essential in the cold spraying of high-performance coatings. The dynamic phase boundary interfaces developed due to CS particle strain result in formation of phases with hard interfaces in the microstructure, which provides a continuous source of strain hardening during impact deformation.
- Graded materials (a graded model of CS coating structure is presented in the previous section) can also be considered as an example of interfaces, which seem to be thicker or less than those discussed above. The graded structure of metallic polycrystalline materials such as Al-, Ti-, Ni-, or Fe-based alloys and its influence on mechanical behavior was analyzed by Dahlberg and Faleskog (2014). The analysis involves submicron thick layers surrounding GBs with a microstructure different from the bulk of the grain. There are two options of this model: (1) the GB layer is softer than the grain interior, and (2) the layer is harder than grain interior. Additionally, the graded model may be applied for interparticle boundaries characterization as it is shown in Section 4.3.1.
- The particle surfaces (Figure 4.75b) which form the interparticle boundaries consist of free areas and areas covered by oxide layers. The free surfaces are known to be completely transparent to dislocations (Hirth and Lothe, 1992), and the dislocations are attracted by the surface. The particle surface areas covered by a thin oxide layer can block the dislocations and raise the local strength (Brugger et al., 2010). In this case, the strain gradient can lead to the cracking of the interface and the relaxation of the constraint.

4.3.2.2 SGP Theory Application

Analysis of all listed types of interfaces occurring at the particles deformation and consolidation at CS (Figure 4.75) demonstrates the presence of plastic gradients which originate from a nonuniform stress and strain distribution in the particle due to its the dynamic loading. It results in some size effects similar to those observed at micro- and nanoindentation, torsion of thin wires, submicron holes or bending of thin films (Fleck and Hutchinson, 1997; Abu Al-Rub and Voyiadjis, 2006), which are described by these authors on the basis of the SGP model. The main idea of the SGP model is taking into account of the interface effects on the plastic flow of the polycrystalline material. Based on this model, Pardoen and Massart (2012) developed a generic approach of interfaces with cohesive zones in the context of an FE model relying on a finite strain implementation of the

SGP theory of Fleck and Hutchinson (2001). The cohesive zones are used to characterize the inter-face with higher order boundary conditions as compared to those of grain interior. These interfaces (GBs, twin boundaries, phase boundaries) are initially constrained, and they will be deformed when the stress at the interface will attain the critical level. Application of the polycrystal SGP model of Pardoen and Massart (2012) and Brugger et al. (2010) to cold-sprayed powdered material descrip-tion at microscale is shown in Figure 4.76.

The main goal of this model is to take into account interface effects on plastic flow of material. The grain interior response is described with SGP theory equations (Fleck and Hutchinson, 2001). Classical J_2 deformation theory and J_2 flow theory of gradient plasticity (Fleck and Hutchinson, 2001) are linked by the fact that they coincide when the deformation involves proportional straining. Hutchinson (2012), used SGP as a template for the flow theory which is constructed to coincide with the deformation theory for proportional straining histories.

As noted by Hutchison (2012), the material inputs are (i) the isotropic elastic properties, (ii) the uniaxial stress $\sigma_0(\varepsilon_p)$, and (iii) a single material length parameter ℓ. The length parameter is assumed to take into account input of strain gradient on an effective plastic strain defined in the classical theory of plasticity. Classical and SGP theories operate with following parameters: u_i is the displacement vector, $\varepsilon_{ij} = (u_{i,j} + u_{j,i})/2$ is the strain tensor, ε'_{ij} is its deviator, σ_{ij} is the symmetric Cauchy stress, s_{ij} is its deviator, and the effective stress is $\sigma_e = \sqrt{s_{ij}s_{ij}/2}$. Throughout, $m_{ij} = 3s_{ij}/(2\sigma_e)$ is a dimensionless deviator tensor codirectional with the deviator stress.

Based on this formulation, the plastic strain tensor may be presented as $\varepsilon^p_{ij} = \varepsilon_p m_{ij}$, where $\varepsilon_p = \sqrt{2\varepsilon^p_{ij}\varepsilon^p_{ij}/3}$. On the basis of elasticity theory Cauchy stress may be presented by the following equation:

$$\sigma_{ij} = 2\mu\varepsilon^{e'}_{ij} + \lambda\varepsilon^e_{kk}\delta_{ij}, \quad \varepsilon^e_{ij} = \varepsilon_{ij} - \varepsilon^p_{ij} \tag{4.30}$$

with $\varepsilon^{e'}_{ij}$ as the deviator of the *elastic strain* ε^e_{ij} and with $\mu = (E/[2(1+v)])$, $\lambda = (E/[3(1-2v)])$ and δ_{ij} as the Kronecker delta. In accordance with the SGP theory considered by Fleck and Hutchinson (2001), the spatial gradient, $\varepsilon_{p,i}$, may be used as the measure of the plastic strain gradients. In this case, a basic equation for determination of gradient enhanced effective plastic strain, E_p, is shown below:

$$E_p = \sqrt{\varepsilon_p^2 + \ell^2\varepsilon_{p,i}\varepsilon_{p,i}} \tag{4.31}$$

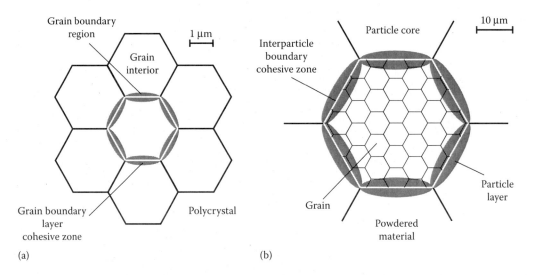

(a)   (b)

FIGURE 4.76 Application of the SGP polycrystal model of Pardoen and Massart (2012) (a), to powdered material description at microscale (b).

Equation 4.31 captures the combined effect of the plastic strain and strain gradients with ℓ Based on Equation 4.30, the total strain energy density of the solid (sum of elastic and plastic strain energies) is presented as

$$U\left(\varepsilon_{ij}^{e'},\varepsilon_p,\varepsilon_{p,i}\right)=\mu\varepsilon_{ij}^{e'}\varepsilon_{ij}^{e'}+\frac{1}{2}\lambda\varepsilon_{kk}^{e2}+U^p\left(E_p\right) \tag{4.32}$$

Here, $U^p(E_p)$ is the energy density defined on the basis of tensile stress–plastic strain curve of the material with equation below:

$$U^p\left(E_p\right)=\int_0^{E_p}\sigma_0\left(\varepsilon_p\right)d\varepsilon_p \tag{4.33}$$

It is assumed (Equation 4.33) that the total plastic strain energy in the presence of strain gradients, E_p, under proportional straining may be taken equal to that at the same strain, $\varepsilon_p^{tension}=E_p$, in the absence of gradients. It allows us to define the gradient contribution to E_p at real structure deformation cases inherent in CS particle impact. As noted by Hutchinson (2012), the gradient contribution to E_p at the nano- and submicroscale accounts for the additional stored GNDs (Figure 4.75). The gradient contribution at the microscale (Figures 4.72 and 4.76) seems to account for GNBs and other structure elements such as cohesive zones in the powdered materials (Figures 4.69 and 4.72). Thus, the SGP tool will allow to define the influence of the *interface* structure features on strength and ductility of cold-sprayed coating materials. The idea that SGP models can be used to describe the influence of the interfaces on plastic flow and, consequently, mechanical properties of the polycrystalline materials have been pursued by many authors, as shown in the review by Pardoen and Massart (2012). However, its application to powdered materials have not been developed yet. The analysis of structure of Al and Cu coatings from the viewpoint of classification of interfaces (Figures 4.75 and 4.76) is presented below.

4.3.2.3 Structure Evolution at Submicro- and Nanoscale

A detailed description at submicro- and nanoscale of microstructural evolution of Al alloys at the CS process is presented in works of Rokni et al. (2014). This study described the microstructure evolution due to CS taking into account development of UFG structures, size and distribution of precipitates, and solute element distribution within the microstructure. The formation of UFG structures in Al alloys during CS has been attributed to the combination of continuous dynamic recovery (DRV) and DRX. In addition, it is well known that SPD can affect the kinetics of precipitation in Al alloys and can result in a higher volume fraction of precipitates. The similar structure evolution processes proceed at CS of other metal powder materials and result in enhanced mechanical properties.

As shown before, the specific deformation features inherent to the CS coating structure may be analyzed both for particle core and particle interface areas. Because the particles experience severe deformation and heat during the impact (see Chapter 2), they have recrystallized fields, resulting in the formation of high angle grain boundaries (HAGBs) and UFG structure (white arrow #1, Figure 4.77). Some low angle grain boundaries (LAGBs) in the micrograph of AA6061 CS coating interface are seen (white arrow #3, Figure 4.77) as well, which is an evidence for the occurrence of DRX and conversion of the LAGBs to HAGBs. The similar data of the DRX process has been reported during the deposition of AA 6061 (Rokni et al., 2014). The results of TEM examination illustrates that in the particle core, the material experiences less, although significant, strain that results in dislocation substructure and presence of subgrains in the microstructure (see Figure 4.78).

The detailed analysis of particle deformation resulting in the UFG structure formation is made in Chapter 2. However, there is a lack of information regarding the evolution of the generated

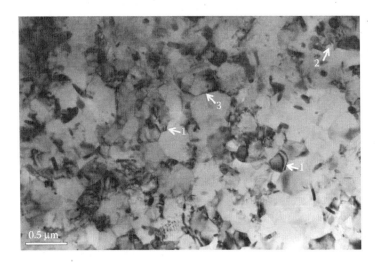

FIGURE 4.77   TEM image of CS AA 6061 coating at the particle/particle interface region.

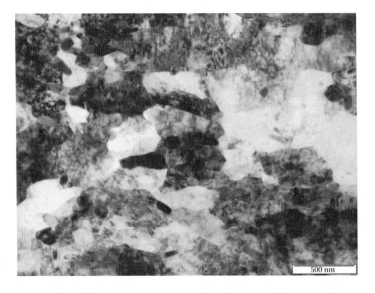

FIGURE 4.78   Dislocation structure in the CS deformed particle core (STEM).

boundaries. Figure 4.77 shows a TEM image of the particle/particle interface region of the CS AA 6061 alloy. Similar to Rokni et al. (2014) three types of boundaries may be selected: (1) polygonized dislocation wall (PDW), (2) partially transformed boundary (PTB), and (3) HAGB. These boundaries are characterized by misorientation angles of <1°, 1–5°, and >15° for PDW, PTB, and HAGBs, respectively. This grain classification was introduced by Chang et al. (2000) during the analysis of GB evolution and UFG formation in 1100 Al alloy deformed by equal channel angular pressing. Chang et al. (2000) found the transformation of LAGBs to HAGBs by an increase in the number of boundary dislocations during the deformation process. The presence of these types of boundaries in the particle/particle interface region of CS AA 6061 coating (Figure 4.78) demonstrates the structure formation process through which UFG structures develop in this case, and indicates about similarity between the particle impact deformation and other SPD processes.

The main conclusions made from the investigation of submicrostructure of CS AA 6061 coatings are the following:

1. Graded structure of the particles (particle/particle boundaries and particle interiors) in CS coating is defined
2. UFG structures are developed during CS due to transformation of the dislocation walls into HAGBs (Rokni et al., 2014)
3. Particle/particle boundaries show a high degree of deformation resulting in the formation of an UFG structure and a low density of LAGBs that may be attributed to DRX effects

Microstructural evolution and nanostructure formation mechanisms in Cu particles during impact differ from those of Al alloys. The mechanisms of the microstructural refinement due to particle plastic deformation may be identified: (i) dislocation movement and accumulation, (ii) deformation twinning, and (iii) shear banding in twin-matrix (T/M) lamellae (Li et al., 2008). Additionally, structure evolution of particle boundary areas may be characterized: (i) adiabatic shear band formation, (ii) DRX and (iii) DRV. Li et al. (2008) have shown that a mixed nanostructure is formed in impact loading of the Cu bulk sample with nanoscale T/M lamellae generation (about 33% of the volume) and nano-sized grains formation ($\approx$67%). In this work, the dynamic plastic deformation was performed with a strain rate of about 10^2 to $2 \times 10^3 \, s^{-1}$ at a temperature of $-100°C$. In the case of the particle impact at the CS process, the strain rate is shown in Chapter 2 to achieve the $10^8 - 10^9 \, s^{-1}$ at ambient temperature. So, it is reasonable to assume the similar processes of structure formation indeed occur during Cu particle impact CS deposition. The results of microstructure evolution study made by authors are shown below. The objective of this study was to examine the microstructure evolution and nanostructure development of CS Cu coatings made by high-pressure and low-pressure CS processes, which is important for achieving coating mechanical properties and technological applications of the CS processes.

An optical microscope (OM, Leica MPS 30) was used to observe the holistic microstructure of the deformed Cu specimens. For OM observations, the samples were etched in a solution containing HNO_3, H_3PO_4 and $H_4C_2O_5$ (with a ratio of 1:1:1) for 5 s. The detailed microstructures of the samples at submicro- and nanoscale were characterized by means of TEM. The cross-sectional thin foils for TEM observations were prepared by means of double-jet electrolytic polishing in an electrolyte consisting of 25% (volume) alcohol, 25% phosphorus acid, and 50% deionized water at about $-10°C$.

The OM images of Cu coating microstructure shown in Figure 4.79 demonstrate the general difference of the high-pressure (Figure 4.79a) and low-pressure (Figure 4.79b) cold spraying processes. The high-pressure CS process results in good interparticle bonding that leads to weak etching capability of interparticle boundaries. The low-pressure CS process does not allow us to achieve the proper interparticle bonding. So, the low-pressure CS interparticle boundaries are etched in high extent. Additionally, the particle shape after high-pressure CS is of mushroom type contrary to the oval shape of the particles after low-pressure CS. Preliminary annealing of the particles for low-pressure CS increases the tendency to uniform strain of the particles. The use of dendrite shape Cu particles leads to uniform deformation and obtaining of the particle oval shape (Figure 4.79c and d). The mentioned specific features of the Cu microstructures demonstrate the much higher effect of strain localization during particle impact with higher particle velocities at high-pressure CS than that at low-pressure CS. It results in different mechanisms of nanostructure formation described below.

When pure Cu with a medium SFE is deformed at high strain rates, deformation twinning may play an important role in accommodating plastic strain as dislocation activity is suppressed. Formation of deformation twins and shear bands is known to be responsible for the structure refinement in the ball milling process, and the high density of nanoscale twins is believed to play a crucial role in the grain refinement in the SMAT process (Li et al., 2008).

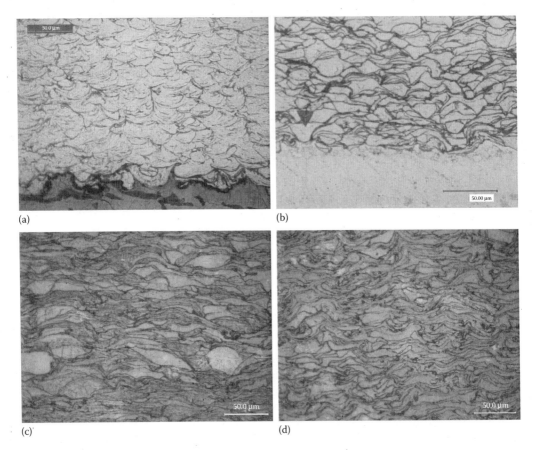

(a)   (b)

(c)   (d)

FIGURE 4.79 OM images of CS Cu coating structure after: high-pressure cold spraying of Cu spherical Cu particles of 25 μm size (a), low-pressure cold spraying of the same particles (b), low-pressure cold spraying of dendrite Cu particles of 45 μm size annealed at 350°C (c), and 550°C (d).

Close SEM observation of the microstructure shown in Figure 4.80 demonstrates some parallel bands in the deformed particle of lens shape. Moreover, some grains of 1–3 μm size are seen in the particle core. The thin layer on the particle surface observed in the micrograph is similar to that shown in Figure 2.42 for SS coating. Thus, the particle inhomogeneous deformation is inherent for high-pressure CS process.

The TEM observation of the area neighboring to particle boundar revealed that numerous deformation twins are found (Figure 4.81a). A selected area diffraction (SAED) pattern (Figure 4.81b and c) from the twin bundles demonstrates a superposition of couple of <001> diffraction patterns which are symmetrical to each other with respect to the {111} plane. As shown by Li et al. (2008), it is indicative for typical nanotwins. The twin lamellas shown in Figure 4.81a have an average width about 30–60 nm and length of 0.5–10 μm.

The elongated *bamboo-like* nano-sized crystallites are seen in some of the bundles areas (Figure 4.82a). They are similar to those identified by Li et al. (2008). The electron diffraction pattern (Figure 4.82b) indicates nano-sized grains with random orientation. A possible mechanism of structure formation in these areas seems to be fragmentation and rotation of the twin lamellas, and DRX due to large shear strains in the areas of strain localization (Xue et al. (2005).

Close TEM observation of CS coating microstructure shown in Figure 4.78 demonstrates the similar features of nanostructure formation (Figure 4.83) only in separate places of the foil. Therefore, a probability of the nanostructure formation during low-pressure CS process seems to be low.

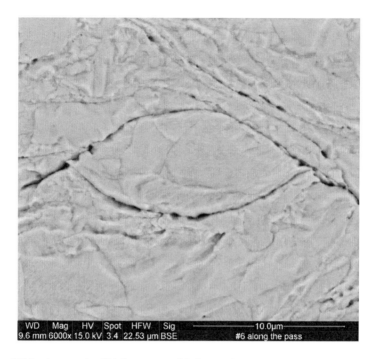

FIGURE 4.80 SEM micrograph of high-pressure CS Cu coating after etching. The parallel bands are seen in the deformed particle of lens shape.

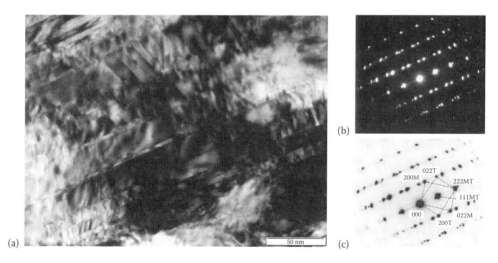

FIGURE 4.81 A close TEM observation of the twin bundles. (a) TEM image of the twins, (b) corresponding SAED pattern, and (c) SAED index of twin bundles.

The results of Cu coating microstructure evolution examination reveal that the impact deformation of Cu particles at the strain rates of $10^8–10^9$ s^{-1} results in formation of submicro and nanostructure that influence on mechanical properties of the coatings. The main mechanisms of Cu microstructural evolution seem to be nanotwinning, nano-sized grain structure formation, twin bundles generation, and DRX. All of them greatly influence the mechanical properties and allow to achieve the anomalous strengthening effects which will be analyzed in the following paragraph.

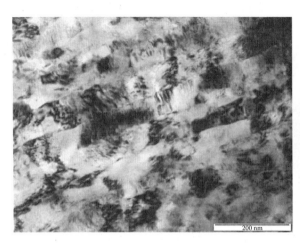
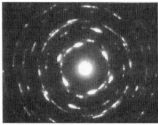

FIGURE 4.82   A close TEM observation of the *bamboo-like* nano-sized crystallites.

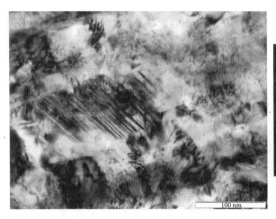

FIGURE 4.83   TEM observation of the twin bundles and nano-sized grains in low-pressure CS coating.

4.3.2.4   Coatings Mechanical Properties

The resulted mechanical properties of the CS Al and Cu-based coatings are defined and analyzed in numerous articles. The mechanical properties have to be examined based on ASTM E8 standard with tensile tests to define both composite strength and ductility. For this reason, we performed the set of examinations by tension tests which were made in the SEM column (Figure 4.82), that allowed us to observe the sample surface topography. The Al and Cu dog bone were tested at quasi-static strain rate of 5×10^{-4} s^{-1}. The Cu tension tested samples are shown in Figure 4.82b and c. A tension fixture (Figure 4.82a, inset) is supplied with extensometer allowing us to register small strains at elastic-plastic region. Tensile bars were machined from as-sprayed deposits in accordance with ASTM E8 geometry.

4.3.2.4.1   Al-Based Composite Coating Properties

To define the coating mechanical properties and the effect of annealing treatment, the bars were wrapped in Al foil envelope and heated in protective (nitrogen) atmosphere to various temperatures.

The tension diagrams of as-sprayed and CS + annealed samples are shown in Figure 4.84. As shown before by analysis of TEM examination results (Figures 4.77 and 4.78), the grain refinement is a significant issue in increase of mechanical properties of the composite coatings. Recent advancements of nanostructuring technologies such as high-energy ball milling with further consolidation

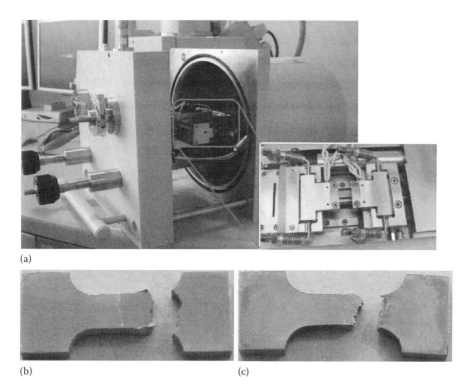

(a)

(b)   (c)

FIGURE 4.84   Tension device in the SEM FEI column (a) and tested tension samples. (b) Cu coating after CS deposition; (c) Cu coating after annealing at 600°C, 1 hour.

with thermomechanical processing (e.g., hot extrusion or other methods of SPD, spark plasma sintering, hot isostatic pressing, etc.) allows us to obtain very fine grain called as ultrafine-grained (UFg) structures with average grain size less than 1 μm. These materials demonstrate the substantial increase of the strength compared to their coarse-grained (CG, average grain size more than 10 μm) counterparts (Oskooie et al., 2015). That is why comparison of the mechanical behavior of cold-sprayed (CS), CS + annealed alloys with UFG and CG Al alloys seems to be very important to define the mechanisms of CS Al alloy composite structure formation.

The results reveal that the mechanical behavior of as-sprayed material is similar to that of UFG Al alloy, while the yield strength is lower. Taking into account that the UFG regions retain their high strength (Oskooie et al., 2015), it would be reasonable to assume that the content of UFG regions in as-sprayed materials would be less than that of UFG ones obtained with powder metallurgy route with severe deformation (Oskooie et al., 2015). Additionally, the possible reason of this behavior seems to be presence of particle interfaces which diminish the effect of UFG structure and decrease the strength of as-sprayed composite. The following heat treatment allows us to considerably improve the particle interface structure due to recrystallization of the interface areas that leads to an increase of a samples total elongation. Annealing of as-sprayed AA6061 bars was made at 300°C and 350°C during 3 hours. One can note that a tensile curve of the AA6061 sample annealed at 350°C lies near to that of CG sample (Figure 4.85) that reveals a similarity of the grain structures of both alloys.

To define the structure formation mechanisms of CS composites and their effect on material ductility, the strengthening curves of as-sprayed and CS + annealed samples are compared with those of Al alloys with bimodal grain structure (Oskooie et al., 2015). The strengthening curves were described with linear approximations $\sigma_s = \sigma_s^0 + C\varepsilon_i$. Approximation parameters are shown in Table 4.5.

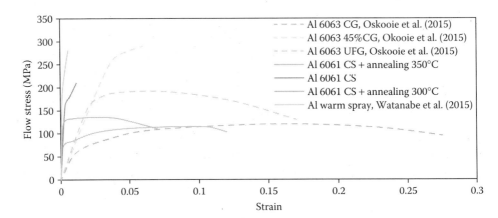

FIGURE 4.85 Tension diagrams of cold-sprayed AA6061 coatings (authors data) and AA6063 UFG composites (Oskooie et al., 2015). For comparison the diagram of warm-sprayed coating (Watanabe et al., 2013) is shown.

TABLE 4.5
Al Alloy Strengthening Curves Approximation Parameters

No.	Sample Grade	Yield Strength, σ_s^0 (MPa)	Strengthening Coefficient, C (MPa)	Elongation to Failure (%)
1	AA6061 CS coating	160	–	1.2
2	AA6061 CS + annealed (300°C) coating	124.3	365.7	3.5
3	AA6061 CS + annealed (350°C) coating	81	415	10
4	AA6063 coarse grained	77.01	352.8	15
5	AA6063 ultrafine grained	233.2	355.1	2.5
6	AA6063 ultrafine + 45% coarse grain	169	352.6	6.5

The presented results allow us to define the following specific features of the materials behavior.

- Yield strength of cold-sprayed and annealed AA6061 coatings are comparable with UFG and bimodal AA 6063 alloys (Oskooie et al., 2015).
- Strengthening coefficients of cold-sprayed and annealed AA6061 coatings depend on annealing temperature due to grain nucleation and growth during annealing.
- Elongation to failure of cold-sprayed and annealed AA6061 coatings are less than that of UFG and bimodal AA6063 alloys because of particle–particle interfaces effect which may be diminished by annealing regime optimization. Creation of bimodal structure similar to that obtained by Oskooie et at. (2015) seems to be effective way of increase of composite coatings performance.
- The total elongation of cold-sprayed and annealed AA6061 coatings is about 10% for coating annealed at 350°C, while it is only 3.5% for coating annealed at 300°C.

4.3.2.4.2   Cu-Based Composite Coating Properties
To analyze the mechanical behavior of cold-sprayed Cu coatings in as-sprayed and annealed state, a comparison of coating and nanocrystalline Cu strengthening curves is performed. The nanocrystalline Cu mechanical behavior was described in work of Dao et al. (2007). They demonstrated that

the nano-twinned Cu structures exhibited unusual both high strength and considerable ductility. The tension diagrams of as-sprayed and annealed CS coatings shown in Figures 4.86 and 4.87. The obtained results reveal that yield strength σ_s^0 of as-sprayed and annealed CS samples does not differ. However, σ_s^0 is higher than that of annealed Cu. While the ductility of as-sprayed material is very low (about 1.5%), annealing of the CS coating at 600°C during 3 hours leads to considerable increase of the total elongation and to strain localization (neck formation). The similar mechanical behavior is observed by Wu et al. (2014), during tension of the gradient-structured (GS) materials obtained by the surface mechanical attrition treatment (SMAT). As shown by Wu et al. (2014), the mechanisms for synergetic strengthening due to SMAT are macroscopic stress gradient and complex stress state caused by the gradient structure under uniaxial applied stress at tension tests. In the work of Wu et al. (2014), the steel plates were annealed at 1173 K for 1 hour to obtain a CG microstructure with a mean grain size around 35 μm. For the purpose of obtaining consistent gradient layers in terms of grain size distribution and layer depth, all samples were processed by SMAT for 5 min on both sides. So, they obtained the integrated samples, which had a CG layer sandwiched between two GS layers obtained by SMAT. One can note that the particle structure of as-sprayed composite (Figures 4.72 and 4.76) seems to be similar to that of GS materials. However, contrary to the SMAT processed steel, the Cu cold-sprayed coating has a gradient structure at the microscale within the Cu particles.

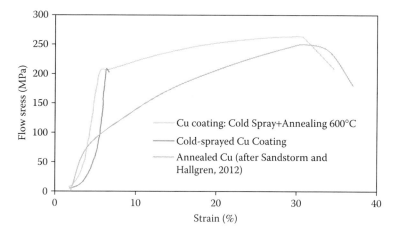

FIGURE 4.86 Tension diagrams of cold-sprayed Cu coatings (authors data) and annealed Cu (after Sandstorm and Hallgren, 2012).

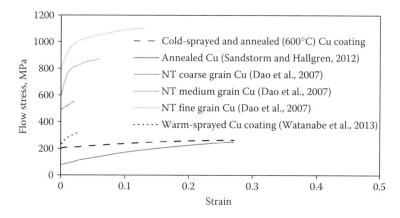

FIGURE 4.87 Strengthening curves of cold-sprayed Cu coatings (authors data)and nano-twinned (NT) Cu (after Dao et al., 2007). For comparison the diagram of warm-sprayed coating (Watanabe et al., 2013) is shown.

As shown in nanostructure analysis of CS coatings, utilization of strain gradient effects at nano, submicro- and microscale is believed to be the mainstream to achieve both high strength and high ductility of the composite coatings. For example, using of nano-twinned structures (Dao et al., 2007) allows us to increase both strength and ductility of the bulk Cu up to $\sigma_s > 1000$ MPa, and total elongation > 15% (Figure 4.87).

The as-sprayed Cu coatings have a very little ductility (Figure 4.86). Fracture at particle–particle interface occurs even at the surfaces parallel to the direction of the tensile load (Figure 4.88a). The main reason of this effect seems to be reaching of interparticle bonds only in local areas of the interface. Considerable increase of the total elongation due to annealing as compared to as-sprayed coatings (Figures 4.86 and 4.87) is observed due to static recrystallization of the CS Cu coatings, development of proper interparticle bonding due to particle interface recrystallization and obtaining multimodal grain structure (Figure 4.88b and c).

In this case, even generation and growth of some pores in the interface field at annealing temperature (Figure 4.88b) does not interfere with ductility increase.

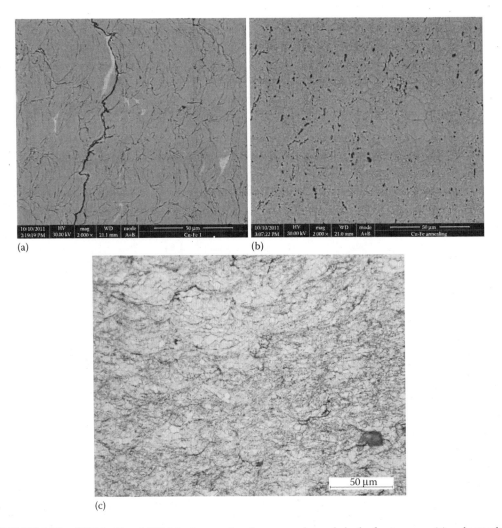

(a)   (b)

(c)

FIGURE 4.88 SEM (a, b) and OM (c) micrographs of as-sprayed sample in the fracture area (a) and annealed structure (b, c) after tension tests. The multimodal grain structure (mixture of micron size and coarse grains) is clearly seen.

TABLE 4.6

Cu Strengthening Curves Approximation Parameters

No.	Sample Grade	Yield Strength, σ_s^0 (MPa)	Strengthening Coefficient, C (MPa)	Elongation to Failure (%)
1	Annealed Cu (Sandstorm and Hallgren, 2012)	91.2	639.2	33
2	Annealed CS Cu coating	210.4	231.6	38
3	Nano-twinned coarse grain Cu (Dao et al., 2007)	490	3000	1.0
4	Nano-twinned middle grain Cu (Dao et al., 2007)	683	3871.4	7.0
5	Nano-twinned fine grain Cu (Dao et al., 2007)	906.7	1875	14
6	Warm-sprayed Cu coating (Watanabe et al., 2013)	230.6	3497.7	1.5

The analysis of linear approximation parameters of the strengthening curves shown in Table 4.6 demonstrates the effect of multimodal grain structure both on strength and ductility of Cu coatings.

The presence of the nano-twins in nano-grains in the coating structure does not considerably influence on the coating mechanical properties. Nevertheless, this effect seems to be seen for warm-sprayed Cu coating (Watanabe et al., 2013), which has the ductility about 3%–5% and yield strength about 230.6 MPa. The strengthening coefficient of the curve is similar to that of nano-twinned Cu (Table 4.6). Thus, one can suggest the similar structure constituents both of the Cu coating and nano-twinned Cu that results in the similar mechanical behavior during tension tests.

4.3.2.4.3 Comparison of Al and Cu Coatings Mechanical Properties in Normalized Parameters

The experimental results of Al and Cu coating tests are shown in Figures 4.75 and 4.76 in normalized parameters:

- Yield strength $\sigma_{s*} = \sigma_s/\sigma_s^o$
- Tensile strength $\sigma_{b*} = \sigma_b/\sigma_b^o$
- Elongation $\varepsilon_{t*} = \varepsilon_t/\varepsilon_t^o$
- Elasticity modulus $E^* = E/E^o$
- Microhardness $HV^* = HV/HV^o$
- Homological temperature $T^* = T/T_m$

The basis of normalization is the mechanical properties σ_s^o, σ_b^o, ε_t^o, E^o, and HV^o of solid Al and Cu and its melting temperature T_m shown in the Table 4.7.

TABLE 4.7

Mechanical Properties of Al and Cu (ASM Handbook, 1990)

Metall	Yield Strength σ_s^o (MPa)	Tensile Strength σ_b^o (MPa)	Elongation at 50 mm Gage ε_t^o (%)	Elasticity Modulus E^o (Gpa)	Microhardness HV^o (Mpa)	Melting Temperature T_m (°C)
Cu	33.3	380	60	125	50	1085
Al	15	50	55	63	120	660.4

In the as-sprayed condition, the Al coating exhibits a brittle behavior (elongation <2%). However, after annealing at recrystallization temperatures (0.4 Tm), the samples demonstrate the ductile behavior (elongation $\varepsilon_t >14\%$), and coating ductility ε_t after annealing at the temperatures 300–500°C is about $\varepsilon_t = 15\%–21\%$ (wrought aluminum $\varepsilon_t = 55\%$, Table 4.7). There are two reasons explaining such a behavior of Al coating (Figure 4.89): (1) presence of pores, and (2) influence of Al oxide films on the particle bonding. The slight fall of the elongation ε_t for annealing temperature of 500°C may be explained by the possible abnormal grain growth at the near to melting temperatures. One can note that dependences of all mechanical strength nondimensional parameters σ_{s*}, σ_{b*}, E^*, and HV^* on the annealing homological temperature only slightly differ one from another (Figure 4.89). This result reveals that the main structure formation processes controlling the Al strength at annealing/sintering are dislocation annihilation and recrystallization.

The dependences of Cu coating mechanical properties (normalized parameters) on temperature (Figure 4.90) demonstrate the similar behavior as compared to those of CS Al coatings (Figure 4.89). However, some specific features of the Cu coating structure formation leads to more strong dependence of mechanical properties on the annealing temperature as compared to those of CS Al coatings.

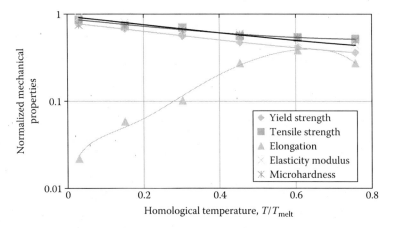

FIGURE 4.89   The dependences of yield strength, tensile strength, elasticity modulus, hardness, and elongation of CS Al coating on the annealing temperatures.

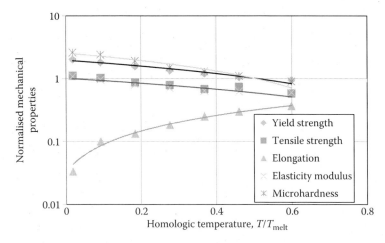

FIGURE 4.90   The dependences of yield strength, tensile strength, elasticity modulus, hardness, and elongation of CS Cu coating on the annealing temperatures.

Based on above analysis of influence of annealing on structure and properties of CS Al and Cu coatings one can conclude that annealed Al and Cu cold spray deposits exhibit yield strength, elastic modulus, ultimate tensile strength, hardness, and elongations similar to those of wrought metals. This indicates that the CS technology may be used both for surface modification at various applications and for spray forming of aluminum components. The primary mechanism of change of the CS coating mechanical properties due to annealing is the reduction in dislocation density and recrystallization of the grains within the particles forming deposit. Additionally, transformation of grain and interparticle boundaries structure due to diffusion processes results in increased ductility compared with as-sprayed coatings. At negligible porosities, microstructures of cold-sprayed coatings are characterized by multimodal structure and presence of oxide films. Annealing experiments and analysis of coatings structure at micro-, submicro- and nanoscale demonstrated kinetics of cold-sprayed coating structure formation may be carefully controlled by annealing regime that is of great importance in a real component manufacturing process.

4.3.3   MECHANICAL BEHAVIOR OF COATING–SUBSTRATE INTERFACE

4.3.3.1   Coating–Substrate Interface Characterization

The adhesion performance of cold-sprayed coatings depends on a variety of factors involved in the interface structure formation process. A multitude of mechanisms controlling the formation of stable and resistant bonding between the substrate and coating lead to difficulties of their identification and definition of structure–property relationships. The prediction and characterization of the adhesion properties for a certain cold-sprayed coating is important for practical applications to ensure its adherence to a given substrate. As shown by Chis and Cojocaru (2005), the process of interaction between substrate and the particles during thermal spraying can be shared out by the following three stages:

1. Development of the physical contact between particles and the substrate at distances in the submicron range
2. The activation of direct contact surfaces and chemical interaction of these materials at the interface in the nanoscale range (10^{-9}–10^{-10} m)
3. A diffusion mass transfer resulting in dimensional changes of both coating and substrate

The interaction time during the particle impact onto substrate in cold spraying is approximately 10^{-8}–10^{-7} s (Papyrin et al., 2007). The cold-sprayed particles undergo high-velocity deformation, heating and rapid cooling. In this short time interval, only localized mass transfer processes can occur at extremely short distances from the surface ($\approx$1–100 nm). For the CS case, the second stage of chemical interaction at the interface is not observed because the particle temperature is low, and the melting due to heat generation at impact may occur only in separate local areas. The mass transfer due to diffusion is recognized well during CS process at the micron range. All these stages contribute the bonding process.

It is well known (in the literature and in practice) that the bonds between the substrate and the particles of the coating involve chemical-type interactions, intermixing and interlocking interactions, and the Van der Waals force interactions. The Van der Waals force interactions are generally somewhat weak, and their contribution to the overall bonding strength of the coating is not significant. The intermixing and interlocking interactions are known to demonstrate a significant input to the overall bonding strength of CS coatings because of severe deformation of both particles and substrate materials (Champagne et al., 2004; Huang et al., 2014). Analysis, understanding, and quantification of these bonding mechanisms will allow to predict the adhesion performance of CS coatings. The mentioned bonding mechanisms depend on mechanical properties both of substrate itself and coating material. From this viewpoint, it seems to be reasonable to analyze the structure and mechanical behavior of cold-sprayed copper coating on aluminum and carbon steel substrates. For the first case, the effect of Cu particles penetration into Al substrate surface is observed (Huang et al., 2014) because the Cu hardness is more than that of aluminum. In the second case, the carbon steel hardness is higher than that of copper. So, penetration of separate Cu particles does not occur, and the effect of substrate surface roughness on structure and properties is known to be valuable. That is why, authors examined the microstructure and adhesion strength of the Cu coating on carbon steel substrate and compared the results with data of Huang et al. (2014) and Champagne et al. (2004).

4.3.3.2   Interface Microstructure

Examination of an interface microstructure of Cu coating on carbon steel substrate in the cross-section AA (Figure 4.71b) and interface fracture topography following the pull-off tests allows us to define the main features of interface structure formation resulting in appropriate mechanical properties. The micrographs of the interface structure of CS Cu coating on carbon steel substrate shown in Figures 4.91 and 4.92 demonstrate an occurrence of the interlocking effect, which depends on initial roughness of the carbon steel substrate. The height of micropeaks formed during the copper particles impingement is higher for substrate with higher roughness (Figure 4.92b and c), and it is clearly seen that the carbon steel substrate surface with higher roughness exhibit a lot of places of local deformation (Figure 4.92c) that results in the development of the interlocking effect.

The detailed view of the micropeaks (higher magnification) is shown in Figure 4.93 at the SEM micrographs. The micrographs exhibit a local plastic flow of the micropeaks which results in formation of substrate material jets. This effect differs considerably from that of copper particles deep penetration into Al alloy substrate (Huang et al., 2014). The thickness of carbon steel jets within the copper coating is about 0.3–0.5 µm that proves the effect of substrate micropeak severe local deformation.

4.3.3.3   Fracture Topography

The fracture morphology of the interface after adhesion tests reflects the features of SPD of both copper particles and steel substrate surface and development of interlocking effects. Because of these effects, sometimes the Cu–steel interface has a higher strength than that of Cu particle–particle interface that results in cohesion failure between Cu particles. The SEM micrograph of the interface with partial cohesion failure is shown in Figure 4.94. The similar features of fracture morphology are found and described in works of Huang et al. (2014), and Hussain et al. (2009) Cu particles impinge with substrate, deform and fill steel substrate surface valleys. The Cu particles severe deformation influence the micropeak thermomechanical conditions that result in its local deformation and formation of the substrate material jets interlocked with deposited Cu particles. The above model of interface structure formation may be examined by numerical simulation and fracture surface examination.

The SEM image of Cu coating–carbon steel interface fracture morphology is shown in Figure 4.95. The spalling zones (area1, Figure 4.95b) and fields of particles plastic deformation

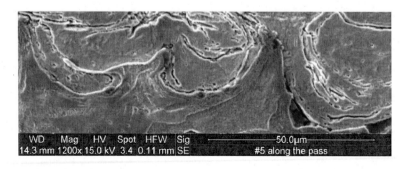

FIGURE 4.91   SEM micrograph of interface of copper coating deposited on carbon steel substrate.

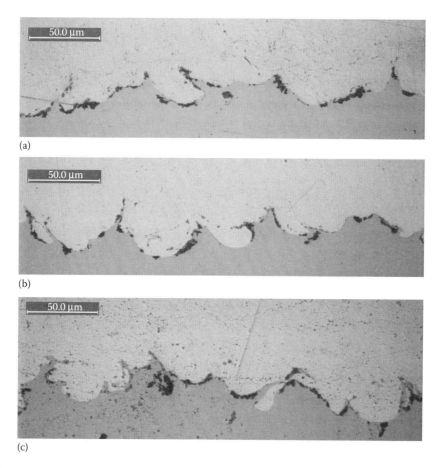

FIGURE 4.92 Optical images of Cu coating–carbon steel interface morphology. (a) Grit blast #0.4 mm, (b) grit blast #0.7 mm, and (c) grit blast #1.0 mm.

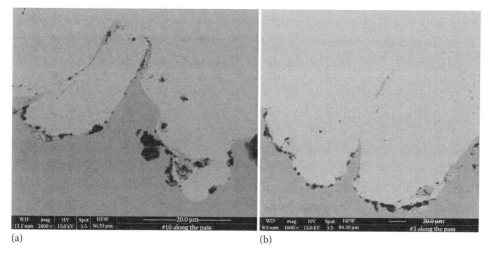

FIGURE 4.93 SEM images of Cu coating–interface morphology. (a) Grit blast #0.7 mm and (b) grit blast #1.0 mm.

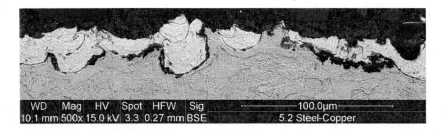

FIGURE 4.94 The SEM micrograph of the interface with partial cohesion failure.

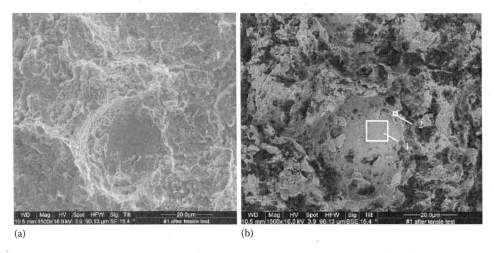

  (a)   (b)

FIGURE 4.95 SEM image of Cu coating–carbon steel interface fracture topography. (a) Fracture surface from Cu coating side and (b) fracture surface from the steel substrate coating side.

during adhesion test 9 area2, Figure 4.95b) are observed at both images. It means that both steel micropeaks and Cu particles undergo plastic deformation at tensile test and exhibit ductile behavior.

Indeed, analysis of Cu coating–carbon steel interface failure demonstrates that at least three modes of fracture zones—spalling zone, metallurgical bond zone, and interlocking zone—input to adhesion strength, and have to be taken into account at the calculation procedure development. To increase the adhesion strength, the main way is to search and optimize the CS technology parameters at which the minimal area of spalling zone and maximal area of interlocking zone may be obtained.

4.3.3.4 Coating–Substrate Interface Bonding Model

Based on the interface structure analysis, it is possible to create interface microstructure model which takes into account input of main microstructure zones to adhesion strength. Modeling results described in Chapter 2 may be used for this purpose. So interface microstructure may be shared by three zones shown in Figure 4.96a: (1) spalling zone, (2) zone of metallurgical bonding, and (3) zone of substrate material jets interlocked with deposited Cu particles. The similar model is developed by Hussain et al. (2009) for description of Cu–Al substrate interface mechanical behavior.

The interlocking effect is reflected at the numerical simulation results (Figure 4.96). However, a real pattern of the interface microstructure is more complex (Figure 4.94) because of stochastic distribution of the particle impacts. Nevertheless, it seems to be reasonable to assume availability of mentioned fields at the interface.

As shown in Figure 4.96 adhesion strength of the Cu coating–steel substrate interface may be defined on the basis of standard theory for longitudinal tensile strength of a long-fiber-reinforced

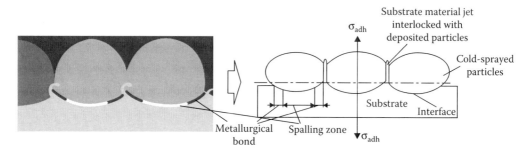

FIGURE 4.96 Interface microstructure model for adhesion strength evaluation. Left side—modeling results of multiparticle impact and right side—microstructure schematics.

composite as stated by Hussain et al. (2009). At this case, the tensile strength of the interface, σ_{adh}, is given by

$$\sigma_{adh} = \sigma_{met} f_{met} + \sigma_{il} f_{il} + \sigma_{sp} f_{sp} \qquad (4.34)$$

where f_{met}, f_{il}, and f_{sp} are the area fractions of the interface corresponding to metallurgical bond, interlocked material and spalling zone, respectively. Taking into account that fracture stress of the spalling areas is much less than that of metallurgical bonding areas, as shown in Section 4.3.1 (Figure 4.56), $\sigma_{sp} \ll \sigma_{met}$,

$$\sigma_{adh} \approx \sigma_{met} f_{met} + \sigma_{il} f_{il} \qquad (4.35)$$

SEM images of fractured surfaces (Figure 4.81) allows us to define area fraction of Cu cohesion failure by determination Cu area with EDX analysis of fractured surface from the steel substrate side, and interlocking area fraction (Figure 4.96) by determination of Fe content on the Cu coating fractured side. Assuming that $\sigma_{met} = \sigma_u^{Cu}$ = and $\sigma_{il} = \sigma_u^{Steel}$, where σ_u is the ultimate strength, it is possible to make estimation of interface adhesion strength and to compare it with experimental results. Results of EDX analysis and adhesion strength calculation are shown in Figure 4.97.

The results reveal that raise of the areas of metallurgical bonding and interlocking results in adhesion strength increase up to 160 MPa and higher (Huang et al., 2014). These parameters are controlled by particle velocity. Experimental data of influence of particle velocity on adhesion strength are shown below.

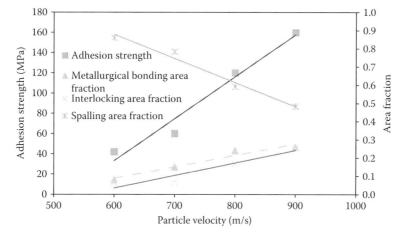

FIGURE 4.97 Calculation results of parameters of Equation 4.34.

4.3.3.5   Experimental Results of Adhesion Strength
Determination: Verification of the Model

Figure 4.98 illustrates the mechanical behavior of Cu–carbon steel substrate interface at the tension tests. It is seen that a shape of the stress–strain curve depends on the cold spraying particle velocity. CS Cu coatings deposited at the particle velocities of 900 and 1000 m/s exhibit ductile properties, while tension diagrams of the samples deposited at lower particle velocities show the classical brittle fracture. Figure 4.99 demonstrates ductile fracture of Cu particle at the interface.

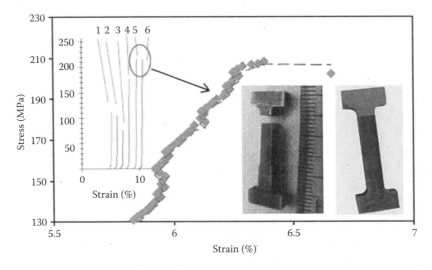

FIGURE 4.98   Fracture part of tension diagram of Cu coating on carbon steel substrate (deposited with Vp = 1000 m/s). Sample images and tension diagrams of samples deposited with Vp 500 m/s (curve#3), 600 m/s (curve#2, 700 m/s (curve#1), 800 m/s (curve#4), 900 m/s (curve#5), and 1000 m/s (curve#6) are shown in insets.

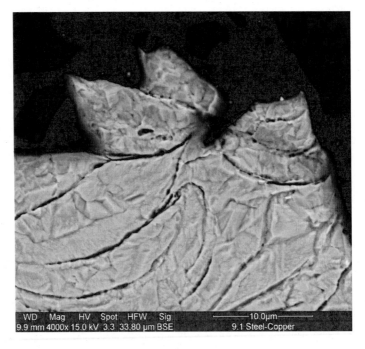

FIGURE 4.99   SEM micrograph of Cu coating fractured area.

Comparison of mechanical behavior of adhesion samples with observation of the carbon steel substrate after fracture (Figures 4.94, 4.95, and 4.99) demonstrates that plastic deformation of both copper particles and interlocked areas of the steel substrate (Figure 4.99) occurs. It can be noted that the interlocking model (Equation 4.34) seems to be valid because only interlocked areas of the coating–substrate interface may exhibit ductile fracture.

The adhesive strength of CS coating depends on type of substrate material, substrate roughness, particle velocity, and other spraying technology parameters. The experimental results of the adhesion strength of Al- and Cu-based coatings on various substrates are shown in Figures 4.100 and 4.101. It is interesting to note that the difference of Al coating adhesion strength on A6061 and 316L substrates is negligible. Figure 4.100 depicts the effect of particle velocity on Al coating adhesion strength (Cavaliere et al., 2014). The particle velocity of 700 m/s is found to be critical to achieve the maximal adhesion strength. A possible explanation of this effect is increase of the residual stresses at the coating–substrate interface formed at the particle velocities more than 700 m/s.

The Cu coating adhesion strength dependences on the particle velocity for various substrates are shown in Figure 4.101. The results reveal that these dependences are the functions of the type of substrate and its mechanical properties. The similar adhesion strength is achieved at lower particle velocities for soft (Al alloy) substrate as compared with that for hardened 316L substrate (Huang et al., 2014). The authors' data of adhesion strength on carbon steel substrate corresponds with the

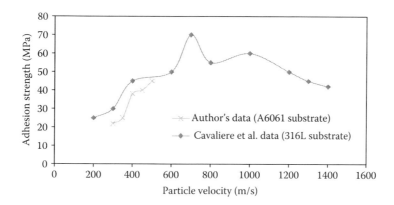

FIGURE 4.100 Adhesion strength of Al coating on A6061 and 316L substrates.

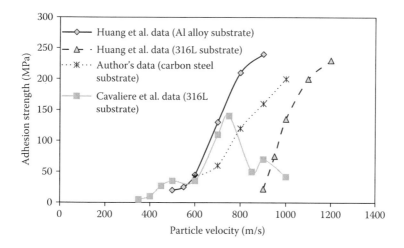

FIGURE 4.101 Adhesion strength of Cu coating on Al alloy, carbon steel, and 316L substrates.

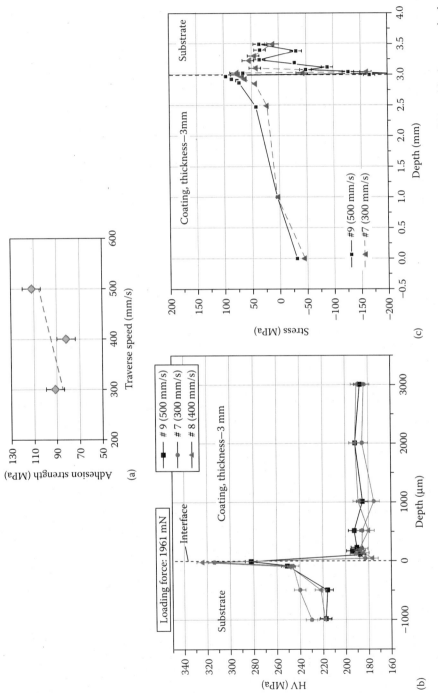

FIGURE 4.102 (a) Cu coating adhesion strength dependence on NTS; (b) Cu coating–substrate microhardness distribution; and (c) residual stresses in the Cu coating and steel substrate.

results of Huang et al. (2014). The carbon steel is softer than 316L SS that results in obtaining of the high adhesion strength at lower particle velocity than that of 316L. These experimental results indicate that the process of substrate surface deformation plays one of the main roles in obtaining coatings with high adhesion strength due to development of interlocking effect.

The authors' results shown in Figure 4.102 reveal the approximate linear dependence of the adhesive strength on the substrate surface roughness. The severe deformation of the substrate micropeaks seems to be the main process leading to the development of the interlocking effect and achieving high adhesive strength. Additionally, the NTS seems to be an effective parameter for CS process optimization. The examination results of the adhesion strength dependence on NTS are shown in Figure 4.102a. It is seen that the increase of NTS results in increase of adhesion strength by 25%–30%. This effect greatly depends on powder feeding rate which controls the particles concentration in the powder laden jet.

Impact deformation of the Cu particles and steel substrate changes the microhardness of both the substrate and the formed coating. The microhardness measurements (Figure 4.102b) show that the carbon steel substrate surface is hardened up to HV 330 MPa as compared with HV 220 MPa for annealed state. The microhardness of Cu coating was relatively uniform and low, about HV 190 MPa as compared to the interface area (about HV 330 MPa) in all measurement points of the coating that implies the presence of a softening processes in the coating during the Cu particle impact and following cooling. The microhardness and residual stress measurement results (Figure 4.102b and c) reveal that NTS barely influence on these coating characteristics. The maximum residual tension stress in the Cu coating near the interface is about 50–80 MPa.

5 Nondestructive Evaluation of Cold Spray Coatings

Roman Gr. Maev and Sergey Titov

CONTENTS

5.1 ULTRASONIC NONDESTRUCTIVE EVALUATION

Cold spray (CS) is a process whereby metallic particles are consolidated to form a coating by means of ballistic impingement upon a suitable substrate. The particles typically ranging in size from 5 to 100 μm are accelerated to high velocity by injection into a stream of high-pressure gas that is subsequently expanded to supersonic velocity through a nozzle. After exiting the nozzle, the particles are impacted onto a substrate, where the solid particles deform and create a bond with the substrate. As the process continues, particles continue to impact and form bonds with the previously consolidated material. The properly established process produces a uniform deposit with a very little porosity and high bond strength. However, because of the complexity of the particle consolidation phenomena and difficulties for controlling all parameters of the spray process various defects may arise in the coatings. Thus, nondestructive evaluation of the coatings and monitoring of the CS process are important problems. This chapter is devoted to application of ultrasonic testing and characterization of the coating deposited by CS technology and ultrasonic in-line monitoring in pulse–echo and passive modes.

5.1.1 METHODS OF ULTRASONIC EVALUATION OF COLD SPRAY COATINGS

Many types of defects inside the coating such as cracks, voids, inclusions, porosity, and delaminations at the coating/substrate interface can be detected by conventional ultrasonic evaluation methods. Detailed description of these methods and their applications can be found in many references including handbooks and monographs (Birks et al., 1991; Cheeke, 2002). However, evaluation of CS coatings has several features, which will be briefly discussed here.

The basic idea of the commonly used pulse–echo method is illustrated in Figure 5.1. The ultrasonic transducer generates the pulse ultrasonic wave that propagates through the layer of immersion liquid and partially passes the upper surface of the coating. If there is a defect inside the coating in a form of material discontinuity the probing ultrasonic wave is reflected back. The amplitude and delay of the received echo depend on the size and depth of the defect. To test a substantial area

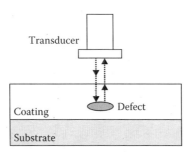

FIGURE 5.1 Detection of defect by pulse–echo method.

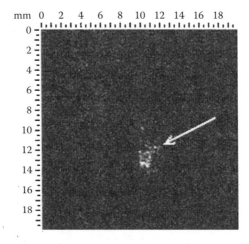

FIGURE 5.2 Acoustic image of inclusions in copper CS coating.

of the coating and construct an acoustic image of the interior of the sample, the mechanical scanning in the transverse plane can be used. Alternatively multielement, array ultrasonic probe can be employed for image formation by means of the electronic scanning of the sample.

Figure 5.2 shows the image defects in the CS copper coating obtained with the scanning acoustic microscope. A focused immersion transducer with a central frequency of 25 MHz was used in this experiment to achieve sufficient spatial resolution. The microstructural analysis of the defect observed in this sample revealed that these ultrasonic reflections are caused by the cluster of inclusions with a diameter of about 50 μm. The overall size of the cluster is approximately 2×1 mm; this estimation is in agreement with the acoustic image. Thus, this example can be considered as a demonstration of high sensitivity and resolution of the ultrasonic pulse–echo method.

5.1.2 SIGNAL ANALYSIS

Besides the defect echo, a small noise uniformly distributed over the overall field of view is presented in the image (Figure 5.2). This noise is caused by scattering the ultrasonic waves on the granular structure of the copper coating. The noise of this kind is often observed in acoustical signals in textured polycrystalline materials (Yang et al., 2007). The granular noise in CS coating is easily distinguished in the data shown in Figure 5.3. The $s(x,t)$ data (Figure 5.3a) is a gray scale image of the received by ultrasonic signal recorded as a function of the position of the transducer x and time t. Bright pixels in this image correspond to a high level of the signal, dark pixels correspond to negative values, and zero level of a signal looks like gray background. For clarity, the waveform obtained at $x = 10$ mm is shown in a separate graph (Figure 5.3b).

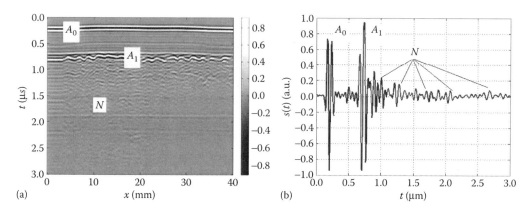

FIGURE 5.3 Ultrasonic echoes measured from substrate side: $s(x,t)$ data (a) and waveform $s(t)$ (b).

The data presented in Figure 5.3 were recorded from the substrate side of the sample perpendicular to the spray passes. The reflections of the ultrasonic probing pulse at the immersion liquid (water)—substrate and substrate—coating interfaces produce the echoes A_0 and A_1, respectively. Inside the coating the ultrasonic wave is scattering on the discontinuities generating multiple noise-like responses N. The intensity and spatial distribution of this granular noise N can be used for assessment of the various parameters of the coating structure. On the other hand, this noise concealing echoes from the internal discontinuities reduces the sensitivity of ultrasonic testing methods in detection of the defects.

The amplitude and delay of the front echo A_0 reflected from the smooth flat surface of the substrate are slowly varying functions of the spatial coordinate x (Figure 5.3a). On the contrary, the echo A_1 changes significantly over x. The approximate period of the delay variation coincides with the distance between the spray passes of 2 mm. Thus, nonuniformities at the interface caused by the particle impact are strong enough to be detected by means of ultrasonic waves. For ultrasonic characterization of the substrate–coating interface, it was proposed to use statistical parameters of the amplitude $A_1(x)$ recorded with a flat ultrasonic transducer at the normal angle of incidence (Maev et al., 2013). To reduce influence of the acoustical contact variations and possible curvature and roughness of the substrate, the normalized coefficient is considered:

$$R = A_1 / A_0 \tag{5.1}$$

The theoretically predicted value of this coefficient can be calculated as follows:

$$R_T = \frac{A_1}{A_0} = \frac{T_{01} T_{10} R_{12}}{R_{01}} \tag{5.2}$$

where T_{01} and T_{10} are the transmission coefficients at the water/copper interface for the probing g and reflected waves, R_{01} and R_{12}, respectively, are the reflection coefficients at the water/copper and copper/steel interfaces, respectively. The reflection and transmission coefficients can be obtained using known relationships (Brekhovskikh and Godin, 1990):

$$T_{01}T_{10} = \frac{4Z_W Z_C}{\left(Z_W + Z_C\right)^2}$$

$$R_{01} = \frac{Z_C - Z_W}{Z_C + Z_W} \tag{5.3}$$

$$R_{12} = \frac{Z_S - Z_C}{Z_S + Z_C}$$

TABLE 5.1

Acoustic Impedances of Materials

Material	Acoustic Impedance Z, 10^6 kg/m²s
Water	1.485
Copper	43.6
Steel	46.02

where Z_W, Z_C, and Z_S are acoustic impedances of water, copper, and steel, respectively. Using (Birks et al., 1991) handbook values of the acoustic impedances of the materials (Table 5.1) the theoretically predicted coefficient was found to be $R_T = 0.0072$.

To establish relationship between the reflection coefficient and properties of the copper/steel interface, samples with a variety of interfacial conditions were produced. The results of the statistical analysis of the measured normalized reflection coefficient R are presented in Figure 5.4 for two samples. One sample has a normal temperature of the substrate (solid curve), whereas the substrate of another sample was preheated up to 500°C (dashed line). It was shown that the probability distribution function of R is close to normal. The deviation of R is much larger than the error of measurement, and the average value of R is larger than its theoretically predicted value R_T. These facts can be linked with the nonuniformity of the interface. The averaged value of R demonstrates correlation with the condition of spraying such as temperature of the substrate and the adhesive strength of the coating. Thus, the experimental results show that the reflection coefficient depends on the parameters of the coating deposition process and it may be used for quantitative characterization of the interface parameters, such as adhesion strength.

It was established that mechanical properties of various materials including CS coatings influence on velocity and attenuation of ultrasonic waves in these media (Papadakis, 1976; Lubrik et al., 2008). Thus, evaluation of the coating structure can be performed by measurement of sound velocity using standard pulse–echo technique. In the experiment, the CS coating samples were cut and their surfaces were polished to provide the sufficient accuracy of measurement. Results obtained for copper coatings prepared at different conditions are presented in Table 5.2. Measured with a 15 MHz, 4 mm transducer, the sound velocity in high pressure sprayed copper is similar to the velocity in wrought copper. However, the attempt to measure sound velocity in the low-pressure

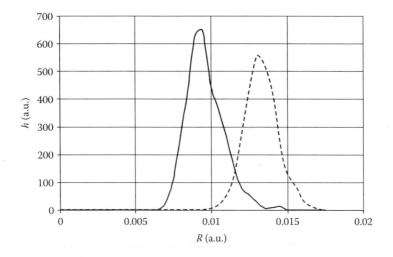

FIGURE 5.4 Histograms of the coefficient R.

TABLE 5.2

Velocity of Longitudinal Waves C_L in Copper Layer

Wrought Copper	High-Pressure CS	Low-Pressure CS	Low-Pressure CS, Heat Treated Sample
4700 m/s	≈4700 m/s	Impossible to measure due to high attenuation	≈3300 m/s

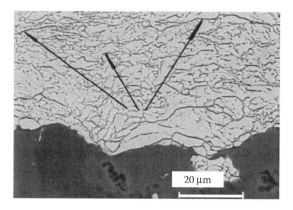

FIGURE 5.5 Scanning electron microscope image of low pressure CS copper coating.

cold sprayed sample was unsuccessful even using 3.5 and 7.5 MHz transducers. Only the front surface echo is present in the recorded waveforms, the back wall echo is absent due to high attenuation of ultrasonic waves. Meanwhile, the sound velocity of the heat-treated low-pressure CS sample is about ≈3300 m/s. In this case, ultrasonic attenuation is much lower and it is possible to measure the sound velocity in the copper coating.

The results of the microstructural analysis show that the structure of the high-pressure samples is uniform, while there are numerous micro cracks inside the low-pressure coating (Figure 5.5). The presence of cracks can explain the elevated ultrasonic attenuation in the low-pressure samples. Annealing of the low-pressure coating results in improved attenuation but still not equal to the wrought copper or high-pressure copper coating sound velocities, presumably due to reduced interparticle bond strength.

The obtained ultrasonic material characterization results of the copper coatings show that the velocity and attenuation coefficient of ultrasonic waves in the sprayed material can be a valuable indication of the coating inspectability and quality.

5.2 ULTRASONIC IN-LINE MONITORING SYSTEM

5.2.1 MEASUREMENT SCHEMATICS AND BASIC RELATIONSHIPS

The scheme of the ultrasonic in-line CS monitoring system is presented in Figures 5.6 and 5.7 (Maev et al., 2011, 2012a, b). This system is based on the portable apparatus equipped with a low-pressure CS gun previously described in this book (Maev and Leshchynsky, 2007). The spraying nozzle is translated along x-axis with velocity V over the top surface of the substrate creating a layer of deposited material. To characterize ultrasonically the CS process, the ultrasonic probe is attached to the opposite surface of the substrate. In the pulse–echo mode, the probe sends ultrasonic waves through a coupling media and the substrate, to the area where the spray process takes place and receives the reflected waves. Immersion liquid, such as water or hard delay line wetted with ultrasonic gel, can be used as a coupling media. In the passive mode, the probe receives the ultrasonic waves generated by

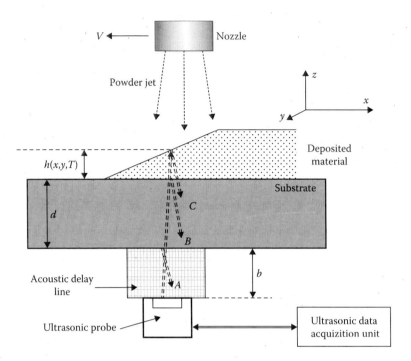

FIGURE 5.6 Scheme of the inline ultrasonic monitoring system.

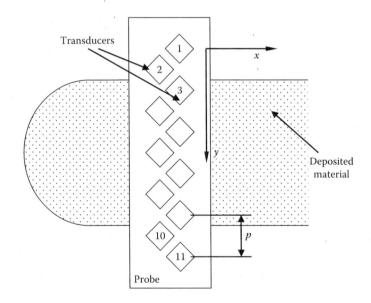

FIGURE 5.7 Arrangement of the ultrasonic transducers. Bottom view.

the particle impacts. An appropriate instrumentation is used for acquisition of the ultrasonic signals, data processing, and representation in coordination with the spray gun operation.

In the pulse–echo mode, the ultrasonic probing wave generated by the probe is reflected from the acoustical discontinuities of the object. The reflected ultrasonic signal received by the same transducer consists of several typical responses. The first pulse A is produced by the reflection of the ultrasonic wave at the interface between delay line and substrate (Figure 5.6). The propagation time of this pulse is

$$t_A = \frac{2b}{C_0} \tag{5.4}$$

where C_0 is the velocity of ultrasound wave in the acoustic delay line and b is its thickness. The pulse B is reflected from the top surface of the substrate. The time delay of this pulse relative to the pulse A is

$$t_B = \frac{2d}{C_S} \tag{5.5}$$

where d and C_S are the substrate thickness and its wave velocity, respectively. If the coating is being sprayed on top of the substrate, the ultrasound wave partially penetrates into the deposited material and the wave being reflected from the top surface produces pulse C. The time delay τ of the pulse C relative to pulse B depends on the thickness of the coating at a particular time of the process and position of the transducer:

$$\tau = \frac{2h(x,y,T)}{C_m} \tag{5.6}$$

where $h(x,y,T)$ is the thickness of the deposited material expressed as a function of time T, spatial longitudinal coordinate x, and transverse coordinate y (Figure 5.6). Thus, it is possible to calculate the profile function of the deposition process $h(x,y,T)$ knowing the measured delay τ, and the sound velocity in the deposited material C_m. Furthermore, by processing these data, it is possible to estimate the deposition rate of the process $\alpha(x,y)$, which can be defined as the speed of the thickness increase at a particular point if the position of the nozzle is unchangeable.

If the nozzle is moving along x-axis with velocity V and passes over the observation point (x,y), the thickness of the coating is simply an integral of the deposition rate:

$$h(x,y,T) = \int_{-\infty}^{T} \alpha(x - V \cdot T, y) dT \tag{5.7}$$

Therefore, using this equation, the deposition rate of the process $\alpha(x,y)$ can be estimated by differentiating of the measured thickness h over time T:

$$\alpha(x - V \cdot T, y) = \frac{dh(x,y,T)}{dT} = \frac{C_m}{2} \frac{d\tau}{dT} \tag{5.8}$$

To achieve sufficient lateral resolution in the transverse direction (along y-axis), it is convenient to have an array probe, whose elements are positioned perpendicular to the movement of the nozzle. In the experiment, an array probe consisting of 11 square elements was used. The elements are located in a zigzag manner as shown in Figure 5.7. The size of the element is 1.2 mm, the pitch of the array in one row is $p = 1.768$ mm. The second row of elements reduces the effective pitch of the array to the value of $p/2 = 0.884$ mm. Because of the shift of the elements 2, 4, 6, 8, and 10 along x-axis relative to the first row at the distance $\Delta x = p/2$, the signals recorded by these elements have an additional time delay $\Delta T = p/(2V) = 0.177$s. In the experiments, the velocity of the nozzle movement was $V = 5$ mm/s, therefore, $\Delta T = 0.177$s. This time delay is compensated numerically at the signal processing stage. The central frequency and relative bandwidth of the elements are 15 MHz and 60%, respectively. For data acquisition, the multichannel ultrasonic pulse–echo flaw detector was used (tessonics.com). All channels record data 40 times per second (25 ms period).

5.2.2   PROCESSING AND ANALYSIS OF THE ULTRASONIC DATA IN PULSE–ECHO MODE

The ultrasonic echoes received by the elements of the array probe are recorded by the monitoring system as a function of the propagation time of the ultrasonic wave t (*fast time*) and the time of the process T (*slow time*). As an example, the data $s(t,T)$ received by the element #6 (Figure 5.7) are presented in Figure 5.8 as a gray scale image (so called B-scan). Bright pixels correspond to high level of the signal, dark pixels correspond to negative values, and zero level of a signal looks like gray background. For clarity, two waveforms obtained at $T = 4$ s and $T = 7$ s are shown in a separate graph (Figure 5.9). In this experiment, the composition (Al + 25% Al_2O_3) with an average particle

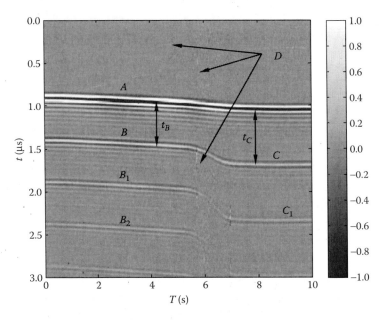

FIGURE 5.8   Grayscale representation of the $s(t,T)$ data.

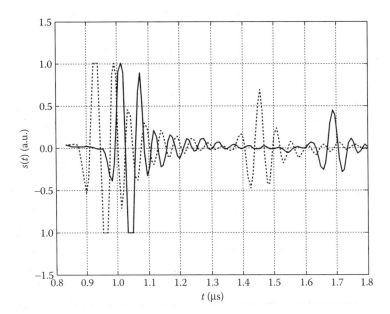

FIGURE 5.9   Waveforms $s(t)$ measured at the time $T = 4$ s (dash line) and $T = 7$ s (solid line).

size of 45 μm was deposited on an aluminum substrate at a speed of the gas flow of 500 m/s and at temperature of the gas of 300°C.

When the nozzle is approaching the area where the probe is attached, the temperature of the substrate and delay line is increased. The sound velocity C_0 in the material of the delay line decreases with increasing temperature. As it follows from (5.4), these causes increase in the time delay t_A, and gradual shifting of the pulse A as seen in Figure 5.8. The pulse B is reflected at the upper surface of the substrate and the position of this pulse also changes with the temperature rise. However, for substrate materials such as steel and aluminum, the temperature coefficient of the ultrasound velocity C_s is much smaller than that of the polymer delay line (Maev, 2008). Thus, the time delay t_B of pulse B relative to pulse A remains practically constant. After pulses B and C, several smaller pulses B_1, B_2, ..., and C_1 are visible (Figure 5.8). These pulses are produced by multiple reflections inside the substrate or coating and are neglected in the future analysis within this work.

In a range of process time $T = 5.5–7$ s, the nozzle passes over the monitoring transducer. Because of the deposition of the coating material, echo C reflected from the top surface of the coating appears. The time delay of this echo t_C grows together with the thickness of the coating. The amplitude of pulse B reflected at substrate/coating interface is small and the interface is not visible in the pictures after the deposition (Figure 5.8). This effect can be explained by the fact that the acoustic impedances of the aluminum substrate and deposited material (Al + 25% Al_2O_3) are very close. Besides of that, the adhesion strength of the coating was measured using the adhesion tests described in detail in monograph (Maev and Leshchynsky, 2007). The measured value of about 40–60 MPa is an evidence of strong bonding and good acoustical contact between the substrate and deposited material. Under these conditions, the reflection coefficient at the interface is small, and the pulse B is not presented in the $s(t,T)$ data after the nozzle passed over the transducer.

Figure 5.10 shows the ultrasonic data recorded for the sample with weak bonding. For this sample, the velocity of the gas flow was reduced that results in low adhesive strength of 15 MPa. In this case the acoustical contact between substrate and coating is not perfect. In the $s(t,T)$ in addition to the pulse C there is the pulse B_m that is produced by the wave reflection at the interface. Thus, the degree of the ultrasonic transmission at the substrate/coating interface can be considered as an indication of the coating bonding.

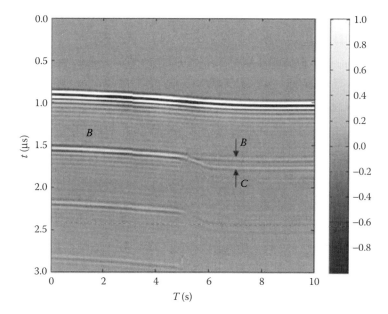

FIGURE 5.10   The $s(t,T)$ data recorded for sample with weak bonding.

5.2.3   ESTIMATION OF THE DEPOSITION RATE

To estimate the deposition rate of the process $\alpha(x,y)$, time delay τ can be found by subtraction of the time delay of the B pulse measured before deposition from the delay of the pulse C relative to the pulse A:

$$\tau = t_C - t_B \qquad\qquad (5.9)$$

The measured delays τ are shown in Figure 5.11 for some elements of the array probe as functions of the position of the nozzle $x = V \cdot T$.

    To determine the thickness of the coating and the deposition rate the sound velocity C_m should be known. Determination of sound velocity in the deposited material was made with separate tests conducted under normal temperature (Lubrik et al., 2008). However, the impacting particles cause the temperature rise in the area of the deposition and changes in the sound velocity as well. Publications devoted to the measurement of the temperature coefficients of the sound velocity in polycrystalline metals and alloys are very rare. Besides that, the velocity and its temperature coefficient depend on the chemical composition and structure of the material. To determine the effect of a temperature change on the coating thickness measurement results, the temperature coefficient of the velocity in the deposited material was estimated experimentally. The velocity was measured in a standard pulse–echo setup as a function of temperature of the sample in a range of 20°C–80°C. The sample was deposited on the substrate using the same parameters of the deposition process. The measured velocity was found to be $C_m = 4500 \pm 60$ m/s at 20°C, and the temperature coefficient was estimated to be in a range of $\xi_m \approx -0.8, \ldots, -1.4$ m/(s·°C). Thus, if the temperature of the samples during the deposition process does not exceed $T^* = 100$°C, decrease of the velocity can be estimated as $\delta C_m = \xi_m (T^* - 20°C) \approx -110$ m/s. Thus, the relative error of the thickness measurement caused by neglecting temperature dependence of the velocity is less than 2.5%.

    The two-dimensional distribution of the deposition rate $\alpha(x,y)$ calculated using the measured delays τ are presented in Figure 5.12 in a form of a waterfall graph. The function $\alpha(x,y)$ has a prominent maximum located close to the center of the plot. This observation confirms the fact that the deposition of the material is efficient in the center part of the powder jet due to higher particle

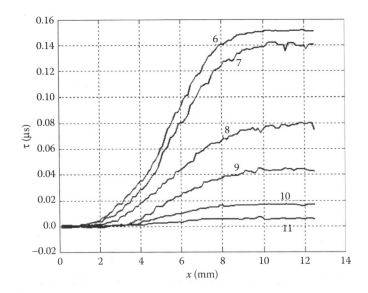

FIGURE 5.11   Delays τ recorded by elements 6–11.

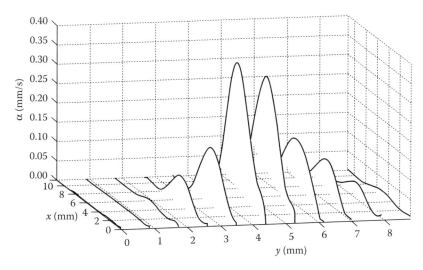

FIGURE 5.12 Spatial distribution of the deposition rate $\alpha(x,y)$.

velocity and concentration. On the other hand, the ultrasonically estimated deposition rate $\alpha(x,y)$ is not symmetrical. The deposition rate is higher for the points located at the right side of the powder spot (see Figure 5.6). It appears that a coating build-up with the left side portion of powder jet changes the angle between the substrate surface and nozzle axis in the x–z plane, which results in the increase of α. Therefore, it is possible to optimize the conditions of coating formation by controlling the nozzle axis angle in the x–z plane.

To validate the proposed method, a cross section of the build-up has been made and the thickness profile $h(y)$ has been measured using an optical microscope (Figure 5.13). The cross section has been made outside of the transient area where the thickness is uniform along x-axis. The deposition rate α and the profile $h(y)$ having a triangle-like shape demonstrate similar nonsymmetrical behavior. The thickness of the coating $h(y)$ estimated by the ultrasonic monitoring system is in a good agreement with the results of the destructive test.

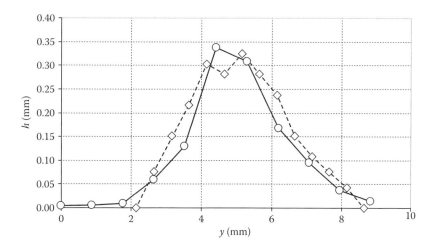

FIGURE 5.13 Coating thickness distribution $h(y)$: diamonds—destructive test, circles—ultrasonic measurement.

5.2.4 Particle Impact Signals

In addition to the typical ultrasonic pulses A, B, C (Figure 5.8), specific noise-like signals D are present in the $s(t,T)$ data during the process time $T = 5.5–7$ s, when the nozzle passes over the transducer the thickness of the coating grows. It was suggested that this noise is generated by particle–substrate impacts (Maev et al., 2012a, b). To prove this statement several experimental observations have been carried out. First of all, this noise is independent from the pulse echoes and it exists before the first pulse A arrives and after all small reverberations in the substrate and coating. Moreover, this noise does not change if the pulse generator of the data acquisition unit is switched off, and the ultrasonic system is in a passive, receiving mode. If the powder feeding is shut down while the deposition apparatus is still working and air jet impinging on the substrate, the received signals are equal to the intrinsic noise of the ultrasonic unit. If the powder feeding is on, the noise level increases to more than 40 dB. Therefore, based on these experimental data it was assumed that the observed noise is solely produced by the particle–substrate impacts.

The described ultrasonic monitoring system can alternatively work in the pulse–echo and passive modes. The typical duration of the active phase is less than 20–30 μs, the rest of the time within the pulse repetition period (100–1000 μs) can be used for receiving of the particle impact signal. Because the process time of CS is substantially larger than the repetition period, the system working in both modes can be considered as simultaneous and continuous.

It was observed that the ultrasonic waves generated by the particle impacts are very wideband. Several contact single ultrasonic transducers in receiving mode were used to establish the frequency range of the particle impact noise. The central frequencies of the transducers were in a range of 1–20 MHz. In the experiment, the powder composition was 75% Al + 25% Al_2O_3, the feeding rate and the temperature of the powder jet were 1.5 g/s and 300°C, respectively. The nozzle was translated with a velocity $V = 5$ mm/s, and the aluminum plate with a thickness $d = 12.8$ mm was used as a substrate.

Using all these wideband transducers the particle impact signals with a high signal-to-noise ratio (>40 dB) were recorded. Figure 5.14a and b shows waveforms received by the 20 and 2.25 MHz transducers, respectively. The frequency content of these signals is entirely determined by the spectral responses of the transducers. It means that the spectrum of the observed signals is wider than the spectral response of any transducer used in the measurements. Thus, the particle impact signals occupy at least a frequency range of 1–20 MHz, and most likely their spectrum spreads well beyond this range.

As it follows from Figure 5.14, it is possible to consider the particle impact signal as a sum of two components: stationary noise for which the parameters do not change over time and superimposed strong pulses with random time positions and amplitudes. To investigate the nature of this phenomenon, the source of the particle impact noise was spatially restricted and the signals were recorded at various positions of the source with respect to the transducer. To localize this source, a steel foil with a hole was placed on the top surface of the substrate (Figure 5.15). There was no acoustic contact between the foil and the substrate; therefore, the ultrasonic waves could only be generated by the particles that hit the substrate through the hole. In the experiment, the 10 MHz contact wideband transducer was attached to the 12.8 mm aluminum substrate. The thickness of the foil and the diameter of the hole were 0.3 and 3 mm, respectively. The recorded waveforms of the CS process (100% Al powder, temperature 20°C) at positions of the hole $\Delta x = 0$ and 16 mm are shown in Figure 5.16.

When the distance between the transducer and the hole is large ($\Delta x = 16$ mm) the received particle impact signals $s(t)$ look like wideband noise for which the power depends very little on the distance Δx (Figure 5.16b). When the hole is located above the transducer ($\Delta x = 0$ mm) sparse strong pulses are observed in addition to the background noise (Figure 5.16a).

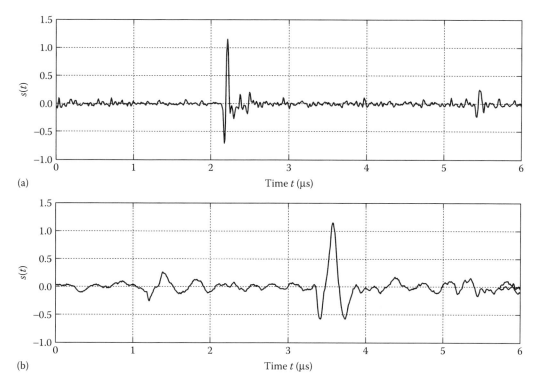

FIGURE 5.14   Waveforms received by 20 MHz (a) and 2.25 MHz (b) transducers.

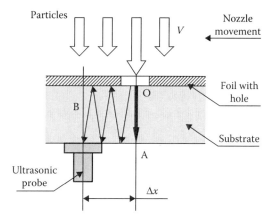

FIGURE 5.15   The experimental setup with localized source of the impact waves.

This specific behavior of the signals can be explained as follows. The particle impact generates a combination of longitudinal, shear, and surface waves (Hutchings, 1979), whose source is located at the point of the incidence. The waves travel in the plate numerous times reflecting at the boundaries. Ray OB is shown in the Figure 5.15 as an example. Because the particles hit the substrate often enough and the reverberations last for a long time, this produces noise-like waves for which the intensity decays very slowly with the distance from the source. The strong pulses can be generated by the bulk, mostly longitudinal waves, which fall at the rear boundary of the

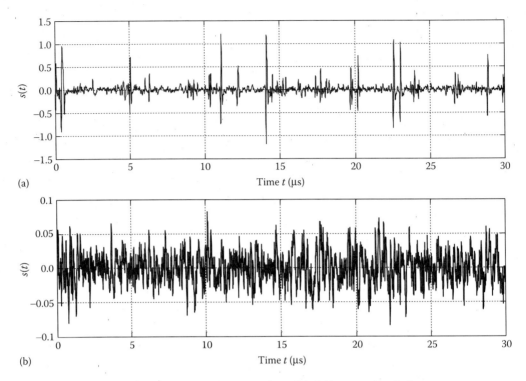

FIGURE 5.16 (a, b) Waveforms $s(t)$ at displacements $\Delta x = 0$ and 16 mm, respectively.

plate at the normal angle (ray OA). The amplitude of these pulses is strong if the source of the wave is above the transducer.

5.2.5 Statistical Analysis of the Particle Impact Signals

To quantitatively evaluate statistical properties of the particle impact signals, the probability distribution g of absolute value of the normalized signal $A_m = |s|$ was assessed for various distances Δx. The distribution g was calculated by counting the number of observations that fall into each of 40 intervals of the variable A_m. The MATLAB® function *histc* was used in the calculation of g. Because the probability distribution g varies in a very large range, the logarithm $w = \log_{10}(g)$ was taken to present the data (Figure 5.17). The probability distributions of the amplitude are almost identical for all distances Δx if the signal value is small $A_m < 0.1$. On the contrary, large amplitudes A_m are only typical for the transducer–hole distance $\Delta x \approx 0$.

To separate the background reverberation noise s_w and primary impact pulses s_h, it is reasonable to use threshold processing. Let ρ be a threshold that divides the particle impact signal $s(t)$ into the weak and strong components:

$$s_w(t) = \begin{cases} |s(t)|, \text{if } |s(t)| < \rho \\ 0, \text{otherwise} \end{cases}$$

$$s_h(t) = \begin{cases} 0, \text{if } |s(t)| < \rho \\ |s(t)| - \rho, \text{otherwise} \end{cases}$$

(5.10)

Then the root mean square (RMS) values of the components can be calculated as follows:

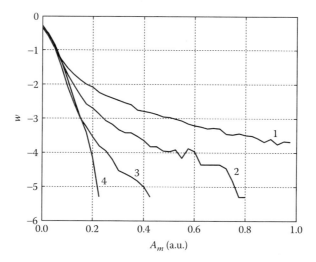

FIGURE 5.17   Logarithm of probability distribution w of amplitude A_m at $\Delta x = 0, 4, 8, 12$ mm, (curves 1,2,3,4, respectively).

$$r_w(\rho) = \left[\frac{1}{t_w} \int_0^{t_0} s_w^2(t)\,dt \right]^{1/2}$$

$$r_h(\rho) = \left[\frac{1}{t_h} \int_0^{t_0} s_h^2(t)\,dt \right]^{1/2} \qquad (5.11)$$

where t_0 is the overall duration of the time window, t_w is the total time when $|s(t)| < \rho$, and t_h is the total time when $|s(t)| \geq \rho$; $t_0 = t_w + t_h$. The squared values of r_w, r_h can be considered as average power of the components.

The RMS r_w and r_h calculated for the various transducer–hole distances Δx and for the threshold $\rho = 0.1$ are presented in Figure 5.18. The RMS of the reverberating noise r_w remains almost constant for all distances Δx. In contrast, the RMS of the primary impact pulses r_h is maximal when the hole

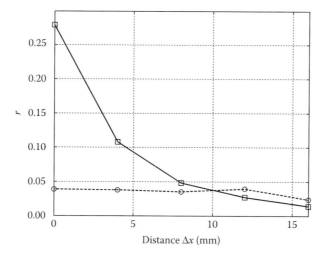

FIGURE 5.18   RMS r_w (dashed line) and r_h (solid line) versus distance Δx ($\rho = 0.10$).

is located above the receiving transducer and it decreases with increasing of Δx. Thus, it is possible to some extent to extract the primary impact responses using an appropriate threshold processing.

The developed processing procedure was applied to analysis of the particle impact signals obtained with the multichannel ultrasonic CS monitoring system (Figures 5.6 and 5.7). In this experiment, the powder composition was 75% Al + 25% Al$_2$O$_3$, and the thickness of the aluminum substrate was $d = 1.5$ mm. The ultrasonic system recorded the pulse–echo multichannel data and the particle impact signals as well. The impact responses were recorded by all 11 elements of the array transducer within the time window $t_0 = 3$ μs. This time window was located on the *fast* time axis t when the pulse–echo signals are absent. Figure 5.19 shows the RMS of these gated $s(t,T)$ data received by seventh element of the array as a function of the nozzle position $x = VT$. The position of the transducer was at $x = 22$ mm (Figure 5.7). Therefore the primary impact pulses are observed in the range approximately $x = 17$–27 mm when the nozzle passes over the transducer.

The RMS of the primary impact pulses $r_h(x)$ was calculated for all array channels, positions x, and various thresholds ρ. It is useful to consider the cumulative output parameter

$$e(x) = \int_0^x r_h(x)\,dx \qquad (5.12)$$

The squared e can be treated as a value that is proportional to the energy of the received ultrasonic particle impact waves. The parameter e calculated for the thresholds ρ = 0.4 is shown in Figure 5.20. This parameter grows as the nozzle passes over the point where the monitoring transducer is attached. It is possible to assume that at properly determined threshold ρ the parameter $e(x)$ is proportional to the thickness of the coating and the RMS $r_h(x)$ is determined by the intensity and spatial distribution of the powder deposition rate.

The parameters e taken at $x = 35$ mm are presented for all channels as a function of the transverse coordinate $y = (i-1)p/2$, where i is the channel number (Figure 5.21). At the position $x = 35$ mm the nozzle is sufficiently far from the measurement point such that further increasing of the parameters e and thickness of the coating should not be expected. The presented function $e(y)$ is normalized on it maximal value. For comparison the normalized thickness of the coating $h_n(y) = h(y)/h_{max}$ directly measured after sectioning of the sample is also presented in Figure 5.21. The results reveal that the appropriate choice of the threshold ρ gives an accurate determination of the CS coating thickness. It is important that definition of the threshold ρ may be made on the base of known width of coating pass that is the function of a powder jet spot. The second feature of application of the passive monitoring method to the coating geometry determination is that the results estimated using the particle impact ultrasonic signals are independent from coating density and speed of sound.

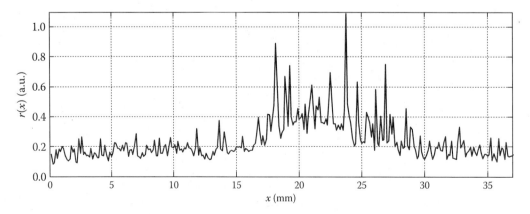

FIGURE 5.19   RMS of the particle impact signals as a function of the nozzle position x (receiving element # 7).

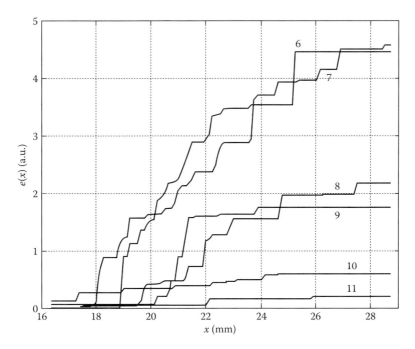

FIGURE 5.20 The cumulative parameter $e(x)$ of the particle impact signals recorded by elements 6, 7, 8, 9, 10, and 11.

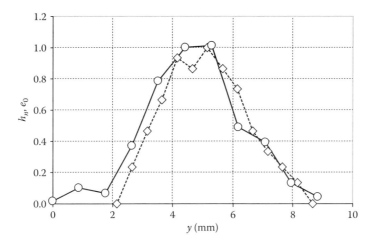

FIGURE 5.21 Normalized thickness distribution $h_n(y)$ of coating measured optically after sample sectioning (diamonds) and normalized parameter $e(y)$ (circles).

5.2.6 CONCLUDING REMARKS

It was shown that the presented multichannel ultrasonic system is a useful tool for the investigation of the particle deposition mechanism and the monitoring of the CS process in real time. In active, pulse–echo mode is possible to observe how the build-up takes place, estimate the spatial distribution of the deposition rate, and measure the profile of the coating. Simultaneously, the system records in passive mode the ultrasonic waves generated by the particle impacts during the CS. It was experimentally established that the particle impact signals occupy a frequency range of 1–20 MHz. The statistical parameters of these signals, their behavior and spatial distribution are related to characteristics of the CS process.

6 Application of Cold Spray

*Victor K. Champagne, Jr., Victor K.
Champagne, III, Dmitry Dzhurinskiy, and Emil Strumban*

CONTENTS

6.1 AIRCRAFT COMPONENTS REPAIR

Victor K. Champagne, Jr

Victor K. Champagne, III

6.1.1 INTRODUCTION

6.1.1.1 Technology Transfer

The aerospace industry is conservative in regard to adopting new technologies, and rightfully so. If a new technology has not been adequately tested and evaluated for a specific application, the consequences could be catastrophic, resulting in the loss of human lives and expensive aircraft. Traditionally, this process takes more than 20 years because of the vast amount of testing and verification required. A significant investment must be made by numerous entities involved including research, engineering, manufacturing, quality control, inspection, logistics, and acquisition. The process involves extensive testing and evaluation of laboratory samples, as well as sub-scale or full-scale components and/or assemblies, some of which require flight testing for flight safety critical components, and progresses to feasibility assessment, and finally demonstration and validation, before it can be applied to production and/or field use. The intent of this chapter is to present the methodology and accompanying technical data for the transition of cold spray (CS) technology into the aerospace industry by the US Army Research Laboratory (ARL), and present several case studies showing the tremendous impact that CS has made for a few select applications, representing three aerospace materials—magnesium, aluminum, and titanium—while attempting to include as much technical data substantiating the advantages and benefits gained by transition of the process. The focus application will involve the restoration of magnesium aerospace components where much data will be presented while the remaining case studies cannot be so inclusive due to page restrictions.

Since the 1980s, the National Aeronautics and Space Administration has used a concept referred to as *Technology Readiness Level or TRL* as a means to assess the maturity of a particular technology and developed a scale to compare technologies. Figure 6.1 summarizes the various stages of the TRL (https://www.nasa.gov/content/technology-readiness-level/). In 1999, the Department of Defense (DoD) adopted a similar TRL concept to assess emerging technologies and incorporated it into their programs to have a formalized methodology for technology transition. The DoD, Technology Readiness Assessment (TRA) Guidance is a document prepared by the Assistant Secretary of Defense for Research and Engineering and is "a systematic, metrics-based process that assesses the maturity of, and the risk associated with, critical technologies to be used in Major Defense Acquisition Programs (Department of Defense, TRA Guidance, 2011).

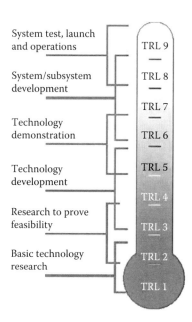

System test, launch
and operations — TRL 9

System/subsystem
development — TRL 8

— TRL 7

Technology
demonstration — TRL 6

Technology
development — TRL 5

TRL 4

Research to prove
feasibility — TRL 3

Basic technology
research — TRL 2

TRL 1

FIGURE 6.1 Technology Readiness Level. (Courtesy of Department of Defense, TRA Guidance, 2011.)

The TRL scale is a measure of maturity of an individual technology for purposes of insertion into operating systems and platforms (Sauser et al., 2006). The aerospace industry utilizes the TRL as a standard and as a basis of comparison to fully assess the level of maturity of a specific technology.

To quantify manufacturing risk in acquisition while increasing the ability to transition new technology, Manufacturing Readiness Levels (MRLs) and assessments of manufacturing readiness have been developed, which correlate to the TRL (Figures 6.1 and 6.2). MRL definitions were developed by a joint DoD/industry working group under the sponsorship of the Joint Defense Manufacturing Technology Panel. The MRL serves as a quantitative means to assess the maturity of a specific technology, component, or system from a manufacturing perspective. The MRL is performed to define the current level of manufacturing maturity, identify the risks and costs associated with maturity shortfalls, and to establish a basis for manufacturing maturation and risk management (Manufacturing Readiness Level (MRL) Deskbook, 2011).

There are 10 MRLs that are correlated to the 9 TRLs in use today throughout the DoD and industry (Figure 6.2). ARL served as the lead technical organization to establish a series of technical multifaceted programs, whereas both the TRL and MRL methodologies were used to assess and transition CS technology into production and field use by the DoD and private industry for aerospace applications. The entire process was accomplished in only 6 years, and culminated with the establishment of the first dedicated CS repair facility in the DoD, located at the Fleet Readiness Center East (FRC-East), Cherry Point, NC and with the official approval allowing CS to be used for the UH-60 Black Hawk helicopter by the original equipment manufacturer (OEM) and the US Army Aviation and Missile Research Development and Engineering Center. Therefore, a TRL 9 and MRL of 10 were achieved. During this time period, ARL also introduced the first military CS specification, Mil-STD-3021 entitled, *Materials Deposition, Cold Spray*, which serves as a standard for the general process.

6.1.2 Components Wear and Repair Methods

The aerospace industry has experienced significant corrosion and wear problems with certain aerospace materials, especially magnesium alloys that are used to fabricate aircraft components. The most severe of these are associated with large and expensive transmission and gearbox housings

MRL 1: Basic manufacturing implications identified

MRL 2: Manufacturing concepts identified

MRL 3: Manufacturing proof of concept developed

MRL 4: Capability to produce the technology in a laboratory environment

MRL 5: Capability to produce prototype components in a production relevant environment

MRL 6: Capability to produce a prototype system or subsystem in a production relevant environment

MRL 7: Capability to produce systems, subsystems, or components in a production representative environment

MRL 8: Pilot line capability demonstrated; ready to begin low rate initial production

MRL 9: Low rate production demonstrated; capability in place to begin full rate production

MRL 10: Full rate production demonstrated and lean production practices in place

FIGURE 6.2 Manufacturing Readiness Level. (Data from Manufacturing Readiness Level (MRL) Deskbook, 2011. With permission.)

for rotorcraft, which have to be removed prematurely from service because of corrosion and/or wear. Additionally, because these parts are produced from castings, there is a need to repair defects associated with the sand casting process, during the production process, such as open porosity and shrinkage that would otherwise render the parts unusable. Many of the parts cannot be reclaimed because there is not an existing technology that can be used for dimensional restoration. This application represents a major aerospace CS application that will serve as an exemplary example of CS transition and implementation.

6.1.2.1 State of the Art and Methodology

It had been determined through extensive testing and evaluation by several major US aerospace manufacturers and the DoD that the deposition of aluminum and its certain alloys, followed by the phenolic resin and a sealant, would have the highest probability of success in reducing corrosion and impact damage for magnesium aerospace components. The technical challenge was to identify a method to deposit the aluminum alloys onto the magnesium and meet all service requirements for bond strength and corrosion resistance without sacrificing the structural integrity of the substrate. The processing of production parts as well as field repair capability was an important attribute in the selection of a viable process. Initially, a review of the current state of the art was conducted and critiqued before CS was down selected as the technology of choice for this application.

6.1.2.1.1 Plasma and High-Velocity Oxy-Fuel Thermal Spray of Al

Plasma and high-velocity oxy-fuel (HVOF) thermal spray have been investigated for deposition of aluminum, but the results have generally been unsatisfactory due to inconsistent coating integrity. Poor adhesion and delamination of the coatings are typically the cause for high rejection rates. Both of these processes involve the use of high thermal energy to melt or partially melt the coating material before it is accelerated onto the surface of the substrate. The tiny molten or partially melted particles rapidly solidify on impact with the substrate and contract forming tensile residual stresses in plasma spray coatings. This is not always the case with the HVOF process because the particles are accelerated at high velocity and have been known to form coatings that are in compression. Regardless, failures occur because the plasma and HVOF processes can generate excessive heat causing the formation of an oxide on the magnesium that is detrimental to adhesion. Besides, the thermal spray pattern is so wide that it would be difficult to apply the coatings to localized areas requiring repair or rebuild.

6.1.2.2 Risk Assessment

Because of the previous and ongoing work being conducted by the US ARL, it was clear that commercially pure aluminum (CP-Al), high-purity aluminum (HP-Al), and other aluminum alloys such as 6061, 7075, and 5056 of various thicknesses could be deposited onto Mg alloys using the CS process depending on the performance requirements, including strength, toughness, and hardness. These coatings have been shown to be very dense and have high adhesion values. Therefore, the technical risk associated with the actual CS deposition process was very low. The principal technical risk was related to the extent to which the application of the aluminum CS coatings would improve the corrosion performance of the Mg alloys and the effect on fatigue properties of the substrate. It was anticipated that a significant amount of parametric optimization work would have to be performed to optimize the corrosion resistance of the coatings. Other risks were associated with the cost of the process. Higher particle velocities and, in many instances, superior coatings are obtained using helium gas that is very expensive and might be cost prohibitive. An issue was raised as to whether acceptable coatings properties could be achieved using less-expensive nitrogen or whether a recycling system was required with helium. If the desired level of corrosion performance could be achieved using nitrogen or by using a helium-recovery system, then there would be little risk associated with application of the coatings onto actual components and the implementation of the technology into a production environment because of the similarity of the CS process with other thermal spray processes, which are used on components of similar geometry and complexity.

6.1.3 CS Process Development

6.1.3.1 CS Program

The US ARL has led the development of CS technology over the past decade and has introduced a process to reclaim aerospace parts, especially those produced from magnesium that shows significant improvement over existing methods. ARL has demonstrated and validated a CS process using aluminum and/or Al alloys as a cost-effective, environmentally acceptable technology to provide surface protection and a repair/rebuild methodology to a variety of magnesium and aluminum aerospace components for use on Army and Navy helicopters and advanced fixed-wing aircraft, which has been adopted worldwide and is being incorporated not only in the US DoD but in private industry as well.

The development and qualification of the CS process to deposit aluminum and its alloys was originally proposed by the Center for Cold Spray Technology at ARL in 2003 for providing dimensional restoration and protection to magnesium components, primarily targeting the aerospace and automotive industries. Initially, ARL concentrated on developing a *nonstructural* repair process that could be qualified quicker and less costly than a *structural* repair process that would require extensive component testing.

Soon multimillion dollar programs were established and executed by ARL, including the Environmental Security Technology Certification Program (ESTCP) that extended from 2005 to 2011, and culminated with the qualification of the CS process for use on the UH-60 Black Hawk, in collaboration with Sikorsky Aircraft Company (Stratford, USA) and the establishment of the first dedicated CS repair facility at the Navy FRC-East, Cherry Point, North Carolina. This program served as an international benchmark for the adaptation of CS for the aerospace industry. The CS process is now viewed as the best possible method for depositing the aluminum alloys to provide dimensional restoration to magnesium components and significantly improving performance and reducing life-cycle costs.

Subsequently, other military services and the private sector followed suit, and a multitude of efforts were undertaken to develop similar repair procedures. Other nationally recognized programs were the National Center for Manufacturing Sciences whose objective was to exploit the use of CS for corrosion control of magnesium. The project participants included all the branches of the armed services: the Army ARL, Navy, Air Force, and Marines, and the cross-industry companies: the Boeing Company (Seattle, USA), Delphi Corporation (Detroit, USA), Ford Motor Company (Dearborn, USA), CenterLine (Windsor) Ltd (Windsor, Canada), and Solidica (Ann Arbor, USA).

6.1.3.2 Current Transition of CS for Magnesium Repair

The deposition of aluminum and its alloys by CS is now becoming an accepted practice to provide dimensional restoration and a degree of corrosion protection to magnesium components (Figure 6.3). It is known that the addition of aluminum to magnesium promotes the formation of better passive films than unalloyed magnesium. Therefore, it is not surprising that the application of cold-sprayed aluminum to magnesium and its alloys constitutes a method to inhibit corrosion in aqueous media. The protective capability of pure aluminum cold sprayed on magnesium has been demonstrated by others (McCune and Ricketts, 2004; Balani et al., 2005; Gärtner et al., 2006; Zheng et al., 2006). In all cases, the corrosion potentials of cold sprayed magnesium coupons approached those of CP-Al. Such polarization behaviors are promising because there is no galvanic protection strategy that is reasonable for magnesium given its strong thermodynamic potential for oxidation. In galvanic corrosion, only small areas surrounding the dissimilar interface require protection, for which CS represents an innovative alternative to the use of washers and insulating bushings.

The various research and development programs concluded that the material characteristics of CS aluminum alloys, especially 6061-Al, would be the best candidate for meeting the prescribed performance requirements while enabling the dimensional restoration for magnesium. The CS process was viewed as the best possible method for depositing the aluminum coatings and would be viewed as part of an overall strategy of replacement of the chromate processes such as Dow 17 and MIL-M-3171 currently in use today, eliminating environmental- and worker-safety issues, while significantly improving performance and reducing life-cycle costs.

This CS process has now been incorporated into production, and has been modified for field repair, across the United States and abroad, and has emerged as a feasible economic alternative over competing technologies. ARL has not only established the first DoD CS repair facility at the FRC-East, North Carolina, but has been involved in the design and implementation of advanced CS systems that have been transitioned into private industry also. In particular, the Office of Secretary of Defense (OSD) Manufacturing Technology Program has sponsored a large program to develop the world's most sophisticated CS repair facility under the technical management of ARL in collaboration with companies including MOOG, MidAmerica, and Sikorsky Aircraft Company, as

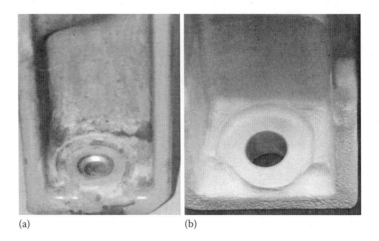

(a)   (b)

FIGURE 6.3 Mg aerospace part (a) before and (b) after being restored by cold spray.

well as other numerous OEMs. ARL has established through cooperative agreements, CS research, prototyping, and production facilities at various locations across the United States in conjunction with other private and government laboratories, research and engineering centers, and private companies. The CS process developed by ARL for the H-60 family (Black Hawk, Seahawk, etc.) of helicopters has been approved by industry, as well as the DoD. The Defense Science Technology Organization (DSTO) of Australia in collaboration with ARL and Rosebank Engineering, through a program funded by the Joint Strike Fighter (JSF), has also been successful at streamlining the process to implement CS for use as a repair technology for rotorcraft parts in Australia.

6.1.3.3   Aluminum and Magnesium Aerospace Alloys

There has been a tremendous motivation in the aerospace industry over the past several decades to make materials lighter and stronger to reduce the weight of aircraft to increase both overall performance and fuel economy. In response to this, there has been significant advancement in the introduction of new materials, especially in the field of aluminum alloy development since the first use of an Al–Cu–Mn casting alloy for the engine crankcase of the Wright Flyer in 1903 (Liu, 2006) The start of aerospace alloy development was marked by Wilm's discovery of precipitation hardening in 1906, which was later followed by advancements for alloys that were higher in strength, toughness, corrosion resistance, and stiffness. The aircraft industry has utilized wrought aluminum alloys that may be strengthened by plastically deforming the crystalline structure of the metal resulting in grain refinement and homogenization as induced by mechanical processing methods such as forging, extrusion, rolling, and upsetting. These alloys are superior to castings in terms of mechanical properties, not only in strength and hardness, but in ductility and fracture toughness too. The wrought alloys are placed into two general categories, specifically, those that are cold worked or nonheat treatable, and include grades such as 5052 and those that are heat treated including 2024, 6061, and 7075. Some of these alloys are susceptible to exfoliation (7075) and all are prone to pitting and crevice corrosion, under the right conditions, so there is a need for a technology to enable the dimensional restoration and repair of these alloys.

The widespread use of magnesium in aircraft occurred during the Vietnam era to reduce weight and increase performance (Navy/Industry Task Group Report, 1995). Magnesium is approximately 35% lighter than aluminum and has exceptional stiffness and damping capacity (Heinrich, 2004). Therefore, magnesium alloys are used for the fabrication of many components on US DoD aircraft, especially for complex components such as transmission and gearbox housings on helicopters and gearbox housings on fixed-wing aircraft, because of their high strength-to-weight characteristics. The more common magnesium alloys include ZE41A and AZ91C for legacy systems, and EV31 for some new aircraft applications. Other magnesium alloys also used are QE22A and WE43B. Magnesium is a very active metal electrochemically and is anodic to all other structural metals. Therefore, it must be protected against galvanic corrosion in mixed-metal systems because it will corrode preferentially when coupled with virtually any other metal in the presence of an electrolyte or corrosive medium (Zhang et al., 2005).

6.1.3.4   Replacement and Sustainment Costs

The US DoD reportedly spends millions of dollars to mitigate magnesium corrosion in their various aircraft fleets (Champagne, 2007). Many expensive magnesium components are decommissioned because the lack of an appropriate repair technique to provide dimensional restoration after corrosion has been removed. The Army and Navy helicopter fleet is comprised of more than 4500 aircrafts, each having numerous components manufactured from magnesium alloys, generally consisting of transmission and gearbox housings. In addition to helicopters, magnesium alloy gearboxes are used extensively in fixed-wing aircraft for the military and are currently designed into the JSF. Add to this the thousands of commercial aircraft around the world, and it is easy to visualize the magnitude of the problem and that thousands of magnesium components currently in service are adversely affected. The cost of each component ranges from several thousand to over $800,000. A Sikorsky Aircraft Company 2009 trade study showed that the annual cost for the H-60 main gearbox corrosion of about 600 helicopters is over US $17 million. In 2001, the FRC-East conducted an extensive review of the cost of corrosion on the main transmission of one type of helicopter and reported that from 1991 to 2000, the total estimated cost for both unscheduled maintenance and module replacement was about $41 million. Most of the corrosion occurs at attachment points where a dissimilar metal is in contact with the coated magnesium component. This includes flanges, mounting pads, tie rods, lugs, and mounting bolts.

6.1.4 CS Repair for Magnesium Aerospace Parts

6.1.4.1 Task Statement

Many of the corrosion problems associated with magnesium helicopter components occur at the contact points between inserts or mating parts, where ferrous metals are located, creating galvanic couples (Vlcek et al., 2005). In addition, magnesium alloys are also very susceptible to surface damage due to impact, which occurs frequently during manufacture and/or overhaul and repair. Scratches from improper handling or tool marks can result in preferential corrosion sites. The DoD and the aerospace industry have expended much effort over the last two decades to develop specific surface treatments to prevent corrosion, increase surface hardness, and combat impact damage for magnesium alloys to prolong equipment service life; however, the means to provide dimensional restoration to large areas on components where deep corrosion has occurred remains a challenge (Champagne et al., 2008).

For OEMs, the standard practice is to hard anodize the surface using a process designated Dow 17, followed by an application of a phenolic resin (one version is designated *Rockhard*). For nonmating surfaces, multiple coats of chromate epoxy polyamide primer followed by multiple coats of epoxy paint are applied. For mating surfaces, no primer or paint is used, but other types of sealant compounds are applied. In repair/overhaul facilities, the standard practice is to use a chromate conversion coating, MIL-M-3171, followed by the Rockhard and the chromate primer and paint for nonmating surfaces and a sealant compound for mating surfaces. It has been demonstrated that CS can be incorporated as a repair technique for dimensional restoration of worn and/or corroded parts in combination with the standard surface treatments and can be used without these surface treatments.

Figure 6.4 shows UH-60 Black Hawk magnesium housings. Figure 6.5 is a schematic of the main transmission housing for the H-60 helicopter showing the areas most susceptible to corrosion. Because of the localized nature of the corrosion, surface treatments intended to mitigate the problem would only have to be applied in these specific areas. A combination of galvanic and crevice corrosion is the primary corrosion mechanism for Mg alloy gearboxes. Water and salt sprays tend to accumulate in the recessed holes around the bolts, and galvanic corrosion occurs between the Cd-plated bolt head and the surrounding Mg alloy, as well as between the shank of the bolt and the ID of the hole. In addition, gearboxes usually have Cd-plated steel studs and bushings press fitted into them, and galvanic corrosion occurs around these areas (see Figures 6.6 and 6.7). In particular, Rosan inserts are frequently used to hold bushings, fittings, and threaded fasteners into housings (see Figure 6.8). These fittings are designed to be press fitted into the housing, where their teeth bite into the Mg alloy and create a high degree of plastic deformation. Any coating or surface treatment that is applied to the Mg at repair facilities must be able to withstand the insertion. When corrosion becomes excessive, material is removed and an Al shim is glued in place of the missing Mg alloy. These shims can be as thick as 0.100 in. and cannot carry load, thereby weakening the structure. Because of the localized nature of the corrosion, surface treatments intended to mitigate the problem would only have to be applied in these specific areas. In summary, the susceptibility of magnesium components to corrosion and damage, even with the most current surface protection schemes, is still very significant as shown by the very high repair/overhaul and maintenance costs.

6.1.4.2 Technical Approach and Test Plan

The critical tasks associated with the qualification of the CS coatings on magnesium components were established in a document known as the Materials Joint Test Protocol (JTP) developed

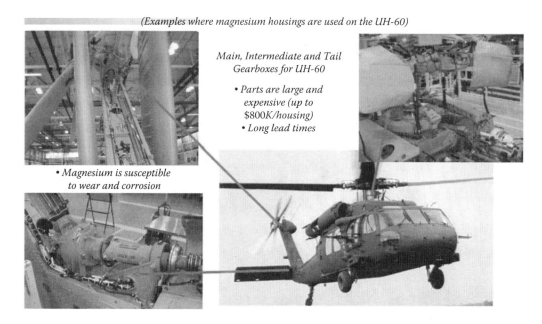

FIGURE 6.4 Black Hawk magnesium housings susceptible to corrosion and wear.

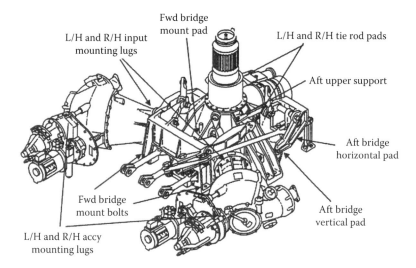

FIGURE 6.5 Schematic of an H-60 Main Transmission Housing showing areas most susceptible to corrosion. Most of the corrosion occurs at attachment points where a dissimilar metal is in contact with the coated magnesium component. This includes flanges, mounting pads, tie rods, lugs, and mounting bolts.

by representatives of the DOD, OEMs, industry, and academia. The quantitative and qualitative performance objectives required to qualify CS to repair magnesium aerospace components are provided in Table 6.1, which also contains a summary of the test results for a *nonstructural* repair for 6061Al CS deposited onto ZE41A-T5 Mg. Some of these test results will be discussed in more detail in a proceeding section and the protocol for a *structural* repair will be presented. The final performance objectives were set by the Naval Air Command and ARL with input from industry including,

FIGURE 6.6 Corrosion around a stud and hole on a helicopter magnesium gearbox housing. (Courtesy Robert Kestler, FRC-E.)

FIGURE 6.7 Corrosion around an insert on a helicopter magnesium gearbox housing. (Courtesy Robert Kestler, FRC-E.)

Sikorsky Aircraft Company, as part of the aforementioned ESTCP Program, which has been established as an aerospace industry standard. The quantitative performance objectives are based on the JTP. The JTP was designed to define the upper and lower technical capabilities of the CS repair process.

The optimal CS process parameters to deposit aluminum and its alloys onto ZE-41A magnesium for nonstructural applications were developed by executing the JTP and from obtaining data from previous efforts in this area. The CS process is significantly different from other established magnesium gearbox coating systems such as Rockhard and Dow 17. Both Rockhard and Dow 17 are thin corrosion-protection coatings that can be considered to have negligible mechanical strength or thickness compared to the magnesium gearbox. CS on the other hand will be applied to damaged areas of the gearboxes with thicknesses ranging from a few mils to approximately one-half of an inch and because of the unique bonding mechanism is capable of carrying a load. A thorough and quantitative evaluation of mechanical performance of the CS repair was therefore

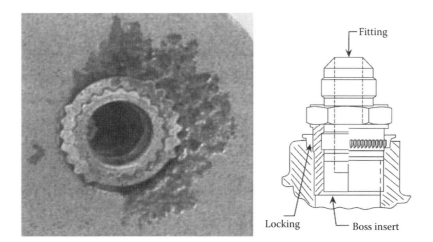

FIGURE 6.8 Rosan insert. Left—corrosion around insert on a helicopter magnesium gearbox housing. Right—cross section of insert with fitting. (Courtesy Robert Kestler, FRC-E.)

deemed necessary during the initial planning stages of the program. Operations such as surface preparation to modeling the blending geometry of the repair site were discussed and researched. Corrosion performance was also a necessary benchmark to establish, so, several quantitative and qualitative corrosion tests relevant to magnesium gearbox castings were also included in the JTP and as performance objectives. Appearance and porosity are included as qualitative performance objectives. Both of these objectives fall under the regime of quality control because they can be evaluated in a production environment with relative ease. As had been done previously with the hard chrome plating replacement projects managed by the Hard Chrome Action Team (HCAT) to qualify the HVOF process for use on aircraft landing gear, a preliminary JTP was prepared and distributed to all potential stakeholders, and then a meeting was held to identify and specify the exact materials tests required to qualify the CS aluminum alloy coatings. Under a contract from the Army, Concurrent Technologies Corporation developed a preliminary JTP titled *JTP for Validation of Corrosion Protection for Magnesium Alloys*. The development of the JTP involved the input from a variety of individuals from Army AMCOM, PEO Aviation, Army RDEC Materials, US Army Corrosion Office, and Sikorsky. The JTP was not directed towards qualification of aluminum alloy coatings but was meant to be generic for qualification of any type of corrosion protection scheme. This preliminary JTP formed the basis of the preliminary JTP that was prepared. This JTP can be used as a basis to qualify other similar applications.

In addition, through the JSF program office, there was international participation from the DSTO, Australia. The JTP was comprised of mechanical, physical, and chemical tests designed specifically for this application. The most important qualification tests were identified as adhesion (uniaxial tension), salt fog exposure (ASTM-B117), and microstructural analysis, which will be the primary focus of this discussion. These screening tests would form the basis of the data required to recommend the use of CS to repair magnesium aircraft components. Additional tests would later be conducted to satisfy specific application requirements for structural repair.

The various stages of the general test plan are shown below and followed by the specific tests of the JTP.

- *CS Systems*: ARL developed a portable high-pressure hybrid CS system specifically for this application, which is now available commercially that can attain gas pressures as high as 1000 psi and gas temperatures of 900°C (Figure 6.9). It can be operated manually or attached to a robot. In addition, a new nozzle was designed and fabricated to prevent clogging that often occurs while spraying aluminum.

- *Selection of Gases*: Initial spray trials using helium gas were very successful in achieving high adhesion and corrosion resistance. CS process parameters were also developed with nitrogen to reduce costs while maintaining satisfactory coating performance.
- *Powders*: Various commercially available aluminum powders were chosen for their flow characteristics, size, geometry, and cost to improve flow and increase deposit efficiency. Aluminum alloys Al-125i, 6061 Al, and 5056 Al were also evaluated.
- *Predictive Modeling*: The predictive model developed at ARL was used to determine starting values of gas temperature and pressure, which resulted in an appropriate particle velocity and surface temperature for optimal coating deposition and minimal damage to the substrate.
- *CS Trials*: A series of CS trials were performed to optimize coverage, adhesion, and coating integrity. Adjustments of the spray parameters were recorded and correlated to test results to improve the overall CS process.
- *Testing and Evaluation*: Materials and coatings characterization and performance-based testing were conducted according to the requirements established in the JTP.

Joint Test Protocol
Mechanical Testing and Materials Characterization

- Adhesion Tensile Bond Test (ASTM C633)
- Residual Stress Analysis (XRD Analysis)
- Flat Tensile Specimens

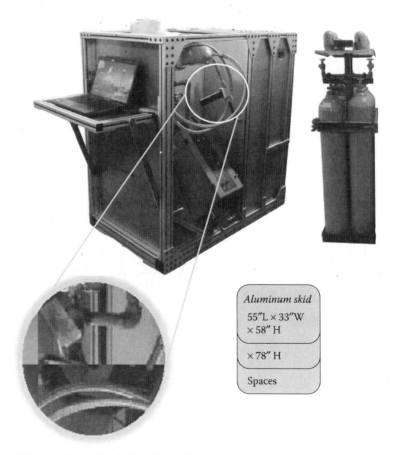

FIGURE 6.9 VRC Gen III HP Hybrid Cold Spray System based on an ARL design.

- RR Moore RB Fatigue surface finished 125RA
- Fretting Fatigue-UTRC
- Impact-ASTM D5420
- Hardness-(Pre/Post 385F-6hrs)
- Microstructural Analysis (Uniformity, Porosity, Grain Refinement)
- ROSAN Insert Test (Sikorsky)
- Triple lug shear test (MIL-SPEC MIL-J-24445A)
- Machinability

Corrosion Testing

- Unscribed ASTM B117
- Scribed ASTM B117
- GM9540 Scribed
- Galvanic Corrosion (G71)
- Crevice Corrosion (G78)
- Beach Corrosion
- G85 SO2

6.1.4.3 Performance Objectives and Test Results

The performance objectives and test results of cold sprayed coatings are summarized in Table 6.1. The performance objectives have been evaluated on the basis of ASTM and other standards. The testing results clearly demonstrate the applicability of cold spray technology for aircraft repair purposes.

6.1.4.4 Structural Repair Testing

Additional mechanical tests were recommended and incorporated into the JTP as a part of a test plan to qualify the CS process for structural repair of magnesium components. Structural CS repair is included for future reference for those wishing to consider CS for structural restoration of components or for additive manufacturing. The following additional tests would be performed to determine whether the CS coating adversely affected the fatigue resistance of the substrate and to test the impact resistance of the coating, both important considerations for structural repair.

- *Fretting Fatigue Testing.* This would be performed using an apparatus similar to the United Technologies Research Center (UTRC) equipment that utilizes standard axial fatigue specimens except with a rectangular cross section in the gage area, where the coatings to be evaluated are applied. Fret pins bear against the center of the gage surface with a constant load and slip against the specimen at an amplitude controlled by an actuator. The fatigue specimens are subjected to cyclical loads that would be expected to result in specimen failure in approximately 105 cycles in the absence of the fretting. The number of cycles to failure is measured and compared to a baseline material.
- *Impact Testing.* Because much of the damage observed on magnesium alloy components in service is due to impact from foreign objects, evaluation of CS-coated specimens would be appropriate. Two types of impact testing would be performed, the first being high-velocity small-particle impact (evaluated by Gravelometry) and the second low-velocity large-item impact simulating dropped tools, handling damage, and so on. The Gravelometry would follow the ASTM D310 protocol, which involves feeding a stream of gravel into an air jet and allowing it to strike the coated specimen at a defined angle. The large-item impact would involve dropping a steel ball bearing from a specific height onto a coated specimen and evaluating the surface damage.

TABLE 6.1
Performance Objectives and Test Results

Performance Objective	Data Requirements	Success Criteria	Results
		Quantitative Performance Objectives	
Deposition rate	Coating thickness measurement to an accuracy of ±0.0005 in.	Ability to deposit coatings at a rate of at least 0.005 in./h with coating quality such that they pass the acceptance criteria specified in the JTP.	Passed Deposition rates were magnitudes higher
Coating Thickness Uniformity	Coating thickness measurement to an accuracy of ±0.0005 in.	Cold spray coating thickness shall be uniform within ±20% for deposition onto various surfaces that simulate Mg alloy components.	Passed Uniform, no waviness
Microstructure	Examined with Optical Microscopy	A uniform microstructure, especially for alloy coatings, must be achieved.	Passed Uniform
Hardness	American Society for Testing and Materials (ASTM) E384 - 10e2	Vicker's Microhardness of as-deposited coatings shall be no less than 50 VHN.	Passed 105 VHN
Fatigue	R.R. Moore High-Speed Rotating Beam Rotating	No debit as compared to baseline noncoated specimens as specified in the JTP.	Passed No debit, credit for 7075Al substrates
Stress/Strain Testing; Ductility	Tensile Testing ASTM E8	Monotonic stress/strain testing shall be conducted in a standard tensile tester. This will evaluate strain tolerance.	Passed 50ksi-UTS 42ksi-YS 7%-EL
Residual Stress	X-ray diffraction (XRD)	Applied coating must be in either a compressive or neutral stress state.	Passed 13 ksi Compression
Adhesion in Tension	ASTM C633	Coating must meet or exceed 8.0 ksi.	Passed >10,000 psi
Shear Adhesion	MIL-SPEC MIL-J-24445A	Coating must meet or exceed 8.0 ksi.	Passed Up to 35,000 psi

(Continued)

TABLE 6.1 (Continued)
Performance Objectives and Test Results

Performance Objective	Data Requirements	Success Criteria	Results
ROSAN Insert	Sikorsky Test	ROSAN insert must be capable of being inserted into a hole containing a cold spray coating without delamination, cracking or spalling.	Passed Coating remained intact
Impact Test	ASTM D5420	Coating is examined for after a weight is dropped at prescribed height.	Passed No spalling
Fretting Fatigue	Conducted by UTRC Epsilon Technology Corp. System (Jackson, USA)	No debit as compared to baseline noncoated specimens as specified in the JTP.	Passed Coating system dependent
Salt Spray Corrosion	ASTM B117	Minimum of 336 hours exposure without penetration of salt spray through coating to the substrate as described in the JTP.	Passed >7000 hours
Cyclic Corrosion	General Motors (GM) 9540 Specification	Minimum of 500 hours exposure without penetration of salt spray through coating to the substrate as described in the JTP.	Passed >500 hours
Qualitative Performance Objectives			
Appearance	Visual Inspection	Coatings are continuous, smooth, adherent, uniform in appearance, free from blisters, pits, nodules, and other apparent defects.	Passed Uniform coating, no defects
Porosity	Examined with Optical Microscopy	Porosity of cold spray coatings should be less than 2%.	Passed <1%
Machinability	Visual Examination	Coating deposited on a rod is machined on a lathe and chip formation is characterized.	Passed Chip comparable to wrought material
Beach Corrosion	Conducted by Navy at Cape Canaveral, FL	No observable penetration or pitting through the coating and into the magnesium.	Passed No pitting
Galvanic Corrosion	ASTM G71-81	No defined criteria. Used for comparison to High-Velocity Oxygen Fuel (HVOF) Al–12 Si baseline specimen.	Passed
Crevice Corrosion	ASTM G78	No observable corrosion product.	Passed

- *Simulated Defect Fatigue Testing.* To obtain quantitative data regarding the effects that a CS would have on the fatigue resistance of the substrate, a special specimen was designed. The specimen consisted of a normal flat fatigue samples that contained a prescribed defect machined into the center into which a CS deposit was applied and subsequently machined flat and parallel to the surface. The data were compared to specimens that also contained a machined defect but without the CS fill. The results revealed that the CS fill did not adversely affect the fatigue resistance of the substrate.

- *Corrosion Tests.* Corrosion tests were performed on selected Mg alloy samples having only the CS aluminum coating and on Mg alloy samples containing a multilayered coating system that would represent actual fielded parts that would include the CS Al coating, Rockhard resin, and sealant once the screening tests were completed.

  - Potentiodynamic polarization measurements in accordance with ASTM G5 and G59 in different electrolyte solutions to characterize performance of the CS Al alloy coatings in terms of open circuit potential, passivation behavior, and pitting potentials.
  - ASTM G85, Annex 4 SO2 salt fog corrosion tests on coated panels
  - Crevice corrosion testing
  - Field corrosion testing. The Navy maintains corrosion test racks on selected aircraft carriers for the purpose of conducting field trials on materials and coatings. Mg alloy panels containing the complete surface finishing of CS Al alloys, Rockhard resin, and sealant would be evaluated in comparison to the current surface-finishing process. An assessment would be made of the extent to which the corrosion life has been increased through the use of the new coating process.

- *Component and/or Flight Testing.* In addition to development and execution of the Materials JTP and in conjunction with the demonstration of the CS coatings onto actual components, lead-the-fleet flight testing was included for structural repair of magnesium components. For this type of application, component rig tests are generally not applicable because they do not simulate the corrosive environment encountered in service, so flight testing is the best way of evaluating the airworthiness of the CS coatings. For identification of candidate components, the Storage, Analysis, Failure Evaluation and Reclamation (SAFER) Program at Corpus Christi Army Depot, which was established by the Army Aviation Engineering Directorate, was involved. Many nonrepairable magnesium alloy components have been placed in the SAFER Program, whose mission is to develop repair and reclamation procedures for high-value components and return them to inventory instead of buying new parts (Figure 6.10). At the FRC-East, CP, the Materials Review Board, which has a similar mission to the SAFER, was utilized to identify candidate components.

6.1.5   CS Systems

6.1.5.1   CS System Employed

The ARL Center for Cold Spray Technology currently maintains six high-pressure stationary CS systems, manufactured by Ktech Corporation CGT Incorporated, Inovati, and Plasma Giken, as well as six portable CS systems. Two of the portable CS systems were manufactured by Dymet Inc., (Obninsk, Russia) two by Centerline, and the remaining systems were designed and produced at ARL. All of the portable systems are considered low pressure with the exception of the ARL Hybrid Gen III system that can operate at low or high pressures and temperatures, manually or robotically controlled. The work performed under this program was conducted primarily with the CGT 4000 system and the VRC Hybrid Gen III ARL system (Figure 6.9).

6.1.5.2   CS Nozzle Design

The conventional CS nozzle that is used is normally fabricated from stainless steel or tungsten carbide. Various nozzle configurations have been designed and tested at ARL, and it is not the intent

FIGURE 6.10   Used aerospace parts in storage at Corpus Christi Army Depot waiting to be restored.

of this chapter to repeat the research of others but to relate those aspects of nozzle design specific to this application. The primary concern was clogging of the nozzle while spraying aluminum, especially CP-Al or HP-Al. Clogging can occur in the throat of the de Laval-type nozzle, where higher temperatures are employed. Aluminum particles tend to stick to the sides of the nozzle interfering with proper gas flow, which adversely affects coating deposition. To mitigate clogging, a plastic nozzle was used with absolute success. ARL and others have conducted studies of various nozzle configurations and materials. Comparisons have been made of nozzles fabricated from ceramic, plastic, carbide, and other metallic materials. ARL has designed a nozzle fabricated from a high-temperature-resistant plastic that can be operated at 400°C with satisfactory results. There are several plastics on the commercial market that can be successfully machined and used as a nozzle material. Such plastics can maintain their properties at high operating temperatures (400°C) and are adequately wear resistant for use with Al for extended periods of time. These plastics tend to be proprietary in nature but can be obtained from the CS equipment manufacturers and typically consist of polybenzimidazole (PBI) plastic, which has the highest mechanical properties of any plastic above 200°C.

6.1.5.3   Selection of Gases

The decision to use helium over nitrogen relies primarily on the added costs involved. However, from a technical standpoint, the velocities that can be achieved by helium and the resultant density of the coatings may well be worth the extra cost, especially for components that are valued at $400,000 each. During operation at ARL, CS system uses 40 scfm of nitrogen at a cost of $0.29/100 scf for a gas-operating cost of $0.116/minute. With helium, it uses 70 scfm at a cost of $11.50/100 scf for a gas-operating cost of $8.05/minute. Therefore, the costs are substantially lower when using nitrogen. To put the gas costs into perspective, the labor rate for operating a CS system is about $1.00/minute and the powder costs are about $2.00/minute. Helium-recycling systems have been designed and put into use that are able to recover approximately 90% of the helium, thereby greatly lowering the operating cost. (Johnson, 2004). For an R&D CS facility that operates intermittently, the cost of the recycling system cannot be justified. But for a production CS facility, the payback time on the cost of the recycling system would be fairly short for these expensive components. Regardless, the results of this work demonstrated that nitrogen can be used as a carrier gas to produce satisfactory aluminum alloy CS coatings depending on the process parameters employed.

6.1.5.4   Powders

Studies have shown that the difference in particle velocities is a function of particle size, with velocities approximately twice as much for helium versus nitrogen for 5 μm particles but only about 30% greater for helium for 25 μm particles. Vlcek et al. (2005) has provided extensive information related to the physical processes that occur during the CS deposition of different materials and why some work better than others. Impact heating, equations of state during impact, dynamic yielding of the particles, and impact pressures were examined, and it was concluded that the materials that are most amenable to CS are those with a face-centered cubic structure, which includes aluminum.

The powders under consideration for the CS repair of magnesium transmission housings were CP-Al, HP-Al, Al-12Si, 6061Al, and 5056 Al. The criteria were based on the requirements associated with this application where galvanic corrosion and corrosion pitting were the primary causes for removing the components from service. In addition, the repair was confined to nonstructural areas of the transmission and gearbox housings. There were advantages and disadvantages associated with each material.

The 5056 Al (composition Al–5Mg–0.1Mn–0.1Cr) was considered because it is compatible with magnesium and has better tensile, yield, elongation, and fatigue strength than any magnesium alloy. It is a work-hardenable alloy and therefore would be more conducive to the CS process, which involves tremendous plastic deformation of each particle deposited than a heat-treatable alloy. The corrosion resistance of 5056 is also among the best of any aluminum alloy. The presence of the manganese is important because it serves to tie up any residual iron contamination that has been shown to degrade corrosion performance. The Al–12Si was chosen as a candidate based on its excellent mechanical properties and resistance to wear and corrosion. It is also used extensively with thermal spray and the powders are commercially available.

The 6061 Al was considered because the powder is commercially available from Valimet Inc. (Stockton, USA) as well as F. J. Brodman & Co., L.L.C., (Harvey, USA) and has high tensile strength (45,000 psi). It is a common aerospace aluminum alloy and therefore would have wide acceptance and have multiple suppliers. However, 6061 Al is a precipitation hardening alloy and obtains its hardness and strength from a sequence of heat-treatment steps including solutionizing, quenching, and subsequent tempering operation. Its major alloying elements are magnesium and silicon.

Initially, concern was expressed over CP-Al and HP-Al as the material of choice for the CS repair of magnesium ZE-41A because it was lower in hardness and strength than other aluminum alloys that were also being considered for this application. It was later selected because the primary reason that the magnesium components were removed from service was due to corrosion and not wear, and the CS repair was to be performed on nonstructural areas of the component, making the strength requirement less of an issue. Additionally, the CP/HP-Al powder could be doped with a certain percentage of hardened particles to impart wear resistance, if required. The cost of CP-Al was attractive at approximately $11.00/lb and was commercially available, while many other alternative alloys originally considered were much more expensive and/or had to be produced as a specialty item. The Al 5056 was an example of a powder that was not commercially available. The stock material used in the melt of the Al 5056 powder was also not available forcing the powder manufacturer to purchase and subsequently mix the raw materials to produce the alloy prior to atomization.

Finally, the spherical particle shape of the CP and the 6061 powder was preferred over that of the Al 5056 flake. Spherical particles form a more densely packed structure when deposited by the CS process resulting in better protection for the magnesium substrate. Figure 6.11 shows micrographs of the two powders and illustrates why the spherical shape is preferred over that of flakes.

6.1.5.5   Predictive Modeling for Process Optimization

The nature of the bond created during particle consolidation and the properties of the material produced by the CS process have been modeled at ARL. These predictive models are used to establish and optimize CS process parameters. Modeling efforts predict the amount of mixing at the interface between the particles and the substrate with concomitant high coating adhesion when

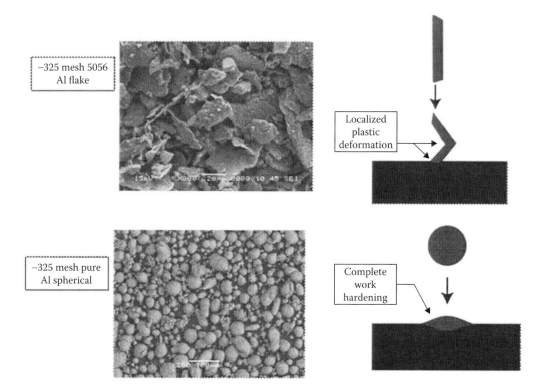

FIGURE 6.11   Differences in powder morphology and the effects of single-particle impact.

the particle velocity reaches a certain minimum value. Compressible, isokinetic flow equations are used to model gas flow within the nozzle. Modified drag and heat transfer coefficients are then used to iteratively calculate the resulting particle velocities and temperatures. An example of this calculation for the aluminum/helium nozzle acceleration is shown in Figure 6.12. The particle size for this calculation is 20 μm, and the initial gas pressure and temperature are 2.75 MPa and 20°C, respectively. Calculated particle velocities are verified experimentally by means of a dual-slit, laser velocimeter. An empirical relationship of the penetration of micrometeorites into spacecraft skin is used to model the interface between the deposit and the substrate, and an empirical relationship between particle characteristics and critical velocity is used to model deposition efficiency. The CS

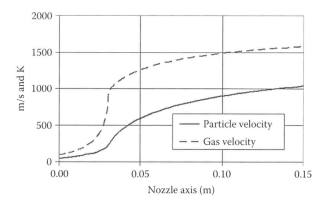

FIGURE 6.12   Theoretical calculation of velocity for a 20 μm Al particle and helium gas along the length of a de Laval nozzle.

material is evaluated for shear strength at the interface and for hardness. Magnified cross sections are examined for density and interface locking. The demonstrated prediction accuracy of these results allows the ability to define operating parameter values and expected coating characteristics prior to CS operation. This prediction algorithm was subsequently used on multiple-coating/substrate combinations with favorable results.

6.1.6 CS TRIALS

6.1.6.1 General Description

A series of CS trials were performed to optimize coverage, adhesion, and coating integrity. Each trial run required the specification of the powder—powder feed rate, gas, gas pressure, and gas temperature. Gas pressure and temperature were specified by model calculations to produce sufficient particle velocity for good deposition. Subsequent coating evaluation prescribed changes to the operating conditions if coating characteristics needed improvement. Besides the limitations imposed by maximum allowable temperature and pressure, two operational restrictions narrowed the available parameter range—these were nozzle fouling and ambient temperature gas (for the portable unit). Low melting point powders can stick to nozzle walls, eventually plugging the nozzle. This is especially true for aluminum powders. Limiting gas heating to lower temperatures can prevent nozzle fouling.

A series of CS trials were carried out to determine optimum spray conditions for the two systems. Operating parameters such as pressure, temperature, stand-off distance, raster speed, and gas were studied. Initially, it was found that helium gas produced superior coatings as compared to nitrogen when using a conventional nozzle fabricated from stainless steel or tungsten carbide. However, fouling occurred when temperatures in excess of 250°C were used with conventional nozzles. It was therefore necessary to conduct similar studies incorporating a plastic nozzle. Higher gas temperatures could be attained without nozzle fouling through the use of plastic PBI nozzles, and denser deposits with significantly increased bond strength were achieved.

It is not possible within the context of this chapter to explain and discuss all of the test methods and accompanying results. However, it is essential to appreciate the tremendous potential that the CS technique has for the repair of aerospace parts that some of the more important tests are presented in some level of detail.

6.1.6.2 Triple Lug Adhesion Tests

The Triple lug shear test method was used as a second test to confirm coating adhesion. Triple lug procedure methodology is prescribed in military specification, MIL-J-24445A. A mechanical drawing of the test coupons is provided as Figure 6.13. A coating with a thickness of greater than 0.125 in. is deposited onto the specimen. Three lugs are machined from the coating. The lugs are sheared from the test specimen using a compressive load frame setup as shown in Figure 6.14. Only one lug is sheared from the specimen at a time. Failure stress is reported based on the load at failure and the surface area of the lug. Control specimens milled from single pieces of cast magnesium were included in the matrix to establish a baseline. All of the 6061 He coatings showed very high adhesion on the magnesium alloys. The average adhesive strength of the 6061 He coating, deposited onto ZE41A-T5 actually exceeded the average strength of the base line ZE41A-T5 samples. Correspondingly, most of the twelve 6061 lugs on the ZE41A-T5 broke off by fracturing material well beneath the coating–substrate interface at strength levels as high as 35,000psi, as shown in Figure 6.15. This serves as substantial proof that that structural repair is possible with the 6061 He CS coating because the weakest point is no longer at the coating interface, but within the Mg substrate.

6.1.6.3 Tensile Testing of Bulk CS Material

The 6061 helium tensile specimens were machined from a sprayed block of CS 6061 as described in Table 6.2. The CS was deposited using the same parameters as for all other 6061 samples in this study. The substrate was a 4 × 4 inch piece of 3/8 in. thick wrought 6061-T6. After deposition,

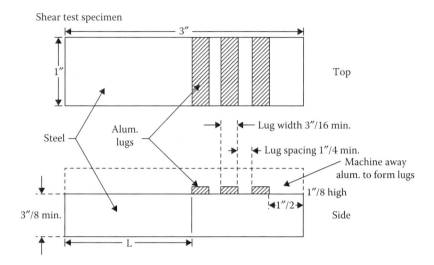

FIGURE 6.13   Triple lug shear test specimen.

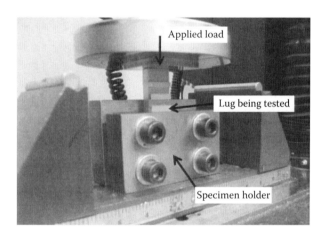

FIGURE 6.14   Triple lug shear test setup.

the substrate was milled away from the deposit. This resulted in a single piece of 6061 CS with an approximate thickness of 0.6 in. Six tensile bars were then machined to the dimensions specified in Table 6.2. In addition to ultimate tensile strength (UTS), the elastic modulus, yield strength (YS), and percent elongation at failure were characterized for the 6061 helium depositions. This additional mechanical data were deemed important to determine the load bearing capability of the CS coatings. The purpose of tensile testing is to determine the tensile strength, yield, modulus, and other tensile materials properties of the substrate/coating combinations, as well as the contribution of the coating to the strength of the substrate. Tensile test results were used to guide fatigue testing. The test was be carried out in accordance with ASTM E8, using a strain rate of 0.05 in./min at room temperature and approximately 50% relative humidity. The results are shown in Table 6.3.

The results listed in Table 6.3, clearly indicate the potential for CS 6061Al as a *structural* material that can be considered for use on other substrate materials than ZE41A-T5 Mg. The UTS of the as-cold sprayed 6061AL was as high as 50 ksi with a corresponding YS of 42 ksi and % elongation (EL) of 3%. A postprocessing procedure of feedstock 6061 Al powder has been developed and can yield CS material with minimum UTS of 45 ksi, YS of 38 ksi, and as much as 7% EL. This has been a major accomplishment for the implementation of CS for *structural* materials repair and for additive manufacturing.

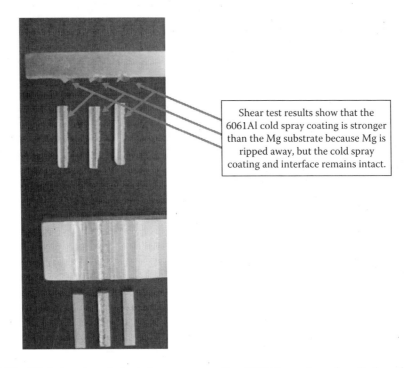

Shear test results show that the 6061Al cold spray coating is stronger than the Mg substrate because Mg is ripped away, but the cold spray coating and interface remains intact.

FIGURE 6.15 Triple lug shear test specimens representing 6061 He coatings deposited on the ZE41A-T5 substrate failed within the Mg substrate, rather than by shearing off at the coating–substrate interface.

TABLE 6.2
6061 Helium Tensile Test Specimen Dimensions (in.)

Block Width	4
Block Length	4
Block Thickness	0.6
Number of Tensile Specimens	6
Tensile Specimen Width	0.25
Tensile Specimen Length	0.125

TABLE 6.3

6061 Al Alloy Grade	Ultimate Tensile Strength (k/square in.)	Yield Stress (k/square in.)	Elongation (%)
Annealed	18	8.0	25
T4, T451	30	16.0	18
T6, T651	42	35.0	10
Cold sprayed - Hi strength	48	42.5	3.5
Cold sprayed - Hi ductility	45	38.0	6.5

6.1.6.4 Microstructural Analysis

Optical and electron microscopies were used to characterize the coating/substrate interface and to access the integrity of the deposit. The density, bond line integrity, microstructure, and inherent material features of the coating were characterized. Acceptable nonstructural repair processing windows for the 6061 helium coating were developed and are proprietary. Although the CS process parameters for the 6061 helium coating are being withheld by ARL, there are exceptional performance test results that are important to relate to illustrate the capability of CS technology. One of the relationships studied with this coating was between pressure, adhesion, and microstructure. Table 6.4 shows the results for bond bars deposited at three different pressures. The pressures were increased in 5 bar increments starting from *a* to *b* and then to *c*. The adhesive strength improves with increasing pressure. The highest pressure samples exhibited only glue failure. Additionally, the highest pressure samples exhibited glue failure at deposition rates exceeding 9 mils/pass. Figures 6.16a–c show optical micrographs of three different 6061 He coatings. The samples were etched with Kroll's Reagent to reveal splat boundaries. Porosity was observed to decrease with increasing pressure, while the temperature was held constant. These results indicate that a direct relationship between particle velocity and porosity exists in this processing regime. As with the CP-Al N2 and the CP-Al He coatings, the preliminary unscribed B117 salt-spray coatings deposited by ARL exceeded 7000 hours of total exposure with no pinhole defects. Testing was stopped prior to observing any failures. This serves to substantiate that the CS coating is dense with no interconnected porosity that extends to the surface. In fact, the porosity level for Figure 6.16c was <0.2%. The microstructure of the as-cold sprayed 6061AL was uniform and consisted primarily of the aluminum matrix, the α-Al(CrMnFe)Si phase at the particle boundaries, and the Mg_2Si phase adjacent to the α phase (Figure 6.17) (Gavras et al. 2007). These observations are in agreement with those by Belsito et al. (2013).

TABLE 6.4
Adhesion Strength (ASTM C633) for 6061 He Coatings Deposited at Various Pressures and Constant Temperature

Slug	Pressure	Deposition Rate (mils/pass)	Max Tensile Adhesion Stress (psi)	Observed Failure Mechanism(s)
1			7802	70% Adhesion + 30% Glue
2	a	3.7	12,753	Mating Bar Surface Failure[a]
3			11,714	60% Adhesion + 40% Glue
4			7246	Mating Bar Surface Failure
5			12,674	100% Adhesion
6	b	3.1	9458	100% Adhesion
7			12,597	100% Adhesion
8			8698	100% Adhesion
9			12,268	100% glue
10	c	4.5	11,938	100% glue
11			13,073	100% glue
12			11,915	100% glue
13			12,842	100% glue
14		9.0	12,674	100% glue
15			12,667	100% glue
16			11,812	100% glue

[a] Mating bar surface failure refers to debonding of the epoxy to the mating bar and does not reflect maximum coating adhesion strength due to the limited strength of the adhesive.

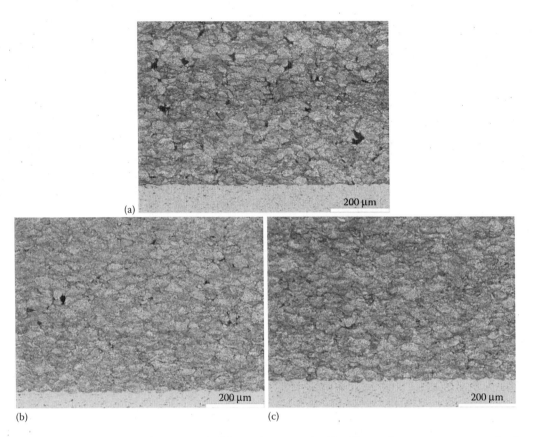

FIGURE 6.16   (a) Optical microscopy of 6061 coating sprayed with a He carrier gas at pressures *a* that is the lowest pressure tested. (b) Optical microscopy of 6061 coating sprayed with a He carrier gas at pressures *b* that represented moderate pressure. (c) Optical microscopy of 6061 coating sprayed with a He carrier gas at pressures *c* that was the highest pressure tested. Substrate is 6061-T6 plate. Temperature held constant.

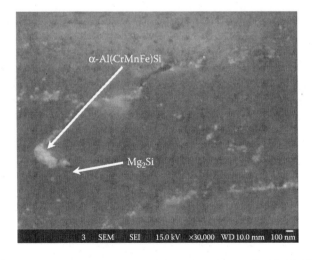

FIGURE 6.17   SEM of as-cold sprayed 6061Al consisting of α-Al(CrMnFe)Si: 0.25% Mg_2Si: 1.68%. (From Gavras A.G. et al., Small fatigue crack growth mechanisms and interfacial stability in cold-spray 6061 aluminum alloys, Worcester Polytechnic Institute Internal Report, 2007. With permission.)

6.1.6.5 Fatigue Testing

Over the past several years, the CS deposition process has become a leading candidate for the fabrication, repair, and enhancement of several cutting-edge defense, aerospace, and commercial components. For applications involving dynamic loading, a comprehensive understanding of the mechanical properties of the CSs was required to insure that performance and lifetime requirements will be achieved. The focus of this section is to summarize the fatigue data from the ARL and its program partners, as well as the data published in the literature. The ARL data for this report comprise of data from three programs. The purpose of fatigue testing was to evaluate the effect of the coating on the fatigue of the underlying material as well as the adhesion between substrate and coating during the course of fatigue. (Coatings may delaminate on failure.) Table 6.5 summarizes each of the program objections as well as the partners. A summary of fatigue results from open literature is also provided.

ESTCP WP-0620 RR Moore Fatigue Testing Parameters

Substrate Materials: AZ91C-T6, ZE41A-T5, and EV31-T6

Coating Systems: CP-Al sprayed with an N_2 carrier gas and 6061 sprayed with a He carrier gas

Material Acquisition: Round bars machined from cast plate. All bars for each material are from same lot

Specimen fabrication: Hourglass 0.25-in. gage diameter (Figure 6.18)

Surface finish: 8 μ in. Ra Precoating, machined after coating

R ratio: R = −1, 1 rpm is equivalent to one cycle

Environment: Laboratory air at ambient temperature

Runout: Defined as 10^7 cycles

Frequency: 167 Hz (10,000 rpm)

Coating Deposition and Grinding: Deposit coatings along the entire gage length and 0.25–0.5 in. along the shoulder. The final coating/thickness will be between 0.012 ± 0.001 in. and 0.015 ± 0.002 in. Use the same surface finish for coated and uncoated specimens.

6.1.6.6 Acceptance Criteria

After data (i.e., cycles to failure) were obtained for all specimens tested under identical parameters (e.g., same coating thickness, same environment) for the coatings, the data were plotted with stress on the vertical axis and cycles-to-failure on the horizontal axis. If the average for the CS coatings at the high or low loads fell on or above the curves for the uncoated, then the CS coatings were considered to have met the acceptance criteria. Spalling or delaminating of coating prior to specimen fracture also constituted failure.

TABLE 6.5

ARL and Partner Programs Included in the Fatigue Study

Program	Main Objective	Partners
Environmental Security Technology Certification Program WP-0620 Supersonic Particle Deposition for Repair of Magnesium Aircraft Components	Depot Level Cold Spray Repair of US Navy Helicopter Gearbox Components at Fleet Reediness Center East (FRCEast)	Naval Air Systems Command (NAVAIR), US Army Aviation and Missile Command (AMCOM), Pennsylvania State University, Sikorsky Aircraft (Sikorsky), United Technologies Research Center (UTRC)
U-60 Sump Repair-OSD Manufacturing Technology	Repair Corrosion Damage on Inside Diameter of U-60 Sump	Sikorsky, MOOG
Australian Defense Science and Technology Organization (DSTO)	Repair Corrosion Damage on Sea Hawk Gearbox Components	DSTO and the Joint Strike Fighter Office

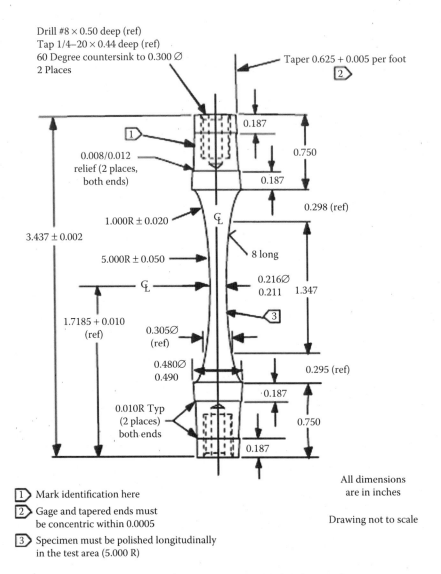

FIGURE 6.18 RR Moore specimen geometry for ESTCP WP-0620 fatigue study.

Two separate methodologies were used for the stress calculations. The first calculation type was used in the previous HCAT WC/Co programs. It will be referred to as the HCAT Approach in this report. The calculation uses only the original substrate diameter to calculate stress (Instron, 2004):

$$S_{HCAT} = \frac{16WL}{\pi c_{substrate}^3}$$

where:

S_{HCAT} = maximum stress (psi)
W = total load on specimen (pounds)
L = moment arm (in.)
$c_{substrate}$ = minimum diameter of the gage (in.)

The HCAT approach assumes that the coatings will bare negligible load compared to the bulk of the specimen. This approach is valid for nonstructural applications where the priority is insuring that the coating does not cause a detrimental fatigue debit. This second approach will be referred to as the two modulus approach. This method was used to examine the structural or load-bearing capability of the coatings. The method has been labeled as the two modulus approach by the authors because it takes into account the elastic modulii differences of the aluminum coatings as compared to the magnesium alloys. This approach also takes into account the moment of inertia of the sample.

It should be noted that a significant load will be transferred through the coating as compared to the substrate due to two separate issues. First, this is a bending beam test, so stresses are intrinsically higher at the surface then along the neutral axis based on the moment of inertia relationship:

$$S \propto \frac{\pi r^3}{4}$$

where S = stress at distance r from the neutral axis.

The magnesium specimen surface is approximately 0.15 in. from the neutral axis, so the top surface of a 0.015 in. coating is approximately 0.165 in. from the neutral axis. Using the above relationship, it can expected that the surface of the CS coating will have a stress level approximately 33% higher than at the surface of the magnesium substrate.

There is a second phenomenon that should be expected to further increase the load through the coating. The elastic modulus of aluminum is higher than magnesium, so greater load should be transferred through the coating. The elastic modulus of the magnesium alloys were taken from the literature. The elastic modulus of the 6061 CS was taken from the tensile testing performed in this study. The elastic modulus of the CP-Al N_2 CS was assumed to be similar to the 6061 CS due to the similar elastic modulii of wrought pure Al versus wrought 6061. The equation for the two modulus approach combines both the moment of inertia relationship and elastic modulus differences:

$$S_{\text{Two Modulus}} = \frac{S_{\text{HCAT}}}{\left(c_{\text{substrate}}/I_{\text{substrate}}\right)/\left(c_{\text{substrate+coating}}/I_{\text{substrate+coating}}\right)}$$

where:

$S_{\text{Two Modulus}}$ = maximum stress (psi)
S_{HCAT} = maximum stress using HCAT method (psi)
$c_{\text{substrate}}$ = radius of substrate in the gage (in.)
$I_{\text{substrate}}$ = moment of inertia at the substrate surface (in.^{4}):

$$I_{\text{substrate}} = \frac{\pi c_{\text{substrate}}^4}{4}$$

$c_{\text{substrate+coating}}$ = radius of specimen at coating surface (in.)
$I_{\text{substrate+coating}}$ = moment of inertia of substrate and coating (in.^{4}):

$$I_{\text{substrate+coating}} = \frac{E_{\text{CS}}}{E_{\text{Mg}}}\left[\frac{\pi c_{\text{substrate}}^4}{4} - \frac{\pi c_{\text{substrate+coating}}^4}{4}\right] + \frac{\pi c_{\text{substrate}}^4}{4}$$

E_{CS} = elastic modulus of the coating (psi)
E_{Mg} = elastic modulus of the magnesium alloy substrate (psi)

6.1.6.7 Summary of ARL Fatigue Results

There was a significant difference in the rotating beam fatigue performance for samples coated with CP-Al N2 versus the 6061 helium (Figure 6.19). A CS coating of CP-Al N2 onto the ZE41A-T5 and AZ91C-T6 did not debit the fatigue performance using the HCAT calculation method (original substrate diameter). Therefore, these coatings met the fatigue performance requirements for nonstructural repair for these two alloys.

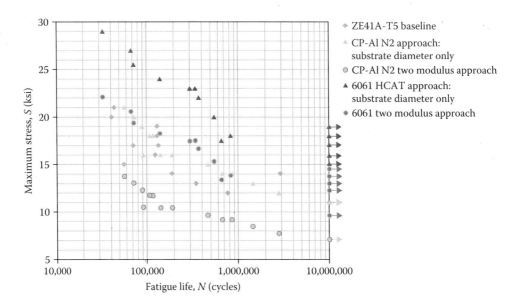

FIGURE 6.19 S-N Curve for 6061 helium and CP-Al N2 cold spray coatings on ZE41A-T5.

Applying the two modulus approach to the CP-Al N2 coating significantly alters the curve as compared the HCAT method. A noticeable debit compared to the baseline magnesium was observed. This shift of the curve indicated that the CP-Al N2 coating cannot be expected to enhance fatigue performance or act as a structural repair material for any of these three magnesium alloys. However, this does not invalidate using these materials for nonstructural or *cosmetic repairs* as noted in the above paragraph.

There was a noticeable fatigue credit compared to the baseline for 6061 helium CS coatings applied to all three alloys using the HCAT calculation method. In fact, the S-N curves showed a significant credit for all three alloys using this calculation methodology. This indicated that these coatings should not debit fatigue performance in a nonstructural repair environment and could potentially increase fatigue life. However, this is certainly dependent on numerous conditions in the specific area of the repair. Applying the Two Modulus calculation to the 6061 coatings shifted the curves back towards the baseline performance for both ZE41A-T5 and AZ91C-T6 substrates. This indicated that there is potential for using this CS coating for structural repairs. The two modulus approach to the 6061 coating on the EV31-T6 substrate indicated a slight fatigue debit. The debit is especially evident for the lower cycle samples. The overall lower fatigue performance of both coatings on EV31-T6 is possibly explained by the higher tensile properties of this alloy compared to the ZE41A-T5 and AZ91C-T6. (Magnesium Electron Inc. (Tamaqua, USA); ASM Handbook)

The focus of this work was nonstructural repairs such as the typical corrosion damage on the H-60 sump and H-53 main gearboxes. However, the 6061 fatigue data are one more indication that structural repair applications using CS could be possible. Further research of Al alloys to repair magnesium and other alloys is suggested. The fatigue performance of CS alloys was significantly dependent on processing parameters. Specifically, the carrier gas and the surface preparation are two key variables that should not be overlooked when designing CS coatings for dynamically stressed applications. Two studies performed by ARL, ESTCP, and DSTO, show that a properly applied CP-Al CS should not debit the fatigue performance of ZE41A-T5 and AZ91C-T6 magnesium alloys, as well as 7075-T651 aluminum alloy. From the literature, two studies show examples of increased fatigue performance for aluminum-based CS coatings on aluminum alloy substrates. It should be noted that ARL's studies focused on nonstructural repair, but data from ARL as well as the literature indicate that a potential exists for structural repair using the CS process.

There are two key papers from the literature that show the benefit of CS on fatigue performance. First, Sansoucy et al. (2007) investigated Al–13Co–26Ce CS deposited with a He carrier gas. The coating–substrate was aluminum alloy 2024-T3. Bending fatigue testing was performed in accordance to ASTM B593-96. Figure 6.20 shows the results of the fatigue testing. The CS outperformed the uncoated baseline 2024-T3 as well as 2024-T3 with an Alcad coating. The authors attributed the increase in fatigue performance to a compressive stress layer generated by the CS as well as the excellent coating adhesion. The second significant study regarding fatigue performance of CS in the literature is by Stoltenhoff and Zimmerman (2012). The application is repair of CH-53 helicopter landing gear shock strut assemblies and C-160 aircraft propeller blades. All of the coating systems and substrates in presented in this paper are aluminum based. A slight debit was reported for this coating–substrate system. A load-cycle graph showing a 7075 CS sprayed with a He carrier gas onto a 7075-T7351 is shown in Figure 6.21. An increase in fatigue performance as compared to an

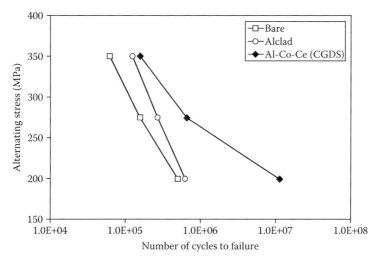

FIGURE 6.20 S-N curve generated by Sansoucy et al. (2007) showing cold spray outperforming bare aluminum alloy 2024-T3 and 2024-T3 coated with Alcad. (From ASM Handbook. *Properties of Pure Metals, Properties and Selection: Nonferrous Alloys and Special-Purpose Materials*, Vol. 2. ASM International, Materials Park, OH, 1990, pp. 1099–1201. With permission.)

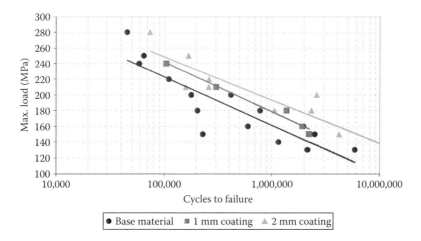

FIGURE 6.21 Load-cycle graph for a cold spray 7075 coating on 7075-T7351. (From Sansoucy E. et al., Mechanical characteristics of Al-Co-Ce coatings produced by the cold spray process. *J. Ther. Spray Technol.*, 16(5-6), 2007, 651–660. With permission.)

uncoated baseline was reported for both coating thicknesses. The better performance of the 2 mm thick coating compared to the 1 mm coating was attributed to compressive stresses (Stoltenhoff and Zimmermann, 2012). Finally, the fatigue resistance for a mixture of 2224 and 2014 aluminum alloy powders cold sprayed onto a 2024 alloy substrate was evaluated. Samples sprayed with a He carrier gas performed in a similar fashion as the baseline. However, the specimens sprayed with N_2 showed a significant debit in fatigue performance. The maximum load for the N_2 samples at given load-cycle changes is 20% lower than the baseline. The authors note that this debit is unacceptable for most applications.

6.1.7 CASE STUDIES

6.1.7.1 Forward Equipment Bay Panel

A common problem on military aircraft is wear due to chafing around fastener holes in skin panels. This chafing results in skin panels that exceed fit tolerances at the fastener locations. The South Dakota School of Mines and Technology Repair Refurbish and Return to Service (R3S) Research Center and the ARL Center for CS in cooperation with the 28th Bomb Wing at Ellsworth AFB SD, Air Force Engineering and Technical Services, Oklahoma City Air Logistics Center, and H. F. Webster Engineering Services developed a repair process for a B1 Bomber Left Upper Aft forward equipment bay (FEB) panel that had chafing damage to the fastener holes resulting in out of tolerance fit for the panel. The locations of these panels are shown in Figure 6.22 on the B-1. This FEB panel is secured to the airframe with 100° tapered flat head TRIDAIR fasteners. The fasteners are designed to be installed flush with the panel for laminar airflow over the skin surface. In service, the chamfer wears causing the fastener holes to become elongated, rendering the panels unserviceable (Figure 6.23). The damage is accelerated by air turbulence on elongated fastener holes. Chafing wear is caused by repeated opening and closing of the panel. Steel fasteners are used to secure the aluminum panes. The panel is made from 2024-T6 aluminum. The CS repair utilized the VRC Gen III ARL system and 6061 Al powder sprayed normal to the chamfered surface of the panel (Figure 6.24). Load transfer, fatigue, and tensile tests were accomplished, along with three-lug shear testing and metallography to characterize the repair. The results demonstrated the capability of CS to provide a permanent repair for this application, restoring the full capability of the panel. Cold sprayed coupons met or exceeded the required bearing loads for the parent material and fastener type for this application. Even when tested to failure (greater than 1.5 times the bearing yield), the CS material did not separate from the panel (Figure 6.25). Fatigue test results revealed that at 15 ksi tensile stress (typical upper end for aircraft skin loads) coupon lasted approximately 500,000 cycles

FIGURE 6.22   FEB panels on B-1. There are eight panels per aircraft, four panels per left- and right-hand sides.

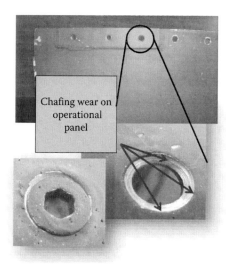

FIGURE 6.23 Wear sites under fastener heads on FEB panels.

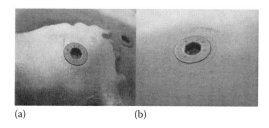

(a) (b)

FIGURE 6.24 (a) Cold-spray-repaired fastener hole as compared to a (b) OEM fastener hole on a new part.

FIGURE 6.25 Even when tested to failure at 5600 lbs, the cold spray material did not separate.

(Figure 6.26). Replacement panels cost as much as $225,000. CS technology has recently been approved as a low-cost (and high return on investment) solution to repair FEB panels. The first panel was repaired and installed on a B-1B for flight testing in August 2012. Cost savings to the B-1 program if this repair is approved for refurbishment of the remaining FEBs is estimated at $9.6 million annually. Further applications of this technology can be applied to all Major Design Series (MDS) aircraft, and is an example of infusing new technologies into current DoD maintenance processes to reduce sustainment cost while maintaining the viability of older weapon systems. If applied to other MDSs across the DOD, cost savings could reach $100 million annually.

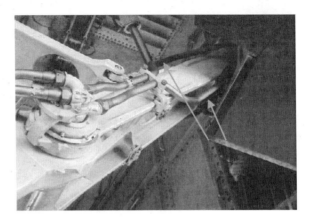

FIGURE 6.26   Wheel well titanium hydraulic lines that chafe during flight.

6.1.7.2   Hydraulic Line Repair

Chafing of titanium hydraulic tubing, Haynes AMS 4944 (Ti3Al2.5V) as a result of vibration and abrasive action is a major maintenance problem on the B-1B aircraft in terms of maintenance man hours (Figures 6.27 and 6.28). A technological solution that reduces the frequency of hydraulic tubing chafing would have broad applicability across the DoD and could be incorporated for use on similar commercial components. CS preventative maintenance of hydraulic lines has been proven to be fairly simple and effective from both an economic and technical standpoint. A feasibility study was accomplished in 2009 that demonstrated the effectiveness of the CS process in preventing hydraulic tube chafing by the application of a CP-titanium coating providing a wear surface in areas known to experience chafing problems. This preventative measure can be performed during programmed depot maintenance or during the high-velocity maintenance process to prevent or reduce occurrences of hydraulic tubing chafing in the field. A characterization study was accomplished that showed that a titanium (Ti) coating could be successfully applied to titanium tubing providing an additional *sacrificial* wear surface (Figure 6.29). The study showed that the Ti coating had adequate deposition efficiency (~70%), bond strength (>12 ksi), density (~99%), and hardness (93HRB). Additionally, the CS deposit was subjected to burst testing >16,000 psi and complete optical and electron microscopies (Figure 6.30) of the microstructure, bond line integrity, and porosity (<1%). An operational wear test was also accomplished to verify the results of the initial feasibility study.

FIGURE 6.27   B-1 spoiler actuator lines that were coated by the cold spray process.

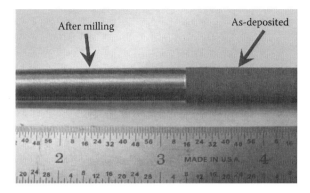

FIGURE 6.28   As-sprayed CP-Ti on the right and machined on the left.

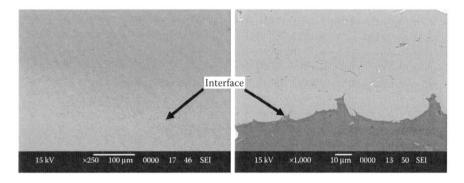

FIGURE 6.29   Scanning electron microscopy of the interface between the cold spray CP-Ti coating and the Ti3Al2.5V hydraulic line showing high density and mechanical mixing.

FIGURE 6.30   AH-64 Apache Helicopter Mast Support.

The test involved two B-1 aircraft from Ellsworth AFB, SD. They had two previously identified hydraulic lines (per aircraft) that have a high incidence of chafing. The hydraulic lines were sprayed with CP-Ti to prevent the chafing, Main Landing Gear Wheel Well (behind the follow-up door) & Wing Spoiler Actuator.

Once the coated lines were installed, an operational wear test was conducted for approximately 5 months. The test involved performing weekly dimensional measurements of the CS coating to determine the effectiveness of the coating, as it related to chafing prevention. The coating was applied to a B-1 nose landing gear (NLG) accumulator hydraulic line (commonly referred to as a curly-Q line), installed, and has been flown for 4 years with no adverse effects and no observed chafing beyond limits and have several thousand flight hours to date.

Since the conclusion of the wear test, proposed logistical and maintenance processes to implement the new process were developed, including the research and development of different coating materials and application parameters based on the outcome of the wear test. The different coating materials will account for varying wear mechanisms on different hydraulic tubing applications. Finally, an indicator layer/mechanism would be developed so that maintenance personnel can identify when the sacrificial layer has been worn through.

6.1.7.3 AH-64 Apache Mast Support Repair

Corrosion and mechanical damage has rendered a number of AH-64 Mast Supports nonserviceable for continued use on the AH-64 Apache Helicopter (Figure 6.30). The US ARL developed a portable CS repair that has been shown to have superior performance in the qualification tests conducted, is inexpensive, can be incorporated into production, and has been modified for field repair, making it a feasible alternative over competing technologies. The goal of this effort was to repair both the corrosion and mechanical damage by blending and machining damaged areas, rebuilding lost material using CS with aluminum powder and finish blending and machining to the original dimensions (Figure 6.31). The mast support is fabricated from Aluminum Alloy 7149 and therefore protective finishes such as a conversion coating, primer and topcoat may be applied to the repaired areas after the CS coating. While it is possible to perform a remediation without rebuilding lost material, the number of times this type of repair could be performed would be limited. With use of CS to rebuild lost material, the component could be remediated as many times as necessary until its safe life has been reached.

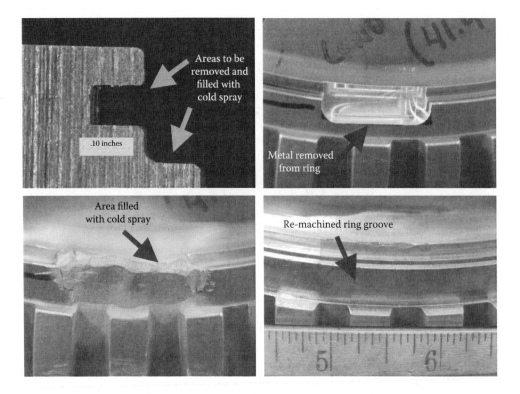

FIGURE 6.31 Step-by-step process to repair the snap ring groove of the AH-64 by cold spray.

6.1.8   Summary and Conclusions

6.1.8.1   CS Implementation into Aerospace

The CS process has matured from an emerging technology to a viable alternative to thermal spray for selected applications (Papyrin, 2001). Research and development efforts of Champagne have shown CS to be a promising cost-effective and environmentally acceptable technology to impart surface protection and restore dimensional tolerances to magnesium, aluminum, and titanium alloy components on helicopters and fixed-wing aircraft (Champagne, 2007; Champagne et al., 2008). Corrosion of magnesium helicopter gearboxes is a major issue both for the DoD and the commercial aircraft industry. The use of CS coatings may be a means to reduce subsequent corrosion damage in gearboxes, especially if it can be combined with improved protection systems, such as Tagnite coatings. CS repairs offer improved performance and permit the reclamation of properties as well as dimensional restoration, and because replacements would be less frequent it would also reduce logistics cost, because fewer gearboxes would be required in the field, and fewer would have to be shipped back and forth between depots and operating bases. There is significant potential for cost reduction by decreasing the number of condemnations and improving operational readiness and safety.

Barbosa et al. (2009) have presented research showing that CS could be used to produce a dense pure titanium coating onto aluminum 7075 substrates, with thickness greater than 300 μm with no deleterious microstructural changes. It was concluded that the CS process could be optimized to produce superior coatings to conventional thermal spray techniques, economically and at higher deposition rates, making it ideal for aerospace applications.

Stoltenhoff and Zimmermann (2012) showed the potential of CS as a repair technology using three different precipitation-hardened aluminum alloys; AA2224, AA6061, and AA7075 for use on CH-53 helicopter landing gear shock struts and C-160 aircraft propeller blades. One of these aerospace applications had already been transferred to production and the other was in the approval process. The fatigue properties of the coating–substrate system were identified as key evaluation criteria for acceptance, and as previously stated, the CS coatings produced at optimized conditions do not adversely affect the component properties. In fact, an improvement of the fatigue resistance was observed for several coated samples, most likely the result of the beneficial compressive stresses induced as a result of the CS process in the coating that helps to avoid crack formation and propagation. The authors note that when service conditions are taken into account, including single events that cause superficial damages like impacts of foreign bodies or defects from tool marks during overhaul and maintenance.

Dr Kumar from the International Advanced Research Centre for powder Metallurgy and New Materials (ARCI) Hyderabad, India has reported using the CS process on a variety of applications to repair gas turbine components including 6061 Aluminum fan casings susceptible to damage where high thermal processes, such as welding could not be tolerated. Another application was the titanium compressor case isogrid fabrication, where titanium CS was used to build up areas to minimize starting stock material and machining waste. This same approach was also successful on other parts using other materials including copper, aluminum, and zinc as well as for other parts including diffusers, combustors, turbines, and exhaust nozzles. Kumar also reported success in the repair of aluminum and magnesium aerospace components such as the S-70 Seahawk and for C-160 Transall propeller blade repair (Kumar and Chavan, 2012).

6.1.8.2   Specification Development

ARL recognized the need to standardize the CS process to facilitate widespread adaptation of the technology and to mitigate the potential misuse of the newly developed repair process for magnesium. The specific concern was the practice of using low-pressure/temperature portable CS systems to repair structural areas on magnesium components. Therefore, two actions were taken; first was to introduce a general CS process specification that could be utilized by the DoD and private industry, throughout the nation, as well as internationally, and the second, to provide the means to educate the user community about CS. In 2008, the Defense Standardization Program (DSP) Office granted approval of the first military specification developed by ARL, MIL-STD-3021, which is a manufacturing process standard, entitled *Materials Deposition, Cold Spray*. The procedures covered by this standard are intended to ensure that CS coating operations, either manual or automated, meet prescribed requirements. ARL will also be introducing another specification in 2014, that covers requirements intended for use in the procurement of powders that will be used to produce coatings utilizing the CS materials deposition process, entitled *Material Powders Utilized for Cold Spray*. These documents are necessary to standardize the feedstock and the CS process to facilitate implementation across major industries.

6.2 CORROSION-PROTECTIVE COATINGS DEPOSITED BY THE LOW-PRESSURE COLD SPRAY PROCESS

Dmitry Dzhurinskiy

Emil Strumban

6.2.1 INTRODUCTION

Materials such as iron, aluminum, magnesium, and their alloys are used in a wide variety of structural, automotive, aircraft, and shipbuilding applications. While these materials are broadly applicable due to their physical characteristics, such as stiffness and high strength-to-weight ratios, they are quite susceptible to corrosion (Nie, 2014).

The total annual cost of damages caused by corrosion is estimated to be around 3%–5% of the gross domestic product of a country (Schmitt et al., 2009).

Therefore, efficiently increasing the corrosion resistance of metal components is one of the most important technological goals of any country.

A generic approach to protecting metals from corrosion is to apply protective coatings, which preserve the desired properties of the metal substrate in corrosive environments. As of today, corrosion protection is one of the most important and largest areas of coating application in terms of its business significance. The presently used in the industry corrosion-protective coatings include barrier coatings, sacrificial coatings, hard coatings as well as conversion and organic coatings (Gray and Luan, 2002). The techniques widely applied for the deposition of coatings on metals include physical and chemical vapor deposition, electrochemical deposition, plasma spraying, and sol–gel process (Davis).

A relatively new coating technology called cold spray has shown encouraging results in the field of corrosion-resistant coatings because the cold spray has certain advantages over conventional thermal spray technologies and allows for tailoring corrosion-resistant properties of the formed coatings (Papyrin et al., 2007). The process will be briefly described below. A more detailed description of the cold spray process can be found in other chapters of this book.

6.2.2 COLD SPRAY TECHNOLOGY

Cold Spray is a low-temperature material-deposition process where particles of diameters between 1 and 100 microns are impacted at high velocity onto a substrate. The particles remaining in solid state are entrained by a high-pressure gas stream (air or nitrogen or helium), which is accelerated through a de Laval supersonic nozzle to speeds ranging from 500 to 1500 m/s. On impact with the substrate, the particles are consolidated forming a coating. Because of low deposition temperature (below the melting point of particles), the cold spray process can be successfully used to form high-density/very low-porosity, low residual stress deposits of a wide range of temperature-sensitive materials, which makes it advantageous comparing to any other thermal spray process (Champagne, 2007).

The prevailing bonding theory in cold spraying is attributed to so-called adiabatic shear instability (Hussain et al., 2009), which occurs at the particle–substrate interface at a certain particle velocity called critical velocity. It is well established that particles impacting on the substrate must exceed the critical velocity to form a deposit instead of being bounced off. The magnitude of the critical velocity depend on particle material parameters, such as density, ultimate strength, and melting point as well as the particle temperature. Particles of different materials have considerably different critical velocities, which define their deposition conditions by the cold spray process (Schmidt et al., 2006). That allows us to use two different types of machines for cold spray coating deposition: (1) High-Pressure Cold Spray machines, which operate utilizing high-pressure (about 20 bar) compressed gas (helium or nitrogen) and (2) Low-Pressure Cold Spray (LPCS) machines, which operate utilizing relatively low pressure (about 5–8 bar) compressed gas (air). Several powder materials, including Al and Al-based alloys, have the critical velocity is in the range of 400–500 m/s and can be deposited by employing LPCS machines. The important advantages of LPCS machines include their portability, mobility, possibility of using inexpensive compressed air as a carrier gas, and high durability of the spray nozzle. All of that allows us to use LPCS machines in both the field conditions as well as in the industrial environment (Maev et al., 2008).

6.2.3   Techniques for Corrosion Behavior Evaluation of Coating/Substrate Systems

Because various corrosion phenomena can be explained in terms of electrochemical reactions, the most used for corrosion behavior evaluation of protective coating/substrate systems are electrochemical methods. Among these methods, most frequently used are polarization, electrochemical impedance spectroscopy, and electrochemical noise analysis techniques.

6.2.3.1   Polarization Technique

Polarization (also called potentiodynamic or potentiostatic) technique, can provide very useful information regarding the corrosion mechanisms, corrosion rate (CR), and vulnerability of specific materials to corrosion in certain environments. Polarization method involves changing the potential of the working electrode and monitoring the current that is produced as a function of time or potential (Uhlig's Corrosion Handbook, 2011). For inert coatings that are not susceptible to corrosion in certain potential ranges, polarization measurements yield the electrical currents due to dissolution of the substrate material through micropores (Rojas and Rodil, 2012). This current value is one of the main characteristics gained from current density–potential measurements for evaluating coating/substrate systems. Other important parameters are the corrosion potential, the passivation range, and the pitting potential. The corrosion potential is the voltage difference between a metal immersed in a given corrosive environment and an appropriate standard reference electrode, which has a stable and well-known electrode potential. The passivation range is a potential region where the sample is protected by a passive film and is characterized by a low anodic current density. Pitting potential is the least positive potential at which pits can form and start propagating. The breakthrough or pitting potential can be determined by an abrupt increase of the anodic current density. This current surge is caused by local breakdowns of the natural protective film due to corrosion reaction. The polarization measurement are often run as a multicycling method where the potential is scanned repeatedly from a starting potential to a reverse potential and back. The procedure is called cyclic voltammetry (Fenker et al., 2014).

6.2.3.2   Electrochemical Impedance Spectroscopy

This electrochemical technique is applied to study an ongoing corrosion process in protective coating/metal substrate systems. The method allows to measure the dielectric properties of materials and structure as a function of frequency by applying a time-varying voltage and registering the current response (Chang and Park, 2010). By measuring the ratio of the two, the frequency-dependent impedance can be recorded. One of the main difficulties with this technique is the proper selection of an equivalent circuit, which should always be based on a physical model of the corroding system. The acquired data are plotted as Nyquist and Bode (impedance versus frequency) plots and the equivalent circuit is used as a basis for the fitting of these plots (Orazem et al., 2006).

6.2.3.3 Open Circuit Potential

The open circuit potential (OCP) is the potential of the working electrode relative to the reference electrode when no potential or current is being applied to the corrosion cell. When a potential is applied relative to the open circuit, the system measures the OCP before turning on the cell, then applies the potential relative to that measurement. The OCP indicates the tendency of a material to electrochemical oxidation/passivation in a corrosive medium. The OCP allows to analyze the electrochemical reaction taking place on the specimen surface without affecting the reaction in any way (Ahmad, 2006). After a period of immersion, OCP usually stabilizes around a stationary value. Nevertheless, the potential may vary with time and oscillate because of changes in the state of the electrode surface due to oxidation or formation of a passive layer (Hwang et al., 2010).

6.2.3.4 Electrochemical Noise Analysis

This is a relatively new technique currently used for corrosion monitoring. The method does not disturb electrochemical processes at electrode surface and is based on the observation of spontaneous current and voltage fluctuations (electrochemical noise) caused by electrochemical processes that induce charge flows between the metal electrode and the electrolyte. The application of this technique is particularly appealing for in situ monitoring of passive samples as no agitation is required. It can be useful for assessing the beginning of localized corrosion or stress corrosion cracking (Smulko et al., 2007; Covac et al., 2010).

6.2.4 CORROSION PROTECTION OF ALUMINUM ALLOYS BY COLD-SPRAY-DEPOSITED COATINGS

Because aluminum and magnesium alloys are very important in aerospace and automotive manufacturing, a number of studies have been conducted on the applicability of cold spray coatings to corrosion protection of such alloys (Lee et al., 2005; DeForce et al., 2007; Spencer et al., 2009; Tao et al., 2009). Deposition of corrosion protective coating on aluminum and magnesium alloys by the LPCS technique presents a special interest due to the possibility of applying the coatings in field conditions and in repair shops (Koivuluoto et al., 2008; Yandouzi et al., 2014; Champagne et al., 2015).

Two examples demonstrating effectiveness of cold-spray-deposited coatings for corrosion protection of high-strength and aluminum (AA2024) and magnesium (AZ31) alloys are given below.

6.2.4.1 High-Strength Aluminum Alloys

High-strength aluminum alloys such as AA2024, containing copper as primary alloying element are used for structural applications in aircraft industry due to their good fatigue resistance and high-strength/weight ratio (Heinz et al., 2000; Nakai and Rto, 2000).

However, it is also known that the copper-rich aluminum alloys are prone to corrosion. In chloride solution, such types of localized corrosion as pitting, galvanic corrosion, and intergranular corrosion, including exfoliation corrosion have been observed (Boag et al., 2010, 2011; Hughes et al., 2011). The localized corrosion of high-strength aluminum components can lead to their sudden failures and result in considerable loses. To protect high-strength aluminum parts from localized and other types of corrosion, protective coatings have to be used (ASM Handbook, 2003).

A systematic study of the dependence between $Al–Al_2O_3–Zn$ powder spray composition and the corrosion protective properties of the LPCS-deposited coatings on high-strength AA2024-T3 Alclad substrates was performed in Dzhurinskiy et al. (2012). The details of this work are presented below.

Coating Materials. The following precursor materials have been used for the LPCS process-based coating deposition: Al metal powder, Zn metal powder, and alumina Al_2O_3 powder. The precursor powder compositions have been varied as shown in Table 6.6.

TABLE 6.6
Specimen Component Contents

Specimen	Composition, vol. (%)		
Designation	Al	α-Al₂O₃	Zn
CP1	100	0	0
CP2	75	25	0
CP3	50	50	0
CP4	30	50	20
CP5	35	40	25

The corrosion resistance of the protective coatings was evaluated using the OCP and potentio-dynamic polarization techniques. The samples were placed in 1M sodium chloride solution at room temperature and subjected to the accelerated corrosion test according to ASTM B117. The CR was calculated in accordance to ASTM G102.

OCP Monitoring. The evolution of the OPS for samples immersed in chloride solution is shown in Figure 6.32a. As can be seen, an abrupt decrease in OCP takes place around 15-min mark,

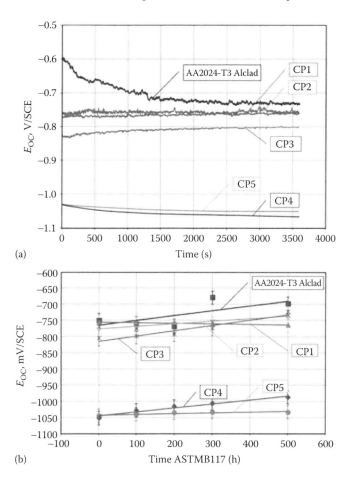

(a)

(b)

FIGURE 6.32 Evolution of the open circuit potential (E_{OC}) for the investigated samples: (a) E_{OC} as a function of immersion time in corrosive media and (b) E_{OC} as a function (linear regression) of accelerated salt fog corrosion test.

indicating that the tendency to corrode is increased. This initial fall is ascribed to the destruction of a virgin aluminum oxide film formed during the cold mill rolling process. The OCP of the Alclad sample gets stabilized at 730 mV/SCE. The corrosion behavior of the examined cold sprayed coatings depends on the powder feedstock composition. The OCP of the cold sprayed coating made of pure aluminum (CP1) does not depend on immersion time in the sodium chloride media. That reveals the presence of a relatively stable protective oxide film on the coating surface. The as-sprayed CP2 and CP3 coatings exhibit the slight change of the OCP towards to the negative value during immersion time, and the potential stabilizes at −762 mV/SCE for CP2, and −801 mV/SCE for CP3, respectively, whereas the Al–Zn–Al$_2$O$_3$ coatings exhibit the decrease of the potential up to −1050 mV/SCE. It should be noticed that the higher is the volume % of Zn in the sprayed powder composition, the more negative OCP potential is observed, which can be linked to the sacrificial behavior of Zn as was demonstrated in (Baker et al., 2000; Guzman et al., 2000).

The OCP oscillation behavior that displays itself as data noise is another characteristic feature of the potential monitoring. The observed oscillation is usually attributed either to the initiation of pitting or to the passive film breakdown. The parameters of the oscillation depend on the surface roughness and electrochemical reaction kinetics (Tao et al., 2010). As can be seen in Figure 6.32a, the amplitude of the OCP oscillation for the Alclad substrate (surface roughness Ra = 0.6 μm) is approximately 8 mV. The amplitude of the OCP oscillation for the Al coating (surface roughness Ra = 12.5 μm) is about 2 mV. That may indicate that the Al coating is dissolving more uniform as compared with AA2024-T3 Alclad substrate surface.

The kinetics of the corrosion process may be evaluated in terms of the dependence of the OCP on the corrosion time. The experimental data of the OCP variation with time $E_{corr} = f(t)$ are approximated by linear functions as shown in Figure 6.32b. It can be seen that the OCP of pure Al coating (CP1) remains stable during the entire test time, while the Al–Al$_2$O$_3$ coatings (CP2 and CP3) become more noble due to permanent oxidation in a corrosion environment that results in a growth of the dense oxide film. The behavior of Al–Zn–Al$_2$O$_3$ coatings (CP4 and CP5) differ from other coating behavior due to the sacrificial effect of Zn and results in higher CR due to providing sacrificial corrosion protection to AA2024-T3 Alclad substrate.

Potentiodynamic Polarization Measurements. The potentiodynamic DC polarization testing was conducted to gain detailed understanding of the coating corrosion properties. The potentiodynamic polarization curves of uncoated AA2024-T3 Alclad substrate and as-sprayed CP1, CP2, and CP3 coating layers are presented in Figure 6.33a. It can be seen that the corrosion potential E_{corr} is decreasing with an increase of alumina content in the CP1, CP2, and CP3 coatings. The corrosion potential is about 720 mV/SCE for the AA2024-T3 Alcald, whereas for the CP1 coating it is −738 mV/SCE that is in agreement with (Guillaumin and Mankowski, 1998; DeForce, 2007). For the CP2 and CP3 coating layers in as-sprayed condition, the corrosion potential becomes about −1080 mV/SCE.

The polarization curves of the examined composite coatings CP2 and CP3, depicted in Figure 6.33a, exhibit two breakdown potentials: the more active E_{b1} is close to −1080 mV/SCE and the nobler E_{b2} is close to −760 mV/SCE, where the region between E_{b1} and E_{b2} is a passivity area in which the current density remains practically constant. Each of these breakdown potentials presents a threshold of the anodic current density sharp growth. According to Maitra et al. (1981), the breakdown potential E_{b2} corresponds to the dissolution potential of an Al matrix, whereas E_{b1} is associated with the dissolution of the aluminum along the boundaries between aluminum and alumina particles. Thereby the corrosion protection efficiency was different with an increase of ceramic particles content in the coatings.

The kinetics of the corrosion potential evolution during an accelerated corrosion test is shown in Figure 6.33b as a function of the emersion time. The obtained results show two types of corrosion potential behavior with time $E_{corr} = f(t)$: (1) E_{corr} increases with time, and (2) E_{corr} decreases with time. This reveals two coating corrosion mechanisms: noble for CP1, CP2, and CP3 (associated with the growth of oxide film), and sacrificial for CP4 and CP5 (associated with the dissolution of the coating layer due to the presence of Zn).

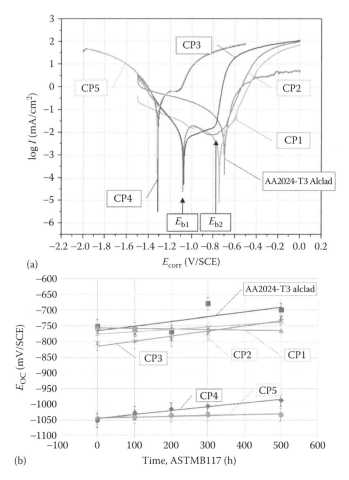

FIGURE 6.33 Evolution of the corrosion potential (E_{corr}) for the investigated samples: (a) potentiodynamic DC polarization curves before corrosion test; (b) $-E_{corr}$ as a function (linear regression) of accelerated salt fog corrosion test.

The behavior of corrosion current (I_{corr}) is shown in Figure 6.34. The I_{corr} is poised at 0.2 μA/cm^2 for the AA2024-T3 Alcald substrate and at 0.8 μA/cm^2 for the CP1 coating. Contrary, the I_{corr} dependence of the CP2 and CP3 coatings decreases from 1.3 to 1.05 μA/cm^2 and from 0.9 to 0.3 μA/cm^2, respectively. Thus, the experimental results show that LPCS-deposited Al powder coating layers have the best corrosion protection properties as compared to other composite coatings studier in this work. Also, the results of DC polarization measurements are in agreement with the OCP measurements, where it was demonstrated that the potential depends on the content of Al$_2$O$_3$ within the coating.

Al–Zn–Al$_2$O$_3$ coatings exhibit significantly higher I_{corr} values than those of AA2024-T3 Alclad substrate (46 μA/cm^2 for CP4 and 50 μA/cm^2 for CP5). The E_{corr} of CP4 and CP5 coatings is significantly lower than that of Alclad substrate, which reveals an active dissolution of the Al–Zn–Al$_2$O$_3$ composite. This behavior confirms that the CP4 and CP5 coatings are providing a sacrificial corrosion protection mechanism. A comparison of the corrosion behavior of LPCS-deposited Al–Al$_2$O$_3$ and Al–Zn–Al$_2$O$_3$ composite coatings shows that the formation of the new interfaces, may significantly change the corrosion mechanism from a noble to sacrificial corrosion protection.

CR Calculation. The CR calculation results depicted in Figure 6.35 reveal that all LPCS-deposited Al–Al$_2$O$_3$ coatings have approximately the same CR. These data may be explained by the fact that galvanic corrosion between Al$_2$O$_3$ and Al particles is unlikely to take place because

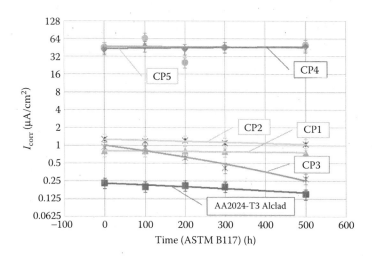

FIGURE 6.34 Evolution of the corrosion current (I_{corr}) as a function (linear regression) of accelerated salt fog corrosion test.

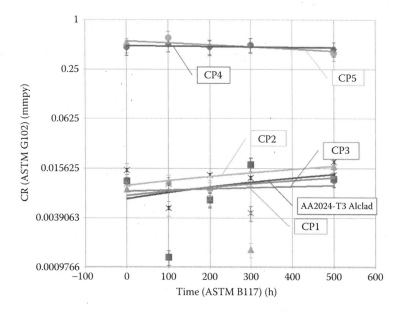

FIGURE 6.35 Evolution of corrosion rate as a function (linear regression) of accelerated salt fog corrosion test environment.

the resistivity of alumina is greater than 10^{14} Ω·cm (DeForce et al., 2007). The higher CRs of the Al–Zn–Al$_2$O$_3$ coatings (CP4, CP5) may be explained by the action of both the Cl$^-$ ions in the NaCl solution and the galvanic coupling effect between the noble Al and the much more active Zn particles resulting in the formation of Zn$_5$(OH)$_8$Cl$_2$·H$_2$O and ZnO corrosion products (Li et al., 2010). It is important to note that the presented CR values should be considered as CR estimations only.

Coating Microstructure Analysis. The analysis of the LPCS-deposited coatings passivation behavior is based on an evaluation of the coating morphology. The coating surface was examined using scanning electron microscopy (SEM) and the chemical composition of the coating layers surface was determined by energy dispersive X-ray spectroscopy (EDX). The images depicted in Figure 6.36 show that the coating morphology is quite different for the as-sprayed coatings. An increase of the reinforcing ceramic phase in the coating leads to an increase of the coating

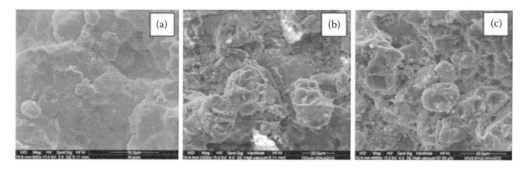

FIGURE 6.36 SEM images of the coating layers topography: (a) Al feedstock powder coating (CP1), (b) 75% Al–25% Al$_2$O$_3$ feedstock powder coating (CP2), and (c) 30% Al–20% Zn–50% Al$_2$O$_3$ feedstock powder coating (CP3).

roughness thereby increasing the number of sites where corrosion attack can be initiated. It was reported (Motzet and Ilmann, 1999; Li et al., 2006) that the coating morphology plays a critical role in corrosion propagation over the sprayed surface, where the higher coating roughness corresponds to lower pitting potential. That has been confirmed by the OCP measurement results shown in Figure 6.32. The observed amplitude of the OCP oscillations is decreasing with the coating roughness increase.

Micrographs of the LPCS-deposited coating surfaces after 500 hour immersion in the corrosion environment in accordance with ASTM B117 are shown in Figure 6.37. The CP1, CP2, and CP3 coating look alike having a similar homogeneously dense scale formed on the surface. However, image analysis revealed that the fraction of the surface area occupied by white rust increases with an increase of alumina content, to approximately 55% for CP1 and 65% for CP2, respectively.

The granular oxide scale around the aluminum particle can be seen at higher magnification. Figure 6.38b shows the region of CP1 coating where aluminum phase islands (dark gray) are found. EDX analysis confirmed that dark grey areas on the image are associated with aluminum, whereas white gray areas correspond to tetrahydroxoaluminate oxide layers. The significant increase of the oxygen content is believed to be associated with the formation of Al$_2$O$_3$·2H$_2$O. According to Takahashi et al. (2010), Pourbaix (1975), Hong and Nagumo (1997), and Sasaki and Burstein (1996), in chloride solution, aluminum ionizes rapidly to the Al^{3+} ion, which in its turn also rapidly hydrolyses: Al^{3+} + 3Cl$^-$ → AlCl$_3$; and AlCl$_3$ + 6H$_2$O → 2Al(OH)$_3$ + 3HCl that is transformed slowly to Al(OH)$_3$ and finally to Al$_2$O$_3$·2H$_2$O, which plays an important role in the passivation of aluminum. Figure 6.38a shows the needle crystal nucleation of the aluminum oxide. Fine-grain lamellas have

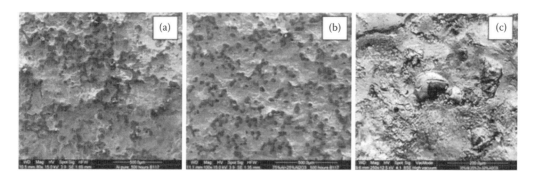

FIGURE 6.37 SEM images of the coating layers topography after immersion in corrosive environment for 500 hours (ASTM B117): (a) Al feedstock powder coating (CP1), (b) 75% Al–25% Al$_2$O$_3$ feedstock powder coating (CP2), and (c) 30% Al–20% Zn–50% Al$_2$O$_3$ feedstock powder coating (CP3).

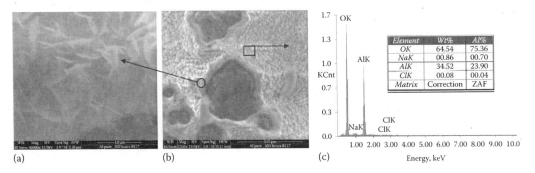

Element	Wt%	At%
OK	64.54	75.36
NaK	00.86	00.70
AlK	34.52	23.90
ClK	00.08	00.04
Matrix	Correction	ZAF

(a)   (b)   (c)

FIGURE 6.38 Magnified SEM images of the coating layers topography: (a) Al_2O_3 nucleation around Al particle, (b) dense passivation layer on the coating surface after 500 hours of salt fog environment, and (c) EDX analysis of the corroded passivation layer composition.

a leaf-like structure with leaves the size of about tens of nanometers. Such needle structure of $Al_2O_3 \cdot 2H_2O$ is known to be typical for aluminum oxide nucleation and its crystal growing (Metzeger and Zahavi, 1979; Tian et al., 2009).

6.2.5 Corrosion Protection of Magnesium Alloys by Cold-Spray-Deposited Coatings

One of the most effective ways to minimize galvanic corrosion of magnesium–aluminum–zinc alloys currently used for efficient weight reduction in the transportation sector is to apply protective coatings (Jia et al., 2006). Of special interest is AZ31 alloy with its high strength-to-weight ratio. However, it is known that AZ31 alloy is prone to corrosion as well as wear, especially in joints with other materials such as aluminum or steel (Kainer and Kaiser, 2003; Pan and Santella, 2012). A study demonstrating the possibility to form LPCS coatings on magnesium–aluminum–zinc alloy components to provide high corrosion resistance was published in Dzhurinskiy et al. (2015). The details of this work is given below.

Coating Materials. The following precursor materials have been used for the LPCS process-based coating deposition: Al metal powder, Zn metal powder, and Al_2O_3 powder. Powder blends were obtained by mechanically mixing the powder composition components for 10 hours. The prepared powder compositions are listed in Table 6.7.

Corrosion-protection coatings were deposited on AZ31 magnesium alloy substrates (chemical composition in wt-%: Al 3.1; Zn 0.73; Mn 0.25; Si 0.02; Cu <0.001; Fe <0.005; Ni <0.001; Ca <0.01; Zr <0.001, and Mg balance). Corrosion behavior of the formed protective coatings was studied by conducting the accelerated salt-spray tests according to ASTM B117. Samples were subjected to salt fog environment for 25, 50, and 194 hours, respectively using the ATLAS CXX advanced cyclic corrosion exposure system.

OCP Monitoring. To evaluate the corrosion performance and estimate the tendency of the formed coatings to electrochemical oxidation, the coated samples were suspended in the 3.5 wt-%

TABLE 6.7
Specimen Component Contents (Zn/Ni Ratio is Constant)

Specimen (Powder Composition-PC)	Composition, vol. (%)			
	Al	α-Al$_2$O$_3$	Zn	Ni
PC A	100	–	–	–
PC B	–	–	75	25
PC C	–	20	60	20

NaCl solution (pH = 7) for 20 min at room temperature. The evolution of OCP for AZ31 as a reference material and AZ31 coated with three different coating compositions (PC A, PC B, and PC C) are shown in Figure 6.39a and b.

As can be seen in Figure 6.39a, after immersion in solution the OCP of AZ31 increases, then slightly drops, after that re-establishes its value, and finally stabilizes at around −1600 mV. Such behavior indicates that initially a corrosion-decelerating passive film is formed, then the dissolution of the film takes place with its subsequent restoration, which is in agreement with the reported data (Jia et al., 2005; Ismail and Virtanen, 2007). Another reason for the cyclic behavior of the OCP could be associated with the variation of pH at the AZ31 surface (Salman et al., 2010). In the conducted experiments, the pH value was recorded before and after each test. The electrolyte solution was observed to change from a neutral solution with a pH about 7.1 to a more basic solution with a maximum pH of 10.5. The pH increase can be attributed to the production of OH^- in the electrolyte, which in its turn conduces to the formation of the corrosion-decelerating Mg hydroxide film in the alkaline environment (Shi and Atrens, 2011).

The OCP of the coated AZ31 samples behaves differently from the OCP of uncoated sample. First of all, the OCP of the coated samples are shifted toward noble potential values, which indicates the presence of the corrosion-inhibiting layer on the surface of the AZ31 sample. The OCP of the coated samples are also dependent on the coating composition. Thus, the OCP of aluminum-coated samples (PC A) shows a substantial decrease toward negative potentials within the first 6 min of immersion in the electrolyte, eventually stabilizing at −1200 mV after 15 min of immersion, which can be related to the distraction of the aluminum oxide initially formed on aluminum coatings (El-Sayed, 2011). The OCPs of the coated samples PC B and PC C display similar behavior, demonstrating a slight shift toward nobler potentials and stabilizing at 1100 mV, after 20 min of immersion in the electrolyte. It can be assumed that by providing a relatively stable OCP, the coating compositions PC B and PC C are able to slow down the corrosion reaction on the surface of AZ31 samples.

The conducted accelerated salt fog corrosion tests showed that the corrosion protection properties of the coating compositions PC A, PC B, and PC C considerably differ with exposure time (Figure 6.39b). The OPC of PC A samples shifts toward noble potentials that can be attributed to the formation of a stable protective aluminum oxide film on the coating surface. The OCP of PC B and PC C samples gradually decrease over the period of 200 hours. The OPC decrease of PC B sample can be associated with a slow corrosion of the coating with formation of a corrosive product $ZnCl_2·4Zn(OH)_2$ (Jiang et al., 2005). Similar behavior was observed for PC C sample.

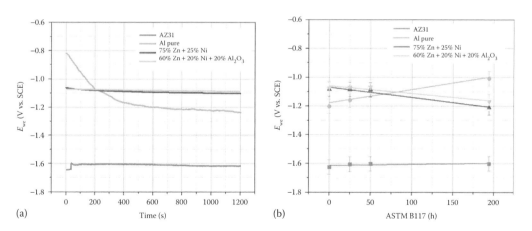

FIGURE 6.39 Evolution of OCP (E_{we}): (a) E_{we} versus sample immersion time in 3.5 wt.% NaCl solution (b) E_{we} versus time (linear regression) for samples exposed to the accelerated salt fog corrosion in accordance with ASTM B117.

However, the CR of the sample was lower that can be attributed to the presence of alumina in the coating composition (Dzhurinskiy et al., 2012). It is important to note that corrosion resistance of all coated samples was considerably higher than the corrosion resistance of bare AZ31 samples.

Potentiodynamic Polarization Measurements. The potentiodynamic polarization curves and kinetics of the corrosion potential evolution during the accelerated corrosion test of the AZ31, PC A, PC B and PC C samples is shown in Figure 6.40.

The corrosion potential and corrosion current density changes for all examined samples are shown in Figure 6.40a. It can be seen that the bare AZ31 samples are subjected to corrosion much more actively than the coated samples. Such behavior is usually associated with the interaction between AZ31 and chloride ions Cl⁻ that are present in the electrolyte solution. It is generally accepted that Cl⁻ ions transform the relatively stable $Mg(OH)_2$ into soluble $MgCl_2$ and, therefore, accelerating the corrosion of magnesium (Song et al., 2003).

The PC A, PC B and PC C samples are much less vulnerable to corrosion than AZ31 alloy. Slow corrosion of PC A coating can be linked to the reaction of the ionized aluminum Al^{3+} with chloride ions Cl⁻ that leads to the formation of aluminum chloride $AlCl_3$, which turns initially into aluminum hydroxide $Al(OH)_3$ after reacting with water molecules and eventually into alumina Al_2O_3 (Breslin et al., 1994; Bonora et al., 2002).

The corrosion mechanism of coating compositions contained Ni and Zn components (PC B, PC C) may be explained by the occurrence of micro-galvanic coupling between the noble Ni and the much more active Zn particles (Muller et al., 2001). Under the influence of the corrosive environment in the presence of Cl⁻ ions, zinc corrodes, forming preferentially $Zn_5(OH)_8Cl_2 \cdot H_2O$ and ZnO, and leaving the top layer of the coating enriched with nickel. That enriched nickel layer acts as a barrier to further corrosion (Abdel, 2008). The results of the E_{corr} as a function of time exposed to the accelerated corrosion test ASTM B117 are shown in Figure 6.40b. There are two types of corrosion behavior can be seen: (i) E_{corr} increases with time for PC A coatings indicating noble type, associated with the growth of the aluminum oxide film, for PC A coating and (ii) E_{corr} decreases indicating sacrificial type, associated with the slow dissolution of the Zn-containing coating layer with time, for PC B and PC C coatings.

Corrosion current I_{corr} behavior for the investigated samples is shown in Figure 6.41. As can be seen, exposing AZ31 to salt fog corrosion environment leads to considerable increase of I_{corr}, which indicates that an active corrosion process takes place (Pardo et al., 2008). The I_{corr} for PC B samples slightly increases with the exposure time to salt fog environment, while for PC A and PC C samples the I_{corr} remains almost constant. It indicates that samples PC A, PC B and PC C are substantially more resistant to corrosion comparing to AZ31 and can be used as corrosion protective coatings.

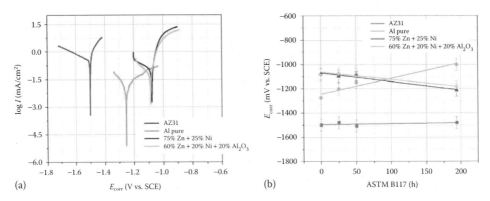

FIGURE 6.40 Evolution of the corrosion potential (E_{corr}) for the investigated samples: (a) potentiodynamic polarization curves and (b) corrosion potential E_{corr} versus time (linear regression) for samples exposed to accelerated salt fog corrosion in accordance with ASTM B117.

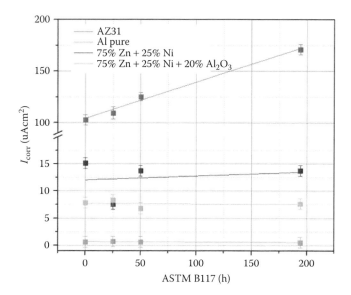

FIGURE 6.41 Evolution of the corrosion current (I_{corr}) versus time (linear regression) for samples exposed to the accelerated salt fog corrosion test.

Corrosion behavior of the investigated samples subjected to the accelerated salt corrosion test was also characterized by the CR (expressed in mm/year) in accordance with ASTM G102 testing procedures.

The estimated CRs as a function of time in the corrosive environment for all samples are shown in Figure 6.42. As can be seen, the CR of the AZ31 samples is high and substantially grows with time spent in the corrosive environment. However, the CR of the PC B samples is lower and growth with time at a much slower rate, while the rate of corrosion for PC C sample remains practically

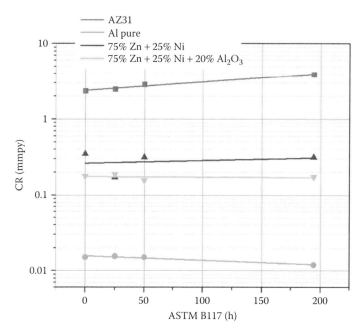

FIGURE 6.42 Corrosion rate versus time (linear regression) for samples subjected to the accelerated salt fog corrosion test in accordance with ASTM B117.

constant with time that can be attributed to the presence of alumina in the coating composition (Dzhurinskiy et al., 2012), which apparently allows to deposit a denser coating structure. A denser coating may prevent the formation of microgalvanic cells between Zn and Ni grains and, therefore, increase corrosion resistance of the PC C samples. The CR for the PC A samples slightly decreases with time in the corrosive environment that can be linked to forming a stable protective aluminum oxide film.

It is important to note that the surface of the PC B and PC C coatings did not contain corrosion-induced blisters, while the surface of the PC A coatings had a number of blisters, which was also reported in DeForce et al. (2007).

Coating Microstructure Analysis. The low porosity level and the absence of cracks at the coating/substrate interface are the major coating characteristics produced by LPCS system (Spenser et al., 2009).

However, as shown in DeForce et al. (2007), cold spray coatings may have defects such as microcracks, micropores, and voids. This fact becomes critical when coatings meet aggressive corrosive corrosion environment. In case of Cl$^-$ and OH$^-$ ions, they can easily react with magnesium substrate and create blister damage of the coating surface. From DeForce et al. (2007), the minimum aluminum cold sprayed coating thickness to provide an effective barrier corrosion protection mechanism was determined to be approximately 400 μm.

The SEM analysis of the coatings as depicted in Figure 6.43 represents typical cold spray microstructures and reveals the presence of closed micropores at the coating interface (less than 1.0 vol.% as measured by image analysis, whereas the rest of the coating was fully dense).

Adhesion strength measurements reveal that the adhesion strength for cold sprayed coating layers with the composition of PC A was 32 ± 3 MPa however for PC B and PC C, there were slight increase in the value up to 37 ± 2 MPa and 40 ± 1 MPa, respectively. The increase in adhesion strength is proportional to an increase of the content of *hard* material within the *soft* coating composite matrix. This is in agreement with Boag et al. (2011) where an increase of Al$_2$O$_3$ in powder composition leads the increase of adhesion strength. Also the following should be noted from adhesion test measurements that in most cases (PC B and PC C), the failure type was cohesive, indicating that the true adhesion strength of those coatings must be greater.

6.2.6 CONCLUDING REMARKS

The following conclusions can be drawn based on the material laid out in this chapter:

- Cold spray process offers unique opportunities to uniformly spray dense composite corrosion-resisting coatings. The use of a portable LPCS machines allows to apply the process in both the field and the industrial environment conditions.

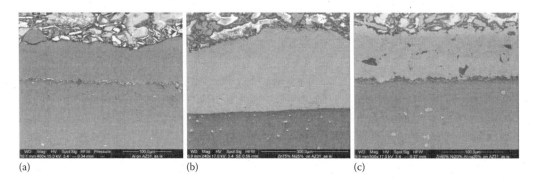

(a)   (b)   (c)

FIGURE 6.43 Cross-sectional microphotographs of the coating layers cold sprayed on AZ31: (a) cold sprayed PC A composition; (b) cold sprayed PC B composition; and (c) cold sprayed PC C composition.

- Electrochemical methods (e.g., polarization measurements, electrochemical impedance spectroscopy, OCP, etc.) can be effectively used for the corrosion behavior evaluation of LPCS-formed protective coatings.
- The results of the accelerated corrosion test demonstrate that the Al6022 high-strength Al alloy can be moat effectively protected by LPCS-deposited Al coating. In case of a composite Al–Zn–Al$_2$O$_3$ coating, the protection properties are gradually degraded via the sacrificial mechanism of the formed Al–Zn galvanic couples.
- The CR in magnesium AZ31 alloy components can be considerably improved by the deposition of the LPCS-deposited protective 75Ni–25Zn and 60Ni–20Zn–20Al$_2$O coatings, due to the microgalvanic actions of Ni–Zn phases that behave as a barrier between the AZ31 alloy and the corrosive environment.

7 Economic Analysis of Cold Spray*

Dennis Helfritch

CONTENTS

7.1 COST PRINCIPLES

Cold spray is employed for scores of applications (Champagne et al., 2014), from the deposition of thin protective coatings (Dzhurinskiy et al., 2012), to parts repair (Champagne et al., 2015) and to the production of free-standing shapes (Cormier et al., 2013). The costs from application to application can vary significantly, but the cost for each is dependent on the same set of operating parameters. Once defined, these parameters can be used to accurately calculate the cost of a finished product or can be used to predict the cost of a potential product. The methods presented here assume that a complete product has not yet been fabricated, and cost estimates will be based on the geometry and composition of the desired product. The methods presented can thus be used to calculate quotation prices for prospective customers.

Manufacturing costs all reside within the following three classic categories:

1. Materials costs
2. Direct labor costs
3. Overhead costs

---

* The research reported in this document was performed in connection with contract/instrument W911QX-14-C-0016 with the US Army Research Laboratory. The views and conclusions contained in this document are those of the authors and should not be interpreted as presenting the official policies or position, either expressed or implied, of the US Army Research Laboratory or the US Government unless so designated by other authorized documents. Citation of manufacturer's or trade names does not constitute an official endorsement or approval of the use thereof. The US Government is authorized to reproduce and distribute reprints for Government purposes notwithstanding any copyright notation hereon.

Gas and powder are materials costs. The salaries paid to workers while engaged directly in manufacturing a specific product are direct labor costs. All other costs, such as utilities, depreciation, and maintenance, are indirect overhead. These costs are interrelated and are subject to task difficulty. In the following sections, we will consider each cost category in detail.

7.1.1 Materials Costs

Cold spray accelerates powder particles suspended in a gas. The gas–particle suspension is accelerated by expansion through a supersonic nozzle. The resulting high-velocity particles impact on a substrate to create a deposition. From this simple description, one can conclude that powder and gas costs and their rates of usage are major contributors to the overall product cost. Materials costs can be determined based on the dimensions of the deposited product and the efficiency of deposition. The product will contain a known volume of deposited powder. Assuming negligible porosity and known density, this volume gives the mass of deposit. The mass of powder needed for the product is therefore that of the deposit plus estimated overspray, divided by the fractional deposition efficiency (DE). The DE can be obtained from a trial spray of the powder, can be estimated from similar applications, or can be iteratively calculated. Once the required mass of powder is determined, the mass of gas needed for powder acceleration can be simply calculated by dividing the powder mass by the selected ratio of powder to gas. If helium is recovered and reused, then this must be acknowledged in the calculation. Given the masses of powder and gas, the costs of these materials are simply the masses multiplied by the cost per mass, such as $/pound. In addition, the time needed for a single part fabrication can be calculated from the mass of the gas needed, the gas pressure and temperature, and the nozzle throat diameter. While this time is not needed for materials cost calculations, it is needed for labor costs below.

7.1.2 Direct Labor

Labor rates, $/man-hour, are known, and the hours needed for product completion are known from the calculation described earlier. Time must be added to take into account initial planning and setup. This time includes one-time initial fixture assembly, robot programming, and operating parameter determination. If multiple pieces are manufactured, then this one-time cost is shared by piece. While rates may vary among workers, typically an average rate for all workers is assumed. The cost of direct labor as described here assumes that the worker is employed in other activities when not operating the cold-spray system. If the worker is paid a salary irrespective of his or her activity, then this labor cost must be part of fixed overhead and must be subjected to utilization considerations, as described next.

7.1.3 Overhead

This cost category can be further divided into the following two subcategories:

1. Variable overhead, such as utilities for direct production, which changes as production changes.
2. Fixed overhead, such as administration, rent, heating and lighting, maintenance, and capital recovery, which remains independent of production.

The cost of direct use of electricity for the production of a product by cold spray can be easily determined. Electricity is used to heat and sometimes to compress the gas used. The usages can be straightforwardly calculated from the temperature, pressure, and flow rate of the gas. Other electrical usages are for robot motion and control systems, but these are negligible in comparison with gas treatment. Once the rate of electricity usage is determined, the total usage in kWh can be calculated by multiplying by the production time. The cost is then obtained by multiplying by the purchased cost of electricity in $/kWh.

Fixed overhead is apportioned to individual jobs depending on the amount of time that particular job requires in relation to all other jobs, and this is where the concept of utilization must be introduced. Utilization is the percentage of time the cold-spray system is used with respect to the total time that is available. Utilization is a measure of how effectively the available resources are being put to use. So the amount of fixed overhead charged to an individual job is equal to (total fixed cost per year/available operating time per year) × (total job time/fractional utilization). It will be seen that utilization is a major cost factor.

The capital recovery factor (CRF) method for the calculation of depreciation takes into account the declining cost of the equipment, as well as the cost lost to interest payments if the purchase money were used instead as a loan. This is similar to a mortgage payment, which consists of principal and interest. This calculation yields a higher fixed cost than straight-line depreciation, unless interest rates are zero. The yearly depreciation cost, calculated by the CRF method is given by

$$\text{Yearly depreciation cost} = \text{CRF}*(\text{capital cost} - \text{salvage cost})$$

$$\text{CRF} = \frac{i(1+i)^n}{(1+i)^n - 1}$$

where i is the fractional interest rate, for example, 5% = 0.05, and n is the years of ownership.

Yearly maintenance is generally estimated as a percentage of capital cost, for example, 5%. Other fixed overhead costs, such as rent and administration, are straightforward, and should be apportioned with respect to fraction of total floor space used and fraction of administrative time devoted to cold spray.

7.1.4 Combined Costs

The cost determination steps described above are easily assembled and carried out in a spreadsheet program. Alternatively, Stier (2014) presents a single equation containing the cost elements described earlier. An example of a typical spreadsheet is shown in Figure 7.1. Values that must be input, such as gas type and labor rate, are shown in the left-hand column. The values shown in Figure 7.1 are typical of actual cold-spray operations. The remaining values, such as flow rate and time for completion, are calculated by the spreadsheet. The costs by category are then also calculated. A bar chart allows for quick assessment of the importance of various cost drivers. For this spreadsheet, the gas flow is calculated from knowledge of the nozzle throat diameter and the gas conditions upstream of the throat. Time for completion, which is needed for the determination of most cost contributions, is simply calculated by dividing total powder mass used by the powder feed rate. The powder mass needed is affected by the volume of the part, the DE, and the overspray fraction. For example, the time needed to complete a single part is given by

$$\text{Powder mass needed} = [(\text{part volume})(\text{metal density})/(\text{fractional DE})][1 + \text{fractional overspray}]$$

$$\text{Time for completion} = (\text{powder mass needed})/(\text{powder feed rate})$$

The gas flow is calculated from the input values of throat diameter, pressure, and temperature with the equation below for sonic conditions at the throat:

$$\text{Mass flow} = Ap\sqrt{\gamma/RT} / \left[1 + \frac{\gamma-1}{2}\right]^{(\gamma+1)/(2(\gamma-1))}$$

where p and T are pressure and temperature upstream of the nozzle, A is the throat area, and γ is the ratio of specific heat at constant volume to the specific heat at constant pressure (=4 for nitrogen and 1.67 for helium).

Input data		Calculated values	
Nitrogen (0) helium (1)	0	Gas flow, NCMH	51.6
Throat diameter	2 mm	% Powder to gas	4.6
Feed rate	3 kg/h	Powder per piece, kg	1.72
Gas temp after heating	500 °C	Time per piece, h	0.57
Compressed gas pressure	4 MPa	Elect per piece, kWh	16.0
Deposit volume	125 cu cm		
Number of pieces	5	Powder	$859
Overspray	10 %	Gas	$37
Material density	10 gm/cc		
Deposition efficiency	80 %	Labor	$590
Unit powder cost	100 $/kg		
Unit nitrogen cost	0.2 $/kg	Administration	$843
Unit helium cost	40 $/kg	Utilities and rent	$562
Helium reuse	0 %	Maintenance	$253
Unit electricity cost	0.15 $/kWh	Depreciation	$303
Prespray set up time	4 h	Production elect	$12
Set-up time per piece	0.20 h		
Hourly labor rate	75 $/h	Cost per piece	$689
Equipment capital cost	900,000 $		
Equipment life	15 years	Total job cost	$3,447
Salvage cost	300,000 $		
Interest rate	4 %		
System utilization rate	70 %		
Maintenance, % of capital	5 %		
Yearly administration	150,000 $		
Yearly rent, utilities	100,000 $		

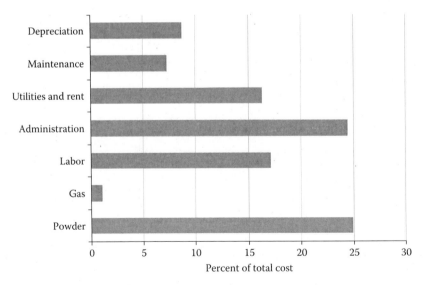

FIGURE 7.1   The cost calculation spreadsheet.

The electricity needed to produce a part is calculated by adding the usages of the gas heater, the gas compressor, and the exhaust fan and then multiplying by the time needed to complete a part. Electricity usage of the compressor and heater can be calculated from the known flow rate and conventional power equations. The ventilation fan power is assumed to be 15 kW for this spreadsheet.

All of the unit costs, cost rates, and fixed costs of the left-hand column must be input. The pre-spray setup time is the time needed for tasks which need to be completed before actual spraying, and include items such as powder purchase and robot programming. The set-up time per piece is the time it takes to remove a completed piece and install a new base for a subsequent piece. The hourly labor rate includes only salary and fringe benefits directly paid to the operator. The spreadsheet assumes only one worker for all tasks. Yearly administrative cost is the cost for supervision, sales, clerical, and so on devoted to the operation of the cold-spray system considered. The yearly rent is similarly prorated, for example where the cold-spray system occupies only part of a building and other production systems occupy the remainder.

The job costs are then calculated, shown in the lower, center column. Powder and gas costs are based on amounts used and on unit costs previously determined. The labor cost is simply the labor rate times the sum of all the time needed to complete the job, including set-up. Overhead values for a single job are prorated based on the fraction of time needed to complete the job divided by the available time per year (here 2000 hours), divided by the fractional utilization (U). For example,

$$\text{Prorated admin} = (\text{yearly admin})\text{total job time}/2220(U)$$

7.2   PARAMETER EFFECTS

Considering Figure 7.1 to be a base case, we can estimate the relative importance of each parameter by its variation. There are often trade-offs that can be made between parameter values that can reduce costs. For example, a more expensive powder may allow the use of nitrogen instead of helium. The effects of major cost-affecting parameters are examined below. All of the calculations made are based on a variation of parameters given by Figure 7.1.

7.2.1   GAS

By far, the largest influence on final cost is the gas used. This can be inferred from the difference in unit price between nitrogen and helium. Adjusting for a higher DE and a lower feed rate, when the nitrogen used in Figure 7.1 is switched to helium, the cost increases from $3447 to $5629. The cost increase is almost entirely due to the difference in unit costs between nitrogen and helium as can be seen from the comparable cost distributions shown in Figure 7.2. On first look, it would seem unreasonable to ever use helium; however, there are quality benefits resulting from helium use that are not evident from a manufacturing point of view. Helium can yield improved bond strength and decreased porosity. Nitrogen will sometimes not produce high enough particle velocity needed to allow hard, refractory particles to deposit.

System recycle of helium can significantly offset the cost increase described in this example. After use for cold spray, helium is captured, purified of particles and air, compressed and sent back to the cold-spray system for reuse. Helium recycle requires that the cold-spray system be gas-tight, such that air contamination be minimized. Solids are filtered from the gas stream and air is removed through pressure swing adsorption or membranes. Air can enter the system through inleakage during operation and between spray runs when the system is opened. Helium is lost through imperfect purification and when opening the system. The recycle system can cost 60%–100% of the basic cold-spray system; however, high helium cost can justify this expense. Figure 7.3 is an example of how helium recycle can affect costs. The figure compares helium cold-spray costs without recycle and

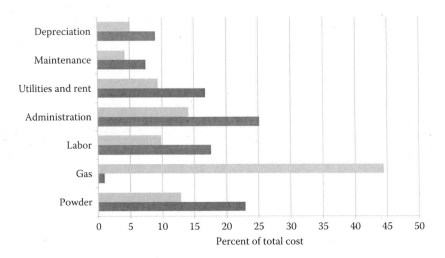

FIGURE 7.2 The cost distribution for nitrogen usage (black) versus helium usage (gray).

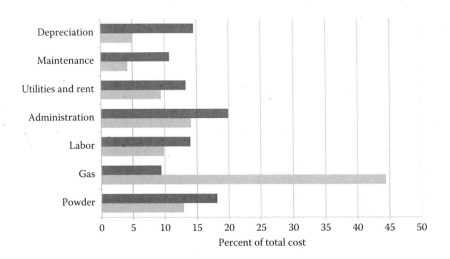

FIGURE 7.3 The effect of helium recycle—without recycle (gray), with recycle (black).

costs with recycle. The bottom-line cost without recycle is $5629 as for Figure 7.2. An additional equipment cost of 80% of the original cold-spray system cost was used for the recycle calculation and 15% air contamination was assumed. The job cost was reduced to $3988, and the helium usage cost percentage was reduced from 45% to 10%.

7.2.2 POWDER COST

While the unit cost for the purchase of gas does not vary significantly from job to job, the cost of powder can vary between $20/kg and $1000/kg. For the case described by Figure 7.1, the effect of only changing the powder unit cost is shown in Figure 7.4. Because production time, hence all other cost contributors, depends on the rate of powder usage, total cost varies linearly with powder unit cost. The total job cost can change tenfold over the possible range of powder unit costs. This result weighs in favor of shopping for the lowest powder cost, but the qualities of the powder can directly influence the quality of the deposit and the DE of the specific operation. Determination of deposit quality and DE among powder candidates must be done by means of test cold-spray runs. Once the

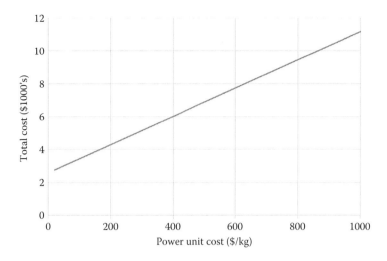

FIGURE 7.4 The effect of powder cost on overall cost, based on the example of Figure 7.1.

results from the test runs are known, the deposition efficiencies and unit costs can be inserted into the cost spreadsheet and job costs determined.

7.2.3 POWDER FEED

An often overlooked but extremely important parameter when attempting to minimize cost is the powder feed rate. This is especially true when operating with helium. Clearly, increasing powder feed rate will shorten the time required for completion, which in turn decreases total gas consumption, labor cost, and prorated overhead. The above example for helium operation that gives a $5629 cost was based on a feed rate of 3 kg/hour, which results in a powder flow equal to 11.6% of the gas mass flow. Increasing the feed rate to 5 kg/hour, without any other changes, would decrease the cost to $4,315. The limit to arbitrary increase of powder feed rate is the carrying capacity of the accelerating gas. It has been shown that particle velocity is decreased by approximately 4% for each doubling of powder loading (Gilmore et al., 1999; Samareh et al., 2009). The decrease in particle velocity in turn adversely affects DE and deposit quality. Thus, powder feed cannot be arbitrarily increased without the risk of inferior deposits. Figure 7.5 shows

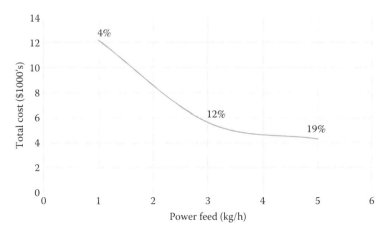

FIGURE 7.5 How powder feed rate affects cost. Percentage of powder to gas indicated.

how dramatically powder feed rate affects cost and can serve to alert cold-spray operators to be aware of the large cost effects of the feeder control. The percentage of powder relative to gas is given at three points along the curve.

7.2.4 Deposition Efficiency

As described earlier, powder characteristics directly affect DE. For a given alloy, particle shape and particle size distribution are the principal powder determinates of DE. Particle density is also important when other alloys are included. Besides powder characteristics, operating parameters, such as gas pressure and temperature, have a large influence on DE. For the case described by Figure 7.1, the effect of DE on job cost is shown in Figure 7.6. The figure clearly shows the importance of maximizing DE. In this case, the time for completion is inversely proportional to DE; hence, the dependence is nonlinear. Powder characteristics and operating parameters can be adjusted to maximize DE. Computer models are sometimes used to predict DE and to determine an optimum set of parameters, but trial cold-spray runs are generally more accurate and preferred.

7.2.5 Utilization

Labor and prorated overhead costs are strongly dependent on the time needed to complete the job. Clearly, labor rates, depreciation cost, and administration yearly cost all directly affect the bottom line job cost, but the actual time that these services are used for the manufacture of the specific job is what assigns their prorated costs to that job. A second factor, related to the job time, is utilization. Utilization is simply the amount of facility time spent in productive utilization, divided by the total time available. A total of 100% utilization means that the system is in full use throughout the year without any idle time. Figure 7.7 shows how utilization affects costs for the case described by Figure 7.1. A larger portion of fixed overhead costs must be assumed by each job as utilization decreases, and in this case job costs can almost double as utilization decreases to below 50%.

7.2.6 Number of Pieces to Be Manufactured

Given a constant, one-time, setup period (robot programming, purchasing, etc.), the cost per piece obviously decreases as this setup cost is shared with many pieces. Again, considering the case

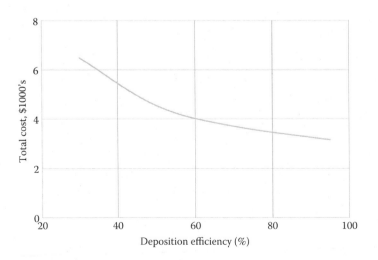

FIGURE 7.6   The effect of DE.

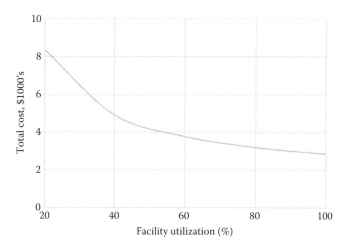

FIGURE 7.7 The effect of utilization.

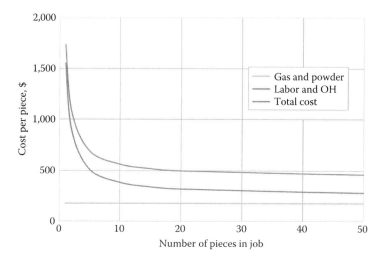

FIGURE 7.8 The effect on cost per piece by the number of pieces made.

described by Figure 7.1, the effect of mass production is shown in Figure 7.8. The powder and gas costs per piece manufactured obviously do not change. Labor and overhead costs do change, because they are affected by the one-time set-up cost. For this example, with 4 hours of upfront setup, the cost per piece does not increase significantly until fewer than 10 pieces are to be produced. Below 10 pieces, the cost per piece increases significantly as upfront costs are shared with fewer pieces.

7.2.7 DETERMINATION OF OPERATING PARAMETERS

Cold-spray parameters are often adjusted to provide maximum particle velocity and DE, which generally results in optimum deposition characteristics and maximum cost. A cost spreadsheet allows the cold sprayer to balance deposit quality with cost. Until deposition models are improved, the characterization of DE and deposit quality is best done by means of trial spray runs. A matrix of variables may be gas type, pressure, temperature, and feed rate. The DE would be measured for each run. Measurement of deposition quality could include cross-sectional examination, bond strength, tensile strength and so on depending on the specific characteristics desired. Costs could be calculated for each run and associated DE. Data generated in this way would then yield how the minimum cost for acceptable quality could be achieved.

7.3 DECISION-MAKING EXAMPLE

An example of cost-effectiveness control is as follows. It is desired to coat five tubes, 10 cm diameter and 1 m length, with 1 mm of tantalum. This yields a deposit volume of 314 cm^3 per tube. The deposit porosity must be less than 2%. A maximum system temperature of 600 °C is chosen. Considering the high expense of tantalum, the gas and pressures that can yield the desired results at lowest cost must be determined. The corresponding cost spreadsheet for nitrogen gas is shown in Figure 7.9. Once the deposition efficiencies and porosities are determined from test runs and the costs are determined from the spreadsheet, Table 7.1 can be generated.

Figure 7.10 shows the porosity and cost values of Table 7.1 for the two gases. It can be seen that nitrogen or helium can yield porosity less than 2% at equivalent costs; however, helium can yield overall lowest porosity at lowest cost. The costs are seen to increase as porosity increases for both gases, which seems counter-intuitive. Although higher pressures and more gas are used to yield lower porosities, the increasing deposition efficiencies result in decreases in powder usages. The decreased cost of powder more than compensates increased gas costs and results in a net cost savings.

Input data		Calculated values	
Nitrogen (0) helium (1)	0	Gas flow, NCMH	48.6
Throat diameter	2 mm	% Powder to gas	4.9
Feed rate	3 kg/h	Powder per piece, kg	11.03
Gas temp after heating	600 °C	Time per piece, h	3.68
Compressed gas pressure	4 MPa	Elect per piece, kWh	139.0
Deposit volume	314 cu cm		
Number of pieces	5	Powder	$44,105
Overspray	10 %	Gas	$223
Material density	16.6 gm/cc		
Deposition efficiency	52 %	Labor	$1,753
Unit powder cost	800 $/kg		
Unit nitrogen cost	0.2 $/kg	Administration	$2,505
Unit helium cost	40 $/kg	Utilities and rent	$1,670
Helium reuse	0 %	Maintenance	$751
Unit electricity cost	0.15 $/kWh	Depreciation	$901
Prespray set up time	4 h	Production elect	$104
Set-up time per piece	0.20 h		
Hourly labor rate	75 $/h	Cost per piece	$10,382
Equipment capital cost	900,000 $		
Equipment life	15 years	Total job cost	$51,909
Salvage cost	300,000 $		
Interest rate	4 %		
System utilization rate	70 %		
Maintenance, % of capital	5 %		
Yearly administration	150,000 $		
Yearly rent, utilities	100,000 $		

FIGURE 7.9 Coating a tube with tantalum example.

TABLE 7.1
Porosity and Cost Changes Resulting from Operational Changes

Gas	Pressure, bar	DE (%)	Porosity (%)	Cost per Tube ($)
N_2	20	35	2.70	15,267
N_2	30	52	2.00	10,382
N_2	40	67	1.20	8130
He	20	81	1.50	9055
He	30	87	0.90	8453
He	40	92	0.40	8011

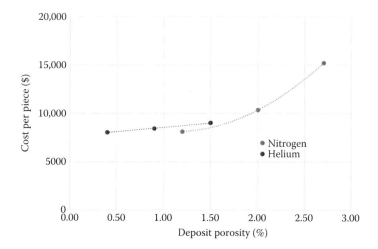

FIGURE 7.10 Production cost versus porosity achieved for nitrogen and helium.

7.4 CONCLUDING REMARKS

A framework for cold-spray cost estimation has been presented. The principle cost categories of materials, labor, and overhead have been defined and broken down into individual components. Methods to determine the cost contribution of each component have been described and examples given. These components have been incorporated into a spreadsheet (Figure 7.1), which utilizes input values, such as DE, powder cost, equipment cost, and coating volume, to calculate the cost of the completed products. The calculations contained in the spreadsheet are straightforward and can be done by readers either by hand calculation or by computer, for example, Microsoft Excel. Additionally, the author may be contacted for assistance

We saw that individual parameters can have a significant cost impact and that the judicious combination of these parameters can result in cost savings. As seen from Figures 7.3 through 7.8, costs can easily range over several multiples, depending on parameter selections. The influence of gas on cost is extreme, due to a two-orders-of-magnitude difference in gas purchase price between nitrogen and helium. Clearly, the use of nitrogen is favored, but at times helium must be used for a desired result. Powder feed rate is often overlooked as a cost driver, but must be carefully understood and controlled to achieve an optimum result at a minimum cost.

There are competing technologies that can be applied to most applications. As cold-spray applications are being developed, it is important to be able to assess the commercial viability of the application with respect to the alternative methods. In addition to quality, a knowledge of relative cost is critical for this assessment.

References

Abdel A. (2008). Protective coating for magnesium alloy. *J. Mater. Sci.*, 43, 2947–2954.

Abramovich G.N. (1963). *The Theory of Turbulent Jets*. MIT Press, Cambridge, MA.

Abu Al-Rub R.K., Voyiadjis G.Z. (2006). A physically based gradient plasticity theory. *Int. J. Plasticity*, 22(4), 654–684.

Ahmad Z. (2006). *Principles of Corrosion Engineering and Corrosion Control*. Butterworth-Heinemann. Elsevier Science Technology.

Akedo J. (2008). Room temperature impact consolidation (RTIC) of fine ceramic powder by aerosol deposition method and applications to microdevice. *J. Therm. Spray Technol.*, 17(2), 181–198.

Alkhimov A.P., Gulidov A.I., Kosarev V.F., Nestrovich N.I. (2000). Specific features of microparticle deformation upon impact on a rigid barrier. *J. Appl. Mech. Tech. Phys.*, 41(1), 204–209.

Alkhimov A.P., Kosarev V.E., Papyrin A.N. (1990). A method of cold gas-dynamic deposition. *Dokl. Akad. Nauk SSSR*, 318, 1062–1065.

Alkhimov A.P., Kosarev V.F., Klinkow S.V. (2001). The features of cold spray nozzle design. *J. Therm. Spray Technol.*, 10, 375–381.

Alkhimov A.P., Kosarev V.F., Papyrin A.N. (1997). Gas-dynamic spraying, study of plane supersonic two phase jet. *J. Appl. Mech. Tech. Phys.*, 38(2), 176–183.

Alkhimov A.P., Kosarev V.F., Papyrin A.N. (1998). Gas-dynamic spraying, experimental study of the spray process. *J. Appl. Mech. Tech. Phys.*, 39(2), 183–188.

Alkhimov A.P., Papyrin A.N., Kosarev V.F., Nesterovich N.I., Shushpanov M.M. (1994). Gas-dynamic spraying method for applying a coating. US Patent 5,302,414.

Anders A. (2004). Fundamentals of pulsed plasmas for materials processing. *Surf. Coat. Technol.*, 183, 301–311.

Anderson J.D. (2004). *Modern Compressive Flow*. McGraw-Hill, Boston, MA, p. 216.

Anderson Jr. J.D. (1982). *Modern Compressible Flow*. McGraw-Hill, New York.

Anderson T.L. (1995). *Fracture Mechanics: Fundamentals and Applications*. CRC Press, Boca Raton, FL.

Ansara I., Dinsdale T., Rand M.H. (1998). *Thermochemical Database for Light Metal Alloys*, Vol. 2. COST, Office for Official Publications of the European Communities, Luxembourg, pp. 23–39.

Antolovich S.D., Armstrong R.W. (2014). Plastic strain localization in metals: Origins and consequences. *Prog. Mater. Sci.*, 59(42–327), 1–160.

Arzt E. (1998). Size effects in materials due to microstructural and dimensional constraints: A comparative review, overview No. 130. *Acta Mater.*, 46(16), 5611–5626.

Arzt E., Dehm G., Gumbsch P., Kraft O., Weiss D. (2001). Interface controlled plasticity in metals: Dispersion hardening and thin film deformation. *Prog. Mater. Sci.*, 46, 283–307.

ASM Handbook. (1990). *Properties of Pure Metals, Properties and Selection: Nonferrous Alloys and Special-Purpose Materials*, Vol. 2. ASM International, Materials Park, OH, pp. 1099–1201.

ASM Handbook. (2003). *Corrosion: Fundamentals, Testing, and Protection*, Vol. 13A. ASM International, Materials Park, OH, pp. 1779–1793.

Assadi H., Gartner F., Stoltenhoff T., Kreye H. (2003). Bonding mechanism in cold gas spraying. *Acta Mater.*, 51, 4379–4394.

ASTM Designation E8-01. (2001). Standard Test Method for Tension Testing of Metallic Materials. (Philadelphia).

Atkinson J.H. (1993). *An Introduction to the Mechanics of Soils and Foundations: Through Critical State Soil Mechanics*. McGraw-Hill, New York, 337pp.

Avitzur B. (1983). *Handbook of Metal-Forming Processes*. John Wiley & Sons, New York.

Bai Y. (1982). Thermo-plastic instability in simple shear. *J. Mech. Phys. Solids*, 30, 195–201.

Baker M.A., Gissler W., Klose S., Trampert M., Weber F. (2000). Morphologies and corrosion properties of PVD Zn-Al coatings. *Surf. Coat. Technol.*, 125, 207.

Balani K., Laha T., Agarwal A., Karthikeyan J., Munroe N. (2005). Effect of carrier gases on microstructural and electrochemical behavior of cold sprayed 1100 aluminum coating. *Surf. Coat. Technol.*, 195(2–3), 272–279.

Barbezat G. (2006). Application of thermal spraying in the automobile industry. *Surf. Coat. Technol.*, 201, 2028–2031.

Barbosa M. et al. (2009). Cold spray deposition of titanium onto aluminium alloys. DIA MUNDIAL DOS MATERIAIS 2009, Menção Honrosa ORDEM DOS ENGENHEIROS.

Bechtold, J.H. (1955). *Acta Metall.*, 3, 252.

Bednarczyk I., Kuc D., Niewielski G. (2008). The structure of FeAl and Fe_3Al-5% Cr intermetallic phase-based alloys after hot deformation processes. *Arch. Mater. Sci. Eng.*, 30(1), 5 8.

Bejan A., Kraus A. (2003). *Heat Transfer Handbook.* Wiley, Weinheim, Germany.

Belsito D., McNally B., Bassett L., Champagne V., Sisson R. (2013). *Proceedings of MS&T 2013 Conference*, Montreal, Canada.

Benson D.J., Nesterenko V.F., Jonsdottir F., Meyers M.A. (1997). Quasistatic and dynamic regimes of granular material deformation under impulse loading. *J. Mech. Phys. Solids*, 45(11/12), 1995–1999.

Bhagat R.B., Amateau M.F., Papyrin A.N., Conway J.C.J., Stutzman B.J., Jones B. (1997). In: Berndt C.C. (Ed.). *Thermal Spray: A United Forum for Scientific and Technological Advances.* ASM International, Materials Park, OH, pp. 361–367.

Bialucki P., Kozerski S. (2006). Study of adhesion of different plasma-sprayed coatings to aluminium. *Surf. Coat. Technol.*, 201, 2061–2064.

Bielousova O. (2013). Structure and properties investigation of composite coatings made by cold gas dynamic spray, PhD thesis, ENISE.

Billig F.S. (1967). Shock-wave shapes around spherical and cylindrical-nosed bodies. *J. Spacecraft*, 4(5), 822–823.

Birks A.S., Green R.E., McIntire P. (1991). *Ultrasonic Testing (Nondestructive Testing) Handbook*, 2nd ed., Vol. 7. American Society for Nondestructive Testing, Columbus, OH.

Boag A., Hughes A.E., Glenn A.M., Muster T.N. (2011). Corrosion of AA2024-T3: Part I. Localized corrosion of isolated IM particles. *Corros. Sci.*, 53(1), 17–26.

Boag A., Taylor R.J., Muster T.H., Goodman N., Hughes A.E. (2010). Stable pit formation on AA2024-T3 in a NaCl environment. *Corros. Sci.*, 52(1), 90–103.

Bonora P.L., Andrei M., Eliezer A., Gutman E.M. (2002). Mechanoelectrochemical behavior and creep corrosion of magnesium alloys. *Corros. Sci.*, 44, 729.

Borchers C., Gartner F., Stoltenhoff, Kreye H. (2003). Microstructural and macroscopic properties of cold sprayed copper coatings. *J. Appl. Phys.*, 93(12), 10064–10070.

Borchers C., Gartner F., Stoltenhoff T., Kreye H. (2005). Formation of persistent dislocation loops by ultra-high strain-rate deformation during cold spraying. *Acta Mater.*, 53, 2991–3000.

Bouayad A., Gerometta Ch., Belkebir A., Ambari A. (2003). Kinetic interactions between solid iron and molten aluminium. *Mater. Sci. Eng.*, A363, 53–61.

Bouche K., Barbier F., Coulet A. (1998). Intermetallic compound layer growth between solid iron and molten aluminium. *Mater. Sci. Eng.*, A249, 167–175.

Brekhovskikh L.M., Godin O.A. (1990). *Acoustics of Layered Media I.* Springer, Berlin, Germany.

Breslin C.B., Treacy G., Cornell W.M. (1994). Studies on the passivation of aluminium in chromate and molybdate solutions. *Corros. Sci.*, 36(7), 1143–1154.

Briggs A., Kolosov O. (2010). *Acoustic Microscopy*, 2nd ed. Oxford University Press, New York, p. 387.

Brugger C., Coulombier M., Massart T.J., Raskin J.-P., Pardoen T. (2010). Strain gradient plasticity analysis of the strength and ductility of thin metallic films using enriched interface model. *Acta Mater.*, 58, 4940–4949.

Buch A. (1999). *Pure Metal Properties. A Scientific-Technical Handbook.* ASM International, Materials Park, OH.

Carlson D.J., Hoglund R.F. (1964). Particle drag and heat transfer in rocket nozzles, *AIAA J.*, 2(11), 1980–1984.

Cavaliere P., Perrone A., Silvello A. (2014). Processing conditions affecting grain size and mechanical properties in nanocomposites produced via cold spray. *J. Therm. Spray Technol.*, 23(7), 1089–1096.

Champagne V. (Ed.). (2007). *The Cold Spray Materials Deposition Process: Fundamentals and Applications.* CRC Press, Cambridge.

Champagne V., Helfritch D., Leyman P., Lempicki R., Grendahl S. (2004). Nozzle design influence on the supersonic particle deposition process. In: *Cold Spray An Emerging Spray Coating Technology*, Akron, OH.

Champagne V., Helfritch D., Leyman P., Lempicki R., Grendahl S. (2005). The effects of gas and metal characteristics on sprayed metal coatings. *Model. Simul. Mater. Sci. Eng.*, 13, 1119–1128.

Champagne V.K, Helfritch D. (2015). Critical Assessment 11: Structural repairs by cold spray. *Mater. Sci. Technol.*, 31(6), 627–634.

Champagne V.K., Helfritch D.J. (2014). Mainstreaming cold spray – push for applications. *Surf. Eng.*, 30(6), 396–403.

Champagne V.K., Leyman P.F., Helfritch D.J. (2008). *Magnesium Repair by Cold Spray.* ARL Technical Report ARL-TR-4438, 34.

Chang B.Y., Park S.M. (2010). Electrochemical impedance spectroscopy. *Annu. Rev. Anal. Chem.*, 3, 207–229.

Chang C.P., Sun P.L., Kao P.W. (2000). Deformation induced grain boundaries in commercially pure aluminium. *Acta Mater.*, 48(13), 3377–3385.

Cheeke J. (2002). *Fundamentals and Applications of Ultrasonic Waves*. CRC Press, Boca Raton, FL, Chapter 14.

Chen L., Batra R.C. (2000). Microstructural effects on shear instability and shear band spacing. *Theor. Appl. Frac. Mech.*, 34, 155–166.

Chis M.C., Cojocaru M.O. (2005). Adhesion prediction on metal thermal spray coatings. *Surf. Eng.*, 21(1), 72–75.

Christian J.W., Mahajan S. (1995). Deformation twinning. *Prog. Mater. Sci.*, 39, 1–157.

Clift R., Grace J.R., Weber M.E. (1987). *Bubbles, Drops and Particles*. Academic Press, New York.

Conzone S.D., Putt D.P., Barlett A.H. (1997). Joining MoSi2 to 316 stainless steel. *J. Mater. Sci.,* 32(13), 3369.

Cormier Y., Dupuis P., Jodoin B., Corbeil A. (2013). Net shape fins for compact heat exchanger produced by cold spray. *J. Therm. Spray Technol.*, 22(7), 1210–1221.

Covac C., Alaux T.J., Marrow E., Covekar A. (2010). Legat, Correlations of electrochemical noise, acoustic emission and complementary monitoring techniques during intergranular stress-corrosion cracking of austenitic stainless steel. *Corros. Sci.*, 52(6), 2015–2025.

Cullison A. (September 2001). Stress relief basics. *Weld. J.*, 80(9), 49.

Dahlberg C.F.O., Faleskog J. (2014). Strain gradient plasticity analysis of the influence of grain size and distribution on the yield strength in polycrystals. *Eur. J. Mech.*, A, 44, 1–16.

Dalley H.S. (1970). Powder feed device for flame spray guns. US Patent 3,501,097.

Dao M., Lu L., Asaro R.J., De Hosson J.T.M., Ma E. (2007). Toward a quantitative understanding of mechanical behavior of nanocrystalline metals. *Acta Mater.*, 55(12), 4041–4065.

Daridon L., Oussouaddi O., Ahzi S. (2004). Influence of the material constitutive models on the adiabatic shear band spacing: MTS, power law and Johnson-Cook models. *Int. J. Solids Struct.*, 41, 3109–3124.

Date H., Kobayakawa S., Naka M. (1999). Microstructure and bonding strength of impact-welded aluminium–stainless steel joints. *J. Mater. Process. Technol.*, 85(1–3), 166.

Davis J.R. (2004). *Handbook of Thermal Spray Technology*. ASM International, Materials Park, OH.

Davis J.R. (Ed.). *Surface Engineering for Corrosion and Wear Resistance*. ASM International, Materials Park, OH, p. 209.

DeForce B., Eden T., Potter J., Champagne V., Leyman P., Helfritch D. (2007). Application of aluminum coatings for the corrosion protection of magnesium by cold spray. *TRI Service Corrosion Conference,* Denver, CO.

Denner S.G., Jones R.D. (1977). Kinetic interaction between al (liquid) and iron/steel (solid) for conditions applicable to hot dip aluminizing. *Met. Technol.*, 4(3), 167–174.

Department of Defense, Technology Readiness Assessment (TRA) Guidance, Prepared by the Assistant Secretary of Defense for Research and Engineering (ASD(R&E)), April 2011.

DiLellio J.A., Olmstead W.E. (1997). Temporal evolution of shear band thickness. *J. Mech. Phys. Solids*, 45, 345–359.

DiLellio J.A., Olmstead W.E. (2003). Numerical solution of shear localization in Johnson–Cook materials. *Mech. Mater.*, 35, 571–580.

Dodd B., Bai Y. (1984). Width of adiabatic shear bands. *Mater. Sci. Technol.*, 1, 38–40.

Dodd B., Bai Y. (1989). Width of adiabatic shear bands formed under combined stresses. *Mater. Sci. Technol.*, 5(6), 557–559.

Drucker D.C. (1967). *Introduction to Mechanics of Deformable Solids*. McGraw-Hill, New York.

Dykhuizen R.C., Neiser R.A. (2003). Optimizing the cold spray process. *Proceedings of the International Thermal Spray Conference*. ASM International, Materials Park, OH, 5–8 May, pp. 19–26.

Dykhuizen R.C., Smith M.F. (1989). Investigations into the plasma spray process. *Surf. Coat. Technol.*, 37(4), 349–358.

Dykhuizen R.C., Smith M.F. (1998). Gas dynamic spray principles of cold spray. *J. Therm. Spray Technol.*, 7(2), 205–212.

Dykhuizen R.C., Smith M.F., Gilmore D.L., Nelser R.A., Jiang X., Sampath S. (1999). Impact of high velocity cold spray particles. *J. Therm. Spray Technol.*, 8(4), 559–564.

Dzhurinskiy D., Maeva E., Leshchinsky Ev., Maev R.Gr. (2012). Corrosion protection of light alloys using low pressure cold spray. *J Therm Spray Tech.*, 2(12), pp. 304–313.

Earvolino P.A., Fine M.E., Weertman J.R., Parameswaran V.R. (1992). Processing an Al Al3Zr0.25Ti0.75 metal-matrix composite by conventional melting, casting and rolling. *Scr. Metall. Mater.*, 26(6), 945–948.

Edwards M. (1996). Properties of metals at high rates of strain. *Mater. Sci. Technol.*, 22(4), 453–462.

Eggeler G., Auer W., Kaesche H. (1986). On the influence of silicon on the growth of the alloy layer during hot deep aluminizing. *J. Mater. Sci.,* 21(9), 3348–3350.

El-Sayed S.M. (2011). Corrosion and corrosion inhibition of aluminum in Arabian Gulf seawater and sodium chloride solutions by 3-amino-5-mercapto-1,2,4-triazole. *Int. J. Electrochem. Sci.*, 6, 1479–1492.

Elmoursi A., Van Steenkiste T., Gorkiewicz D., Gillispie B. (2005). Fracture study of aluminum composite coatings produced by the kinetic spray method. *Surf. Coat. Technol.*, 194(1), 103–110.

Eskin D. (2005). Modeling dilute gas-particle flows in horizontal channels with different Li W-Y wall roughness. *Chem. Eng. Sci.*, 60, 655–663.

Eskin D., Voropaev S., Dorokhov I. (2003). Effect of particle size distribution on wall friction in high-velocity gas-dynamic apparatuses. *Theor. Found. Chem. Eng.*, 37(2), 122–130.

Eskin D., Voropaev S. (2001). Engineering estimations of opposed jet milling efficiency. *Miner. Eng.*, 14, 1161–1175.

Eskin D., Voropaev S. (2004). An engineering model of particulate friction in accelerating nozzles. *Powder Technol.*, 145, 203–212.

Eskin D., Voropaev S., Vasilkov O. (1999). Simulation of jet milling. *Powder Technol.*, 105, 257–265.

Estrin Y., Toth L.S., Molinari A., Brechet Y. (1998). A dislocation based model for all hardenibg stages in large strain deformation. *Acta Mater.*, 46(15), 5509–5522.

Evans J.F., Kirkwood D.H., Beech J. (1991). Modeling of copper solidification process. In: Rappaz M., Ozgu M.R., Mahin K.W. (Eds.). *Modeling of Casting, Welding and Advanced Solidification Processes V*. TMS, Warrendale, PA, p. 533.

Evstratov V.A. (1995). *Theory of Metal Forming*, Russian High School, Moscow, Russia, 248pp.

Fabel A. (1976). Powder feed device for flame spray guns, US Patent 3,976,33.

Fenker M., Balzer M., Kappl H. (2014). Corrosion protection with hard coatings on steel: Past approaches and current research efforts. *Surf. Coat. Technol.*, 257, 182–205.

Fleck N.A., Hutchinson J.W. (1997). Strain gradient plasticity. *Adv. Appl. Mech.*, 33, 295–361.

Fleck N.A., Hutchinson J.W. (2001). A reformulation of strain gradient plasticity, *J. Mech. Phys. Solids*, 49, 2245–2271.

Flemming R.P., Olmstead W.E., Davis S.H. (2000). Shear localization with an Arrhenius flow law. *J. Appl. Math.*, 60(6), 1867–1886.

FLUENT (1996). 4.4.4 User Guide. Fluent Inc., SimScale GMBH, Munich, Germany.

Frost H.J., Ashby M. F. (1982). *Deformation-Mechanism Maps: The Plasticity and Creep of Metals and Ceramics*. Pergamon Press, Oxford.

Fuchs N.A. (1997). *The Mechanics of Aerosols*. Pergamon Press, New York.

Fukanuma H., Ohno N., Sun B., Huang R. (2006). In-flight particle velocity measurements with DPV-2000 in cold spray. *Surf. Coat. Technol.*, 201, 1935–1941.

Gabel H., Taphorn R.M. (1997). Solid state spray forming of aluminum near-net shapes, *JOM*, 8, 31–33.

Gao C.Y., Zhang L.C. (2010). A constitutive model for dynamic plasticity of FCC metals. *Mater. Sci. Eng. A*, 527, 3138–3143.

Gartner F., Borchers C., Stoltenhoff T., Kreye H. (2003). Numerical and microstructural investigations of the bonding mechanisms in cold spraying. *Proceedings Of the International Thermal Spray Conference*.

Gärtner F., Stoltenhoff T., Schmidt T., Kreye H. (2006). The cold spray process and its potential for industrial applications. *J. Therm. Spray Technol.*, 15(2), 223–232.

Gavras A.G., Lados D.A., Champagne V.K., Singh D. (2007). Small fatigue crack growth mechanisms and interfacial stability in cold-spray 6061 aluminum alloys. Worcester Polytechnic Institute Internal Report.

Gedevanishvili S., Deevi S.C. (2002). Processing of iron aluminides by pressure less sintering through Fe+Al elemental route. *Mater. Sci. Eng. A*, 325, 163–176.

Georgiou E.P., Achanta S., Dosta S., Fernandez J., Matteazzi P., Kusinski J., Piticescu R.R., Celis J.-P. (2013). Structural and tribological properties of supersonic sprayed Fe–Cu–Al–Al$_2$O$_3$ nanostructured CerMet, *Appl. Surf. Sci.*, 275, 142–147.

Gerland M., Presles H.N., Guin J.P., Bertheau D. (2000). Explosive cladding of a thin Ni-film to an aluminium alloy. *Mater. Sci. Eng., A*, A280, 311–319.

Gialanella S. (1995). FeAl alloy disordered by ball milling. *Intermetallics*, 3, 73–76.

Gidaspow D. (1994). *Multiphase Flow and Fluidization: Continuum and Kinetic Theory Descriptions*. Academic Press, Boston, MA.

Gilmore D.L., Dykhuizen R.C., Neiser R.A., Roemer, Smith M.F. (1999). Particle velocity and deposition efficiency of cold spray process. *J. Therm. Spray Technol.*, 8(40), 576–582.

Gleiter H. (1990). Nanocrystalline materials. In: Bunk W.G.J. (Ed.). *Advanced Structural and Functional Materials*, Vol. 33. Springer, New York, pp. 223–315.

Gougcon P., Moreau C., Lacasse V., Lamontage M., Powell I., Bewsher A. (1994). *Adv. Processing Tech. Particulate Mater.*, 6, 199–210.

Gray J.E., Luan B. (2002). Protective coatings on magnesium and its alloys—A critical review. *J. Alloys Compd.*, 336(1–2), 88–113.

Grujicic M., Saylor J.R., Beasley D.E., DeRosset W.S., Helfrich D. (2003). Computational analysis of the interfacial bonding between feed-powder particles and the substrate in the cold-gas dynamic-spray process. *Appl. Surf. Sci.*, 219, 211–227.

Grujicic M., Zhao C.L., DeRosset W.S., Helfritch D. (2004). Adiabatic shear instability based mechanism for particles/substrate bonding in the cold-gas dynamic-spray process. *Mater. Design*, 25, 681–688.

Guillaumin V., Mankowski G. (1998). Localized corrosion of 2024 T351 aluminium alloy in chloride media. *Corros. Sci.*, 41, 421–438.

Guzman M., Adami W., Gissler S., Klose, De Rossi S. (2000). Vapour deposited Zn-Cr alloy coatings for enhanced manufacturing and corrosion resistance of steel sheets. *Surf. Coat. Technol.*, 125, 218.

Hackette C.M., Settles G.S. (1995). The influence of nozzle design on HVOF spray particle velocity and temperature. In: Berndt C.C., Sampath S. (Eds.). *Thermal Spray Science & Technology.* ASM International, Materials Park, OH, pp. 135–140.

Hall A.C., Cook D.J., Neiser R.A., Roemer T.J., Hirschfeld D.A. (2006). The effect of a simple annealing heat treatment on the mechanical properties of cold-sprayed aluminum. *J. Therm. Spray Technol.*, 15(2), 233–238.

Hammerschmidt M., Kreye H. (1981). Microstructure and bonding mechanism in explosive welding. In: Meyers M.A., Murr L.E. (Eds.). *Shock Waves and High Strain-Rate Phenomena in Metals.* Plenum Press, New York, 961–973.

Hansen N., Jensen D.J. (2011). Deformed metals—Structure, recrystallisation and strength. *Mater. Sci. Technol.*, 27(8), 1229–1240.

Han T., Zhao Z.B., Gillispie B., Smith J R. (2004). A Fundamental study of kinetic spray process, *ITSC 2004 Thermal Spray Solutions Advances in Technology and Application.* Osaka, Japan.

Heath G.R., Dumola R.J. (1998). Fundamentals and application. In: Coddet C. (Ed.). *Thermal Spray: Meeting the Challenges of 21st Century.* ASM International, Materials Park, OH, pp. 1495–1500.

Heinrich K. (2004). The cold spray process and its potential for industrial applications. Presentation at *Cold Spray 2004*, September 27–28, 2004, Akron, OH. Sponsored by Thermal Spray Society of ASM International.

Heinrich P., Kreye H., Stoltenhoff T. (January 6, 2005). Laval nozzle for thermal and kinetic spraying. U.S. Patent 0001075 A1.

Heinz A., Haszler A., Keidel C., Bendictus R., Mille W.S. (2000). Recent development in aluminium alloys for aerospace applications. *Mater. Sci. Eng. A*, 280(1), 102–107.

Henderson C.B. (1976). Drag coefficients of spheres in continuum and rarefied Flows, *AIAA J.* 14(6), 707–721.

Hibbitt, Karlsson, Soersen. (2002). ABAQUS/Explicit 6.9-1 manual, Pawtucket, RI.

Higgis R. (1971). Engineering Metallurgy, p.1 Applied Physical Metallurgy, London, 467.

Hinze J.O. (1975). *Turbulence.* McGraw-Hill, New York.

Hirth J.P., Lothe J. (1992). *Theory of Dislocations.* Krieger, Malabar, FL.

Holmberg K., Laukkanen A., Ghabchi A., Rombouts M., Turunen E., Waudby R., Suhonen T., Valtonen K., Sarlin E. (2014). Computational modelling based wear resistance analysis of thick composite coatings. *Tribol. Int.*, 72, 13–30.

Holmberg K., Laukkanen A., Ronkainen H., Wallin K., Varjus S. (2003). A model for stresses, crack generation and fracture toughness calculation in scratched TiN-coated steel surfaces. *Wear*, 254, 278–291.

Holmberg K., Matthews A. (2009). *Coatings Tribology: Properties, Mechanisms, Techniques and Applications in Surface Engineering.* Elsevier Tribology and Interface Engineering Series No. 56, 2nd ed. Elsevier, Amsterdam, the Netherlands.

Hong T., Nagumo M. (1997). Effect of surface roughness on early stages of pitting corrosion of type 301 stainless steel. *Corros. Sci.*, 39, 665.

http://www.a-tkt.co.kr/home/html/index.php# (in Korean)

http://www.impact-innovations.com/en/coldgas/cg_index_en.html

http://www.inovati.com/KMequipment/KM%20systems/systems-specs.php

http://www.licenz.ru/eng/tech_dymet.html

http://www.medicoat.com/thermal-spray-systems/cold-spray/

http://www.oerlikon.com/metco/en/

http://www.plasma.co.jp/en/

http://www.supersonicspray.com/en/cold_spray?pg=SE20000

Huang R., Fukanuma H. (2012). Study of the influence of particle velocity on adhesive strength of cold spray deposits. *J. Therm. Spray Technol.*, 21(3–4), 541–549.

Huang R., Ma W., Fukanuma H. (2014). Development of ultra-strong adhesive strength coatings using cold spray. *Surf. Coat. Technol.*, 258, 832–841.

Hughes A.E., Boag A., Glenn A.M., Muster T.N., Ryan C., Luo C., Thompson G.E. (2011). Corrosion of AA2024-T3 Part II co-operative corrosion. *Corros. Sci.*, 53(1), 27–39.

Hu N., Molinari J.F. (2004). Shear bands in dense metallic granular materials. *J. Mech. Phys. Solids*, 52, 499–531.

Hussain T., McCartney D.G., Shipway P.H., Zhang D. (2009). Bonding mechanisms in cold spraying: The contributions of metallurgical and mechanical components. *J. Therm. Spray Technol.*, 18(3), 364–379.

Hutchings I.M. (1979). Energy absorbed by elastic waves during plastic impact. *J. Phys. D: Appl. Phys.*, 12, 1819–1824.

Hutchinson J.W. (2012). Generalizing J2 flow theory: Fundamental issues in strain gradient plasticity. *Acta Mech. Sin.*, 28(4), 1078–1086.

Hwang I.J., Hwang D.Y., Kim Y.M., Yoo B., Shin D.H. (2010). *Alloys J. Compd.*, 504, 527–530.

Idrissi H., Renard K., Ryelandt L., Schryvers D., Jacques P.J. (2010). On the mechanism of twin formation in Fe–Mn–C TWIP steels. *Acta Mater.*, 58, 2464–2476.

Instron, Model Moore R.R. (2004). High Speed Rotating Beam Fatigue Testing Machine, Operating Instructions, 000058-02-0604-EN., Instron Corporation.

Ismail K.M., Virtanen S. (2007). Electrochemical behavior of magnesium alloy AZ31 in 0.5M KOH solution. *Electrochem. Solid State*, 10(3), 9–11.

Jeandin M., Rolland G., Descurninges L.L., Berger M.H. (2014). Which powders for cold spray? *Surf. Eng.*, 30(5), 291–298.

Jenkins J.T. (1992). Boundary conditions for rapid granular flow: Flat, frictional walls. *J. Appl Mech.*, 59, 120–134.

Jen T.C., Li L., Cui W., Chen Q., Zhang X. (2005). Numerical investigations on cold gas dynamic spray process with nano- and microsize particles. *Int. J. Heat Mass Transfer*, 48, 4384–4396.

Jia J.X., Atrens A., Song G. (2005). Simulation of galvanic corrosion of magnesium coupled to a steel fastener in NaCl solution. *Mater. Corros.*, 56, 486–474.

Jia J.X., Song G.L., Atrens A. (2006). Influence of geometry on galvanic corrosion of AZ91D coupled to steel. *Corros. Sci.*, 48, 2133–53.

Jiang Y.F., Zhai C.Q., Liu L.F. (2005). Zn-Ni coatings pulse-plated on magnesium alloy. *Surf. Coat. Technol.*, 191, 393–399.

Jindal V., Sravastava V.C., Das A., Ghosh R.N. (2006). Reactive diffusion in the roll bonded iron-aluminium system. *Mater. Lett.*, 60, 1758–1761.

Jodoin B., Raletz F., Vardelle M. (2005). Cold spray modeling and validation using an optical diagnostic method. *Surf. Coat. Technol.*, 200(14–15), 4424–4432.

Johnson A. (2004). Helium recycle—A viable industrial option for cold spray. *Cold Spray Conference*. ASM International, Akron, OH.

Johnson G.R., Cook W.H. (1983). A constitutive model and data for metals subjected to large strains, high strain rates, and high temperatures. In *Proceedings of the 7th International Symposium on Ballistics. Organized under the Auspices of the Royal Institution of Engineers (Klvl), Division for Military Engineering in cooperation with the American Defense Preparedness Association.* The Hague, the Netherlands. pp. 541–547.

Jordan J., Deevi S.C. (2003). Vacancy formation and effects in FeAl. *Intermetallics*, 11, 507–28.

Joslin D.L., Easton D.S., Liu C.T., Babu S.S., David S.A. (1995). Processing of Fe₃Al and FeAl alloys by reaction synthesis. *Intermetallics*, 3, 467–81.

Jozwiak S., Karczewski K., Bojar Z. (2010). Kinetics of reactions in FeAl synthesis studied by the DTA technique and Jma model. *Intermetallics*, 18, 1332–1337.

JSME (1993). *Heat Transfer Handbook*. Japan Society of Mechanical Engineers, Tokyo, Japan, p. 44 (in Japanese).

Kainer K.U., Kaiser F. (2003). *Magnesium Alloys and Technology*. Wiley-VCH GmbH, Weinheim, Germany.

Kalla G. (1996). Soudage au laser CO_2 de produits en acier de construction pouvant avoir jusqu'a 20 mm d'epaisseur (CO2-laser beam welding of structural steel with a thickness up to 20 mm), *Revue de Metallurgie. Cahiers D'Informations Techniques*, 93(10), 1303.

Kang K., Yoon S., Ji Y., Lee C.H. (2008). Oxidation dependency of critical velocity for aluminum feedstock deposition in kinetic spraying process. *Mater. Sci. Eng.*, A, 486, 300–307.

Kannatey E., Asibu Jr. (1997). Milestone developments in welding and joining processes, *J. Manuf. Sci. Eng.* ASME 119(4), 801–810.

Karthikeyan J. (2005). The cold spray process has the potential to reduce costs and improve quality in both coatings and freeform fabrication of near-net-shape parts, *Adv. Mater. Process.*, p. 35.

Karthikeyan J., Kay C.M, Lindeman J., Lima R.S., Berndt C.C. (2000). In: Berndt C.C. (Ed.). *Thermal Spray: Surface Engineering via Applied Research*. ASM International, Materials Park, OH, pp. 255–262.

Karthikeyan J., Kay C.M., Lindemann, Lima R.S., Berndt C.C. (2001). New surfaces for a new millenium In: Berndt C.C., Khor K.A., Lugscheider E.E. (Eds.). *Thermal Spray J 2001*. ASM International, Materials Park, OH, pp. 383–387.

Kashirin A.I., Klyuev O.F., Buzdygar T.V. (2002). Apparatus for gas-dynamic coating. US Patent 6,402,050.

Kay A., Karthikeyan J. (2003). Advanced cold spray system. US Patent 6,502,767.

Kay A., Karthikeyan J. (2004). Cold spray system nozzle. US Patent 6,722,584.

Kim H.J., Lee C.H., Hwang S.H. (2005). Fabrication of WC–Co coatings by cold spray deposition. *Surf. Coat. Technol.*, 191, 335–340.

Kiselev S.P., Kiselev V.P. (2002). Superdeep penetration of particles into a metal target. *Int. J. Impact Eng.*, 27, 135–152.

Kitron A., Elperin T., Tamir A. (1981). Monte Carlo analysis of wall erosion and direct contact heat transfer by impinging two-phase lets. *J. Thermophys.*, 3, 112–122.

Kittel C. (2004). *Introduction to Solid State Physics*, 8th ed. Wiley, UK.

Klepaczko J.R. (1988). A general approach to rate sensitivity and constitutive modeling of FCC and BCC metals. In: Balkema A.A. (Ed.). *Impact: Effects of Fast Transient Loadings*, Rotterdam, the Netherlands, pp. 3–17.

Klinkov S.V., Kosarev V.F., Rein M. (2005). Cold spray deposition: Significance of particle impact phenomena. *Aerosp. Sci. Technol.*, 9, 582–591.

Klinkov V., Kosarev V.F., Sova A.A., Smurov I. (2009). Calculation of particle parameters for cold spraying of metal-ceramic mixtures. *J. Therm. Spray Technol.*, 18(5–6), 944–956.

Kochs U.F., Argon A.S., Ashby M.F. (1975). *Thermodynamics and Kinetics of Slip*. Pergamon Press, Oxford, UK.

Koivuluoto H., Lagerbom J., Kylmalahti M., Vuoristo P. (2008). Microstructure and mechanical properties of low-pressure cold-sprayed (LPCS) coatings. *J. Therm. Spray Technol.*, 17(5–6), 721–727.

Kong F.Y., Li M., Li D.B., Xu Y., Zhang Y.X., Li G.H. (2012). Synthesis and characterization of V_2O_3 nanocrystals by plasma hydrogen reduction. *J. Cryst. Growth.*, 346, 22–26.

Korpiola K., Hirvonen J.P., Laas L., Rossi F. (1997). The influence of nozzle design on HVOF exit gas velocity and coating microstructure. *J. Therm. Spray Technol.*, 6(4), 469–474.

Kosarev V.F., Klinkov S.V., Alkhimov A.P., Papyrin A.N. (2003). On some aspects of gas dynamics of the cold spray process. *J. Therm. Spray Technol.*, 12(2), 265–281.

Kouzeli M., Mortensen A. (2002). Size dependent strengthening in particle reinforced aluminium. *Acta Mater.*, 50, 39–51.

Kowaisky K.A., Marantz D.R., Smith M.F., Oberkampf W.L. (1990). Thermal spray research and applications. In: Bernecki T.F. (Ed.). *ASM International*. Materials Park, OH, pp. 587–592.

Krautkramer J., Krautkramer H. (1983). *Ultrasonic Testing of Materials*, 3rd ed., pp. 580–587.

Krautkramer J., Krautkramer H. (1990). *Ultrasonic Testing of Materials*. Springer-Verlag, Berlin, Germany.

Kreye H., Stoltenhoff T. (2000). Cold spray-study of process and coating characteristics. In. C.C. Berndt (Ed.). *Thermal Spray: Surface Engineering via Applied Research*. ASM International, Materials Park, OH, pp. 419–422.

Kumar K., Van Swygenhoven H., Suresh S. (2003). Mechanical behavior of nanocrystalline metals and alloys. *Acta Mater.*, 51, 5743–5774.

Kumar S., Chavan N.M. (2011). *Cold Spray Coating Technology: Activities at ARCI,* briefing. International Advanced Research Centre for Powder Metallurgy and New Materials, Hyderabad, India.

Kuroda S., Watanabe M., Kim K.H., Katanoda H. (2011). Current status and future prospects of warm spray technology. *J. Therm. Spray Technol.*, 20(4), 653–676.

Kuruvilla A.K., Bhanuprasad V.V., Prasad K.S., Mahajan Y.R. (1989). Effect of different reinforcements on composite-strengthening in aluminium. *Bull. Mater. Sci.*, 12(5), 495–505.

Kussin J., Sommerfeld M. (2002). Experimental studies on particle behaviour and turbulence modification in horizontal channel flow with different wall roughness. *Exp. Fluids*, 33, 143–159.

Lee H., Jung S.H., Lee S.Y., Ho Y., Kyung Y., Ko H. (2005). Fundamental study of cold gas dynamic spray process. *Appl. Surf. Sci.*, 252, 1891–1898.

Lee W-S., Liu C-Y., Chen T-H. (2008). Adiabatic shearing behaviour of different steels under extreme high shear loading. *J. Nucl. Mater.*, 374, 313–319.

Leshchynsky V., Papyrin A., Bielousova O., Yadroitseva I., Smurov I. (2011). Impact particle behaviour in cold spray of composite coatings. *Proceeding of OTSC*, Hamburg, Germany, pp. 1076–1081.

Li C., Liao H. (2006). Effect of annealing treatment on the microstructure and properties of cold-sprayed Cu coating. *J. Therm. Spray Technol.*, 15(2), 206–211.

Li C.-J., Li W.-Y. (2003). Deposition characteristics of titanium coating in cold spraying. *Surf. Coat. Technol.*, 167, 278–283.

Li C.J., Ohmori A., Harada Y. (1996). Formation of an amorphous phase in thermally sprayed WC-Co. *J. Therm. Spray Technol.*, 5(1), 69–73.

Li C.J., Wang V.V., Zhang Q., Yang G.-J., Li W.-Y., Liao H.L. (2010). Influence of spray materials and their surface oxidation on the critical velocity in cold spraying. *J. Therm. Spray Technol.*, 19, 95–101.

Li H., Li X., Sun M., Wang H., Huang G. (2010). Corrosion resistance of cold-sprayed ZN-50AL coatings in seawater. *J. Chin. Soc. Corros. Prot.*, 30, 62–66.

Li J.R., Yu J.L., Wei Z.G. (2003). Influence of specimen geometry on adiabatic shear instability of tungsten heavy alloys. *Int. J. Impact Eng.*, 28, 303–314.

Lima R.S., Karthikeyan J., Kay C.M., Lindemann J., Berndt C.C. (2002). Microstructural characteristics of cold-sprayed nanostructured WC-Co coatings. *Thin Solid Films*, 416(1–2), 129–135.

Liu J. (2006). Advanced aluminum and hybrid aerostructures for future aircraft. *Mater. Sci. Forum*, 519–521, 1233–1238. (Online available since 2006/Jul/15 at www.scientific.net.)

Li W., Li D.Y. (2006). Influence of surface morphology on corrosion and electronic behavior. *Acta Mater.*, 54, 445–452.

Li W., Zhang C., Li C., Liao H. (2009). Modeling aspects of high velocity impact of particles in cold spraying by explicit finite element analysis. *J. Therm. Spray Technol.*, 18(5–6), 921–933.

Li W.-Y., Gao W. (2009). Some aspects on 3-D numerical modeling of high velocity impact of particles in cold spraying by explicit finite element analysis. *App. Surf. Sci.*, 255, 7878–7892.

Li W.-Y., Liao H., Li C.-J., Li G., Coddet C., Wang X. (2006). On high velocity impact of microsized metallic particles in cold spraying. *Appl. Surf. Sci.*, 253, 2852–2862.

Li W.-Y., Liao H., Wang H.-T., Li C.-J., Zhang G., Coddet C. (2006). Optimal design of a convergent-barrel cold spray nozzle by numerical method. *Appl. Surf. Sci.*, 253(2), 708–713.

Li W.-Y., Li C.-J. (2005). Optimal design of a novel cold spray gun nozzle at a limited space. *J. Therm. Spray Technol.*, 14(3), 391–396.

Li W.-Y., Li C.-J., Liao H. (2010). Significant influence of particle surface oxidation on deposition efficiency, interface microstructure and adhesive strength of cold-sprayed copper coatings. *Appl. Surf. Sci.*, 256, 4953–4958.

Li X., Dunn P.F., Brach R.M. (1999). Experimental and numerical studies on the normal impact of microspheres with surfaces. *J. Aerosol Sci.*, 30(4), 439–449.

Li Y.S., Tao N.R., Lu K. (2008). Microstructural evolution and nanostructure formation in copper during dynamic plastic deformation at cryogenic temperatures. *Acta Mater.*, 56, 230–241.

Li Z., Zhu J., Zhang C. (2005). Numerical simulations of ultrafine powder coating systems. *Powder Technol.*, 150, 155–167.

Louge M.Y., Mastorakios E., Jenkins J.T. (1991). The role of particle collisions in pneumatic transport. *J. Fluid Mech.*, 231, 345–359.

Lubrik M., Maev R., Leshchynsky V. (2008). Young's modulus of metal=matrix composites made by low pressure gas dynamic spray. *J. Mater. Sci.*, 43, 4953–4961.

Maev R., Leshchynsky V. (2006). Air gas dynamic spraying of powder mixtures: Theory and application. *J. Therm. Spray Technol.*, 15(2), 198–205.

Maev R., Titov C., Leshchynsky V. (2012a). Passive and pulse-echo ultrasonic monitoring of cold spray process. *J. Therm. Spray Technol.*, 21(3–4), 620–627.

Maev R., Titov C., Leshchynsky V. (2012b). *Passive and Pulse-Echo Ultrasonic Monitoring of Cold Spray Conference and Exposition*. ASM International, Houston, TX, pp. 357–362.

Maev R., Titov C., Leshchynsky V., Dzhurinskiy D., Lubric M. (2011). In situ monitoring of particle consolidation during low pressure cold spray by ultrasonic technique. *J. Therm Spray Technol.*, 20(4), 845–851.

Maev R., Titov S., Doyle D., Hatton C. (2013). Ultrasonic evaluation of anticorrosive copper cold spray coating of steel nuclear waste containers. *Proceedings of the 52th Annual Conference of the British Institute of Non-Destructive Testing NDT*, Telford, UK.

Maev R.G., Leshchynsky V. (2007). *Introduction to Low Pressure Gas Dynamic Spray: Physics & Technology*. Wiley-VCH Verlag GmbH, Weinheim, Germany.

Maev R.G., Strumban E., Leshchynsky V., Beneteau M. (2004). Supersonic induced mechanical alloy technology and coatings for automotive and aerospace applications. *Cold Spray: An Emerging Spray Coating Technology*, September 27–28, 2004. ASM International, Akron, OH.

Maev R.Gr. (2008). *Acoustic Microscopy: Fundamental and Applications*. Wiley-VCH, Weinheim, Germany.

Maev R.Gr., Leshchinsky Ev. (2006). Low pressure gas dynamic spray: Shear localization during particle shock consolidation. *International Thermal Spray Conference Proceedings*, Seattle, WA (CD Proceeding).

Maev R.Gr., Leshchynsky V. (2006). Compaction and bonding alloyed powder by air gas dynamic spray technology. *Surf. Coat. Technol.* (to be submitted).

Maev R.Gr., Leshchynsky V., Papyrin A. (2006). Structure formation of Ni-based composite coatings during low pressure gas dynamic spraying. *International Thermal Spray Conference Proceedings*, Seattle, WA (CD Proceeding).

Magnesium Electron Inc., Data Sheet: 452A, www.magnesium-elektron.com, Manchester, NJ, p. 4.

Maitra S., English G.C. (1981). Mechanism of localized corrosion of 7075 alloy plate. *Metall. Trans. A*, 12A, 535.

Manufacturing Readiness Level (MRL) Deskbook, Version 2.0, May, 2011. Prepared by the OSD Manufacturing Technology Program in collaboration with The Joint Service/Industry MRL Working Group.

Marchand A., Duffy J. (1988). An experimental study of the formation process of adiabatic shear bands in a structural steel. *J. Mech. Phys. Solids*, 36, 25.

Martin L.P., Rosen M. (1997). Correlation between surface area reduction and ultrasonic velocity in sintered zinc oxide powders. *J. Am. Ceram. Soc.*, 80, 839–846.

Mason David L., Rao K. (1984). Thermal spray coatings—New materials, processes and applications. In: Longo F.N. (Ed.). *American Society for Metal*. Materials Park, OH, pp. 51–63.

Massalski T.B. (Ed.). (1990). *Binary Alloy Phase Diagrams*. ASM International, Materials Park, OH, p. 147.

Matuo K. (1994). *Compressible Fluid Dynamics*. Rikougakusha, Tokyo, Japan, p. 83 (in Japanese).

Mazumder J., Choi J., Nagarathnam K., Koch J., Hetzner D. (1997). The direct metal deposition of H13 tool steel for 3-D components. *IOM*, 49(5), 555–560.

McCune R., Ricketts M. (2004). Selective galvanizing by cold spray processing. *Cold Spray 2004*, ASM International, Akron, OH.

McCune R.C, Donlon W.T., Popoola O.O., Cartwright E.L. (2000). Characterization of copper layers produced by cold gas-dynamic spraying. *J. Ther. Spray Technol.*, 9(1), 73–81.

McCune R.C. (2003). Potential applications of cold spray technologies in automotive manufacturing. In: Moreau C., Marple, B. (Eds.). *Thermal Spray 2003: Advancing the Science and Applying the Technology*. ASM International, Materials Park, OH, pp. 63–70.

McCune R.C., Donlon W.T., Cartwright E.L., Papyrin A.N., Rybicki E.E., Shadley J.R. (1996). Thermal spray: practical solutions for engineering problems. In: Berndt C.C. (Ed.). ASM International, Materials Park, OH, pp. 397–403.

Mebtoul M., Large J., Guidon P. (1996). High velocity impact of particles on a target—An experimental study. *J. Mineral Process.*, 44–45, 77–91.

Metzeger M., Zahavi J. (1979). *Passivity of Metals*. The Electrochemical Society, Princeton, NJ, p. 960.

Meyers M.A. (1994). *Dynamic Behavior of Materials*, 1st ed. Wiley, New York, p. 299.

Meyers M.A., Benson D.J., Olevsky E.A. (1999). Shock consolidation: Microstructurally-based analysis and computational modeling. *Acta Mater.*, 47(7), 2089–2108.

Meyers M.A., Mishra A., Benson D.J. (2006). Mechanical properties of nanocrystalline materials. *Prog. Mater. Sci.*, 51, 427–556.

Meyers M.A., Nesterenko V.E, LaSalvia J.C., Xue Q. (2001). Shear localization in dynamic deformation of materials: Microstructural evolution and self-organization. *Mater. Sci. Eng.*, 317(1–2), 204–225.

Milewski J.O., Lewis G.K., Thoma D.J., Keel G.I., Nemec R.B., Reinert R.A. (1998). Directed light fabrication of a solid metal hemisphere using 5-axis powder deposition. *J. Mater. Process. Technol.*, 75, 165.

Molak R.M., Araki H., Watanabe M., Katanoda H. (2014). Warm spray forming of Ti-6Al-4V. *J. Therm. Spray Technol.*, 24(1–2), 197–212.

Molinari A. (1997). Collective behaviour and spacing of adiabatic shear bands. *J. Mech. Phys. Solids*, 45(9), 1551–1575.

Molinari A., Ravichandran G. (2005). Constitutive modeling of high-strain-rate deformation in metals based on the evolution of an effective microstructural length. *Mech. Mater.*, 37, 737–752.

Molinari J.F., Ortiz M. (2002). A study of solid-particle erosion of metallic targets. *Int. J. Impact Eng.*, 27, 347–358.

Morgan R., Fox P., Pattison J., Sutcliffe C., O'Neill W. (2004). Analysis of cold gas dynamically sprayed aluminium deposits. *Mater. Lett.*, 58, 1317–1320.

Moridi A., Hassani-Gangaraj S.M., Guagliano M., Dao M. (2014). Cold spray coating: Review of material systems and future perspectives. *Surf. Eng.*, 36(6), 369–395.

Motzet H., Ilmann H.P. (1999). Synthesis and characterization of sulfite-containing AFM phases in the system $CaO-Al_2O_3-SO_2-H_2O$. *Cem. Concr. Res.*, 29, 1005–1011.

Muehlberger E., de la Vega. P. (1989). Powder feeder. US Patent 4,808,042.

Murr E., Staudhammer K.P., Meyers M.A. (Eds.). (1986). *Metallurgical Applications of Shock-Wave and High-Strain-Rate Phenomena*. Marcel Dekker, New York.

Murty S.V.S., Nageswara R.B., Kashyap B.P. (2000). Instability criteria for hot deformation of materials. *Int. Mater. Reviews.*, 45(1), 15–26.

Müller C., Sarret M., Benballa M. (2001). Some peculiarities in the codeposition of zinc-nickel alloys. *Electrochim. Acta.*, 46, 2811.

Nakai M., Rto T. (2000). New aspect of development of high strength aluminum alloys for aerospace applications. *Mater. Sci. Eng. A*, 285(1–2), 62–68.

Narayanasamy R., Ramesh T., Pandey K.S. (2005). Workability studies on cold upsetting of Al-Al$_2$O$_3$ composite material. *Mater. Design*, 27(7), 566–575.

NASA Public Communications Office. https://www.nasa.gov/content/technology-readiness-level/.

Navy/Industry Task Group Report, Impact of anticipated OSHA hexavalent chromium worker exposure standard on navy manufacturing and repair operations, October 1995.

Needleman A., Sevillano J.G. (2003). Preface to the viewpoint set on: Geometrically necessary dislocations and size dependent plasticity. *Scripta Mater.*, 48, 109–111.

Nesterenko V.F. (1995). Dynamic loading of porous materials: Potential and restrictions for novel materials applications. In: Murr L.E., Staudhammer K.P., Meyers M.A. (Eds.). *Metallurgical and Materials Applications of Shock-Wave and High-Strain-Rate Phenomena. Proceedings of the International Conference*, August 6–10, 1995, El Paso, TX, Elsevier, Amsterdam, the Netherlands, pp. 3–13.

Nie J.F. (2014). *Physical Metallurgy of Light Alloys*. Physical Metallurgy, Elsevier, Amsterdam, the Netherlands, pp. 2009–2156.

Niranatlumpong P., Koiprasert H. (2006). Improved corrosion resistance of thermally sprayed coating via surface grinding and electroplating techniques. *Surf. Coat. Technol.*, 201, 737–743.

Oesterle B., Petitjean A. (1993). Simulation of particle-to-particle interactions in gas-solid flows. *Int. J. Multiphase Flow*, 19, 199–211.

Ohsaki S., Kato S., Tsuji N., Ohkubo T., Hono K. (2007). Bulk mechanical alloying of Cu–Ag and Cu/Zr two-phase microstructures by accumulative roll-bonding process. *Acta Mater.*, 55, 2885–2895.

Olsson G.B. (1997). Designing a new material world. *Science*, 288(5468), 993–998.

Orazem M.E., Pebere N., Tribollet B. (2006). Enhanced graphical representation of electrochemical impedance data. *J. Electrochem. Soc.*, 153(4), B129–B136.

Oskooie M.S., Asgharzadeh H., Kim H.S. (2015). Microstructure, plastic deformation and strengthening mechanisms of an Al–Mg–Si alloy with a bimodal grain structure. *J. Alloys Comp.*, 632, 540–548.

Pan T., Santella M. (2012). Corrosion behavior of mixed-metal joint of magnesium to mild steel by ultrasonic spot welding. SAE Technical Paper 2012-01-0472.

Papadakis E.P. (1976). Ultrasonic velocity and attenuation: Measurement method with scientific and industrial application. In: Mason W.P., Thurston R.N. (Eds.). *Physical Acoustics. Principles and Methods*, Vol. XII. Academic Press, New York, pp. 227–374.

Papyrin A. (2001). Cold spray technology. *Adv. Mater. Process.*, 159(9), 49–51.

Papyrin A.N., Kosarev V.F., Klinkov S.V., Alkhimov A.P. (2002). On the interaction of high speed particles with a substrate under cold spraying. In: Lugscheider E.F., Berndt C.C. (Eds.). *Proceeding of the International Thermal Spray Conference*, Dusseldorf, Germany, DVS-Verlag, p. 380.

Papyrin A.N., Kosarev V.F., Klinkov S.V., Alkhimov A.P., Fomin V.M. (2007). *Cold Spray Technology*. Elsevier, 336pp.

Pardo A., Merino M.C., Coy A.E., Arrabal R. (2008). Corrosion behaviour of magnesium/aluminium alloys in 3.5 wt.% NaCl. *Corros. Sci.*, 50, 823.

Pardoen T., Massart T.J. (2012). Interface controlled plastic flow modelled by strain gradient plasticity theory. *C. R. Mecanique*, 340, 247–260.

Peker A., Johnson W.L. (1993). A highly processable metallic glass—Zr41·2Ti13·8Cu12·5Ni10·0Be22·5. *Appl. Phys. Lett.*, 63, 2342–2344.

Pinkerton F.R., Van Steenkiste T.H., Moleski J.J. (October 15, 2002). Magnetostrictive composite coating. US Patent 6,465,039.

Pourbaix M. (1975). *Atlas of Electrochemical Equilibria in Aqueous Solutions*. NACE, Houston, TX.

Product catalogue Carbide Jet System (CJS), OZU-Mashinenbau GmbH, Castrop-Rauxel, Germany, 1995.

Product catalogue DIAMOND Jet, SULZER METCO (US) Inc., New York, 1995.

Raabe D., Choi P.-P., Li Y., Kostka A., Sauvage X., Lecouturier F., Hono K., Kirchheim R., Pippan R., Embury D. (2010). Metallic composites processed via extreme deformation: Toward the limits of strength in bulk materials. *MRS Bull.*, 35(12), 982–991.

Raabe D. et al. (2010). Metallic composites processed via extreme deformation: Toward the limits of strength in bulk materials. *MRS Bull.*, 35, 982–991.

Raghukandan K. (2003). Analysis of the explosive cladding of cu–low carbon steel plates. *J. Mat. Proc. Techn.*, 139(1–3), 573–577.

Rajan T.P.D., Pillai R.M.B., Pai C. (1998). Reinforcement coatings and interfaces in aluminium MMCs. *J. Mater. Sci.*, 33, 3491–3503.

Rajaratnam N. (1976). *Turbulent Jets*, Elsevier, Amsterdam, the Netherlands.

Raletz F., Ezo'o G., Vardelle M., Ducos M. (2003). Characterization of cold-sprayed nickel-base coatings. In: Marple B.R., Moreau C. (Eds.). *Thermal Spray 2003: Advancing the Science & Applying the Technology*. Orlando, FL, ASM International, Materials Park, OH, pp. 45–50.

Ranz W.E., Marshall T. (1952). Evaporation from drops. *Chem. Eng. Prog.*, 48, 141–146.

Raybould D. (1981). The properties of stainless steel compacted dynamically to produce cold interparticle welding. *J. Mater. Sci.*, 16(3), 589–598.

Rice R.W. (1996). Evaluating porosity parameters for porosity-property relations. *J. Am. Ceram. Soc.*, 76, 1801–1805.

Rocheville C.F. (Augest 13, 1963). Device for treating the surface of a workpiece. US Patent 3,100,724.

Rojas P. N., Rodil S. E. (2012). Corrosion behaviour of amorphous niobium oxide coatings. *Int. J. Electrochem. Sci.*, 7, 1443–1458.

Rokni M., Widener C., Champagne V. (2014). Microstructural evolution of 6061 aluminum gas-atomized powder and high-pressure cold-sprayed deposition. *J. Therm. Spray Technol.*, 23(3), 514–524.

Rokni M.R., Widener C.A., Crawford G.A. (2014). Microstructural evolution of 7075 Al gas atomized powder and high-pressure cold sprayed deposition. *Surf. Coat. Technol.*, 251, 254–263.

Rotolico A.J., Romero E., Lyons J.E. (1983). Device for the controlled feeding of powder material. US Patent 4,381,898.

Rudinger G. (1980). *Fundamentals of Gas-Particle Flow, Handbook of Powder Technology*, Vol. 2. Elsevier, Amsterdam, the Netherlands.

Sakaaki K., Tajima T., Li H., Shinkai S., Shimizu Y., Nagano J. (2004). Influence of substrate conditions and traverse speed on the cold sprayed coatings. ITSC.

Sakaki K., Akashi T., Hosono T. (2014). Influence of cross sectional shape of cold spray nozzle. *Proceedings of the JSME/ASME International Conference on Materials and Processing*, June 9–13, Detroit, MI, ICMP2014-4961.

Sakaki K., Huruhashi N., Tamaki K., Shimizu Y. (2002). Effect of nozzle geometry on cold spray process. In: Lugscheider E., Berndt C.C. (Eds.). *International Thermal. Spray Conference*, March 4–6, Essen, Germany, DVS Deutscher Verband für Schweißen, pp. 385–389.

Sakaki K., Shimizu Y. (2001). Effect of the increase in the entrance convergent section length of the gun nozzle on the high-velocity oxygen fuel and cold spray process. *J. Ther. Spray Technol.*, 10(3), 487–496.

Sakaki K., Shimizu Y., Gouda Y. (1999). *J. Japan Inst. Metals*, 63(2), 269–276 (in Japanese).

Sakaki K., Shimizu Y., Gouda Y., Devasenapathi A. (1998). Thermal Spray: Meeting the Challenges of 21st Century. In: Coddet C. (Ed.). ASM International, Materials Park, OH, pp. 445–450.

Sakaki K., Shimizu Y., Gouda Y., Minamida T. (1998). *J. Japan Therm. Spraying Soc.*, 35(3), 195–203.

Sakaki K., Shimizu Y., Saitoh N., Gouda Y. (1997). *J. Japan Therm. Spraying Soc.*, 34(1), 1–9 (in Japanese).

Sakaki K., Shinkai S., Ebara N., Shimizu Y. (2006). Effect of geometry of the gun nozzle, the increase in the entrance convergent section length and powder injection position on cold sprayed titanium coatings. *Mater. Sci. Forum*, 534–536, 413–16.

Sakaki K., Takeda K., Takada K., Hosono T., Shimizu Y. (2008). Influence of the expansion ratio of the gun nozzle and gas pressure on properties of cold sprayed copper coatings. *Proceedings of the 3rd JSME/ASME International Conference on Materials Processing*, IL, MSEC_ICM&P2008-72015.

Salman S.A., Ichino R., Okido M. (2010). A comparative electrochemical study of AZ31 and AZ91 magnesium alloy. *Int. J. Corros.*, 2010, Article ID 412129.

Samareh B., Stier O., Lu V., Dolatabadi A. (2009). Assessment of CFD modeling via flow visualization in cold spray process. *J. Therm. Spray Technol.*, 18(5–6), 934–943.

Sandstrom R., Hallgren J. (2012). The role of creep in stress strain curves for copper. *J. Nucl. Mater.*, 422(1–3), 51–57.

Sansoucy E., Kim G.E., Moran A.L., Jodoin B. (2007). Mechanical characteristics of Al-Co-Ce coatings produced by the cold spray process. *J. Ther. Spray Technol.*, 16(5-6), 651–660.

Sasaki G.T. (1996). Burstein, the generation of surface roughness during slurry erosion-corrosion and its effect on the pitting potential. *Corros. Sci.*, 38, 2111.

Sathiya P., Jaleel M.Y.A. (2011). Influence of shielded gas mixtures on bead profile and microstructural characteristics of super austenitic stainless steel weldments by laser welding. *Int. J. Adv. Manuf. Technol.*, 54, 525–535.

Sauser B. et al. (April 7-8, 2006). *From TRL to SRL: The Concept of Systems Readiness Levels Conference on Systems Engineering Research*, Los Angeles, CA.

Schinella A.A. (1971). Powder feeder. US Patent 3,618,828.

Schlichting H. (1979). *Boundary Layer Theory*. McGraw-Hill, New York.

Schmidt J., Dorner H., Tenckhoff E. (1990). Manufacture of complex parts by shape welding. *J. Nucl. Mater.*, 171(1), 120–127.

Schmidt T., Assadi H., Gartner F., Richter H., Stoltenhoff T., Kreye H., Klassen T. (2009). Fro particle acceleration to impact and bonding in cold spraying. *J. Therm. Spray Technol.*, 11, 794–808.

Schmidt T., Gartner F., Assadi H., Kreye H. (2006). Development of a generalized parameter window for cold spray deposition. *Acta Mater.*, 54, 729–742.

Schmidt T., Gartner F., Kreye H. (2003). High strain rate deformation phenomena in explosive powder compaction and cold gas spraying. In: Marple B.R., Moreau C. (Eds.). *Proceedings of the ITSC*, Orlando, FL, ASM International, Materials Park, OH, pp. 9–17.

Schmitt G., Schütze M., Hays G. (2009). Global needs for knowledge dissemination, research and development in materials deterioration and corrosion control. World Corrosion Organization, New York, May 2009.

Schoenfeld E., Wright T.W. (2003). A failure criterion based on material instability. *Int. J. Solids Struct.*, 40(4), 3021–3037.

Schuh C., Hufnagel T., Ramamurty U. (2007). Mechanical behavior of amorphous alloys. *Acta Mater.*, 55, 4067–4109.

Senior R.C., Grace J.R. (1998). Integrated particle collision and turbulent diffusion model for dilute gas-solid suspensions. *Powder Technol.*, 96(1), 48–78.

Seo D., Ogawa K., Sakaguchi K., Miyamoto N., Tsuzuki Y. (2012). Parameter study influencing thermal conductivity of annealed pure copper coatings deposited by selective cold spray processes. *Surf. Coat. Technol.*, 206(8–9), 2316–2324.

Sharma M.M., Eden T.J., Golesich B.T. (2015). Effect of surface preparation on the microstructure, adhesion, and tensile properties of cold-sprayed aluminum coatings on AA2024 substrates. *J. Therm. Spray Technol.*, 24(3), 410–422.

Shi Z., Atrens A. (2011). An innovative specimen configuration for the study of Mg corrosion. *Corros. Sci.*, 53, 226–246.

Simmons J.W., Wilson RD. (1996). Joining of high-nitrogen stainless steel by capacitor discharge welding. *Weld. J., (Miami, Fla)*, 75(6), 185–190.

Smulko J.M., Darowicki K., Zieliński A. (2007). On electrochemical noise analysis for monitoring of uniform corrosion rate. *IEEE Trans. Instrum. Meas.*, 56(5), 2018–2023.

Sommerfeld M. (2003). Analysis of collision effects for turbulent gas-particle flow in a horizontal channel: Part I. Particle transport. *Int. J. Multiphase Flow*, 29, 675–699.

Sommerfeld M., Kussin J. (2003). Analysis of collision effects for turbulent gas-particle flow in a horizontal channel: Part II. Integral properties and validation. *Int. J. Multiphase Flow*, 29, 701–718.

Song G.L., Atrens A. (2003). Understanding magnesium corrosion—A framework for improved alloy performance. *Adv. Eng. Mater.*, 5, 837–858.

Soni, P.R. (1999). *Mechanical Alloying: Fundamentals and Applications*. Cambridge International Science Publishing, Cambridge

Spencer K., Fabijanic D.M., Zhang M.-X. (2009). The use of Al–Al₂O₃ cold spray coatings to improve the surface properties of magnesium alloys. *Surf. Coat. Technol.*, 204, 336–344.

Steenkiste T. Van, Smith J.R. (2004). Evaluation of coatings produced via kinetic and cold spray processes. *J. Therm. Spray Technol.*, 13(2), 274–282.

Steinhauser S., Wielage B. (1997). Composite coatings: Manufacture, properties, and applications. *Surf. Eng.*, 13(4), 289–294.

Stier O. (2014). Fundamental cost analysis of cold spray. *J. Therm. Spray Technol.*, 23(1–2), 131–139.

Stoltenhoff T., Borchers C., Gartner F., Kreye H. (2006). Microstructures and key properties of cold-sprayed and thermally sprayed copper coatings. *Surf. Coat. Technol.*, 200, 4947–4960.

Stoltenhoff T., Kreye H., Richter H.J. (2002). An analysis of the cold spray process and its coating. *J. Therm. Spray Technol.*, 11(4), 542–550.

Stoltenhoff T., Zimmermann F. (2010). *LOXPlate® Coatings for Aluminum Aerospace Components Exposed to High Dynamic Stresses*. Praxair Surface Technologies GmbH, Ratingen, Germany.

Stoltenhoff Th., Zimmermann F. (2012). *LOXPlate® Coatings for Aluminum Aerospace Components Exposed to High Dynamic Stresses*. Praxair Surface Technologies GmbH, Ratingen, Germany.

Straffelini G., Pellizzari M., Molinari A. (2004). Influence of load and temperature on the dry sliding behavior of Al-based metal matrix-composites. *Wear*, 256, 754–763.

Streeter V.L. (1961). *Handbook of Fluid Dynamics*. McGraw-Hill, New York.

Suegama P.H., Espallargas N., Guilemany J.M., Fernández J., Benedettia A.V. (2006). Electrochemical and structural characterization of heat-treated Cr_3C_2–NiCr coatings. *J. Electrochem. Soc.*, 153(10), B434–B445.

Sun Z., Karppi R. (1996). Application of electron beam welding for the joining of dissimilar metals: An overview. *J. Mater. Process. Technol.*, 59, 257–267.

Surinach S. et al. (1996). Hermoanalytical characterization of a nanograined Fe-40 Al alloy. *Mater. Sci. Forum*, 225–227, 395–400.

Suryanarayana C. (2001). Mechanical alloying and milling. *Prog. Mater. Sci.*, 46, 1–184.

Takahashi H., Sakairi M., Kikuchi T. (2010). Three-dimensional microstructure fabrication with aluminum anodizing, laser irradiation, and electrodeposition. *Mod. Asp. Electrochem.*, 46, 59–174.

Tanaka T., Tsuji Y. (1991). Numerical simulation of gas-phase flow in a vertical pipe: On the effect of interparticle collision. In: Stock D.E. et al. (Eds.). *Gas-Solid Flows*, Vol. 121. ASME FED, pp. 123–128.

Tao Y., Xiong T., Sun C., Jin H., Du H., Li T. (2009). *Appl. Surf. Sci.*, 256, 261–266.

Tao Y., Xiong T., Sun C., Kong L., Cui X., Li T. (2010). Microstructure and corrosion performance of a cold sprayed aluminium coating on AZ91D magnesium alloy. *Corros. Sci.*, 52, 3191–3197.

Tapphorn R.M., Gabel H. (2004). Powder fluidizing devices and portable powder-deposition apparatus for coating and spray forming. US Patent 6,715,640.

The leaflet of Kinetics 4000R Cold spray system by Cold Gas Technology GmbH, 05 May 2006.

Thope M.L., Richer H.J. (1992). Thermal Spray: International Advances in Coating Technology. In: Bernt C.C. (Ed.). ASM International, Materials Park, OH, pp. 137–147.

Tian W., Wang Y., Zhang T., Yang Y. (2009). Sliding wear and electrochemical corrosion behavior of plasma sprayed nanocomposite Al_2O_3-13%TiO_2 coatings. *Mater. Chem. Phys.*, 118, 37–45.

Tjong S.C., Zhu S.M., Ho N.J., Ku J.S. (1995). Microstructural characteristics and creep rupture behavior of electron beam and laser welded AISI 316L stainless steel. *J. Nucl. Mater.*, 227(1–2), 24.

Triesch O., Bohnet M. (2001). Measurement and CFD prediction of velocity and concentration profiles in a decelerated gas-solid flow. *Powder Technol.*, 115, 101–113.

Tsirkunov Y.M. (2001). Gas-particle flows around bodies—Key problems, modeling and numerical analysis. In: Michaelides E. (Ed.). *Proceedings of the 4th International Conference on Multiphase Flow*, New Orleans, LA (CD-Rom: Proc. ICMF'2001, paper 609).

Tsuji Y., Morikawa Y., Schiomi H. (1984). LDV measurements of an air-solid two-phase flow in a vertical pipe. *J. Fluid Mech.*, 139, 417.

Tucker, Jr. R.C. (2013). *ASM Handbook Vol. 5A. Thermal Spray Technology*. ASM International, Materials Park, OH, pp. 55–56.

Turgutlu, A., Al-Hassani, S.T.S., Akyurt, M. (1995). Experimental investigation of deformation and jetting during impact spot welding. *Int. J. Impact Eng.*, 16(5–6), 135–152.

Turgutlu A., Al-Hassani S.T.S., Akyurt, M. (1997). Assessment of bond interface in impact spot welding. *Int. J. Impact Eng.*, 19(9-10), 755–767.

Underwood E.E. (1985). *Metals Handbook*, 9th edn., ASM International, Metals Park, OH, pp. 123–134.

Van Steenkiste T.H. (2003). Method of producing a coating using a kinetic spray process with large particles and nozzles for the same. US Patent 6,623,796.

Van Steenkiste T.H. (2004). Low pressure powder injection method and system for a kinetic spray process. US Patent No 6,811,812.

Van Steenkiste T.H. et al. (1999). Kinetic spray coatings. *Surf. Coat. Technol.*, 111, 62–71.

Van Steenkiste T.H., Smith J.R. (2003). Evaluation of coatings produced via kinetic and cold spray processes. *Proceedings of the International Thermal Spray Conference*, May 5–8, Orlando, FL, ASM International, Materials Park, OH, pp. 53–61.

Van Steenkiste T.H., Smith J.R., Teets R.E. (2002). Aluminum coatings via kinetic spray with relatively large powder particles. *Surf. Coat. Technol.*, 154, 237–252.

Van Steenkiste T.H., Smith J.R., Teets R.E., Moleski J.J., Gorkiewicz D.W. (2000). Kinetic spray coating method and apparatus. US Patent 6,139,913.

Villafuerte J. (2005). Cold spray: A new technology. *Weld. J.*, 84(5), 24–29.

Vlcek J., Gimeno L., Huber H., Lugscheider E. (2003). A systematic approach to material eligibility for the cold spray process. In: Marple B.R., Moreau C. (Eds.). *Thermal Spray 2003: Advancing the Science & Applying the Technology*, May 5–8, 2003, Orlando, FL, ASM International, Materials Park, OH, pp. 37–44.

Vlcek J., Gimeno L., Huber H., Lugscheider E. (2005). A systematic approach to material eligibility for the cold spray process. *J. Therm. Spray Technol.*, 14, 125–133.

Vlcek J., Huber H., Voggenreiter H., Fischer A., Lugscheider E., Hallen H., Pache G. (2001). Kinetic powder compaction applying the cold spray process—A study on parameters. In: Berndt C.C., Khor K.A., Lugscheider E.F. (Eds.). *Thermal Spray 2001: New Surfaces for a New Millennium*. ASM International, Materials Park, OH.

Vlcek J., Huber H., Voggenreiter H., Lugscheider E. (2002). Melting upon particle impact in the cold spray process, Presentation at *Materials Week 2002*, September 2002, International Congress on Advanced Materials. Their Processes and Applications, *Deutsche Gesellschaff fur Materialkunde* (DGM), Munich, Germany.

Volkov A., Tsirkunov Yu., Oesterle B. (2005). Numerical simulation of a supersonic gas-solid flow over a plunt body: The role of inter-particle collisions and two-way coupling effects. *Int. J. Multiphase Flow*, 31, 1244–1275.

Walter J.W. (1992). Numerical experiments on adiabatic shear band formation in one dimension. *Int. J. Plasticity*, 8, 657–693.

Wang H.M., Chen Y.L., Yu L.G. (2000). "In situ" weld-alloying/laser beam welding of SiCp/6061 Al MMC, *Mater. Sc. and Eng. A*, 293(1–2), 1–6.

Wang X.-B. (2006). Temperature distribution in adiabatic shear band for ductile metal based on Johnson-Cook and gradient plasticity model. *Trans. Nonferrous Met. Soc. China*, 16, 333–338.

Watanabe M., Brauns C., Komatsu M., Kuroda S., Gärtner F., Klassen T., Katanoda H. (2013). Effect of nitrogen flow rate on microstructures and mechanical properties of metallic coatings by warm spray deposition. *Surf. Coat. Technol.*, 232, 587–599.

Wei Q., Cheng S., Ramesh K.T., Ma E. (2004). Effect of nanocrystalline and ultrafine grain sizes on the strain rate sensitivity and activation volume: Fcc versus bcc metals. *Mater. Sci. Eng. A*, 381, 71–79.

Weinert H. (2013). Exfoliation based technology of large scale manufacturing molybdenum disulphide graphene-like nanoparticle mixtures. *Arch. Civil Mech. Eng.*, 13(2), 144–149.

Winston Revie R. (Ed.). (2011). *Uhlig's Corrosion Handbook*, 3rd ed. Wiley, p. 1296.

Wright J.T. (2002). *The Physics and Mathematics of Adiabatic Shear Bands*. Cambridge University Press.

Wright T.W. (1987). Steady shearing in a viscoplastic solid. *J. Mech. Phys. Solids*, 35, 269–282.

Wright T.W. (1995). Scaling laws for adiabatic shear bands. *Int. J. Solid Struct.*, 32(17/18), 2745–2750.

Wright T.W. (2002). *The Physics and Mathematics of Adiabatic Shear Bands*. Cambridge University Press, New York, 241pp.

Wright T.W., Batra R.C. (1985). The initiation and growth of adiabatic shear bands. *Int. J. Plasticity*, 1, 205–212.

Wu J., Fang H., Yoon S., Kim H.J., Lee C. (2005). Measurement of particle velocity and characterization of deposition in aluminium alloy kinetic spraying process. *Appl. Surf. Sci.*, 252, 1368–1377.

Wu J.W., Fang H.Y., Lee C.H., Yoon S.H., Kim H.J. (2006). Critical and maximum velocities in kinetic spraying. *Proceedings of the 2006 International Thermal Spray Conference*, Seattle, WA.

Wu X.L., Jiang P., Chen L., Zhang J.F., Yuan F.P., Zhu Y.T. (2014). Synergetic strengthening by gradient structure. *Mater. Res. Lett.*, 2(4), 185–191.

Xu Y., Zhang J., Bai Y., Meyers M.A. (2008). Shear localization in dynamic deformation: Microstructural evolution. *Metall. Mater. Trans. A*, 39A, 811–843.

Xue Q., Liao X.Z., Zhu Y.T., Gray III G.T. (2005). Formation mechanisms of nanostructures in stainless steel during high-strain-rate severe plastic deformation. *Mater. Sci. Eng.*, A, 410–411, 252–256.

Yandouzi B.M., Jodoin P., Richer, Jodoin B. (2007). Pulsed-gas dynamic spraying: Process analysis, development and selected coating examples. *Surf. Coat. Technol.*, 201, 7544–7551.

Yandouzi M., Gaydos S., Guo D., Ghelichi R., Jodoin B. (2014). Aircraft skin restoration and evaluation. *J. Therm. Spray Technol.*, 23(8), 1281–1290.

Yang L., Ghoshal G., Turner J.A. (2007). Ultrasonic scattering in textured polycrystalline materials. In: Kundu T. (Ed.). *Advanced Ultrasonic Methods for Material and Structure Inspection*. ISTE, London, pp. 189–235.

Yilbas B.S., Sami M., Nickel J., Coban A., Said S.A.M. (1998). Introduction into the electron beam welding of austenitic 321-type stainless steel. *J. Mater. Process. Technol.*, 82, 13.

Yin S., Wang X., Li W., Xu B. (2009). Numerical effects of interactions between particles on coating formation in cold spraying. *J. Therm. Spray Technol.*, 18(4), 686–693.

Yin S., Wang X., Li W., Xu B. (2010). Numerical study on the effect of substrate angle on particles impact velocity and normal velocity component in cold gas dynamic spraying based on CFD. *J. Therm. Spray Technol.*, 19, 1155–1162.

Zerilli J., Armstrong R.W. (1997). Dislocation mechanics based analysis of material dynamics behaviour: Enhanced ductility, deformation twinning, shock deformation, shear instability, dynamic recovery. *J. Phys.*, 7, c3-637–642.

Zhang D., Shipway P.H., McCartney D.G. (2003). Particle-substrate interactions in cold gas dynamic spraying. In: Marple B.R., Moreau C. (Eds.). *Thermal Spray 2003: Advancing the Science & Applying the Technology*. ASM International, Orlando, FL, pp. 45–52.

Zhang D., Shipway P.H., McCartney D.G. (2005). Cold gas dynamic spraying of aluminum: The role of substrate characteristics in deposit Formation. *J. Therm. Spray Technol.*, 14, 109–116.

Zhang H., Wang G., Luo Y., Nakaga T. (2001). Rapid hard tooling by plasma spraying for injection molding and sheet metal forming. *Thin Solid Films*, 390, 7.

Zhang H., Xu J., Wang G. (2003). Fundamental study on plasma deposition manufacturing. *Surf. Coat. Technol.*, 171(1–3), 112–118.

Zhao Z.B. et al. (2005). US Patent Application 20,050,040,260.

Zheng W., Derushie C., Lo J., Essadigi E. (2006). Corrosion protection of joining areas in magnesium die cast and sheet products. *Mater. Sci. Forum*, 546–549, 523–528.

Zhou F., Wright T.W., Ramesh K.T. (2006). A numerical methodology for investigating the formation of adiabatic shear bands. *J. Mech. Phys. Solids*, 54, 904–926.

Zhu L., Ruan H., Li X., Dao M., Gao H., Lu J. (2011). Modeling grain size dependent optimal twin spacing for achieving ultimate high strength and related high ductility in nanotwinned metals. *Acta Mater.*, 59, 5544–5557.

Zimmerly C.A., Inal O.T., Richman R.H. (1994). Explosive welding of a near-equiatomic nickel-titanium alloy to low-carbon steel. *Mater. Sci. Eng. A*, A188(1–2), 251–254.

Zukas J.A. (1990). *High-Velocity Impact Dynamics*. Wiley, New York.

Index